Integration of Nature and Technology for Smart Cities

Anil Ahuja

Integration of Nature and Technology for Smart Cities

Third Edition

Springer

Anil Ahuja, PE, RCDD, LEED BD+C, CxA
President, C.C. Johnson & Malhotra, P.C.(CCJM)
Chicago, IL, USA

ISBN 978-3-319-25713-6 ISBN 978-3-319-25715-0 (eBook)
DOI 10.1007/978-3-319-25715-0

Library of Congress Control Number: 2015960460

Springer Cham Heidelberg New York Dordrecht London

Printed on acid-free paper

Springer International Publishing AG Switzerland is part of Springer Science+Business Media (www.springer.com)

Preface

This is the third edition of the book *Integrated M/E Design: Building Systems Engineering* published by Anil Ahuja in 1997. As you can see, the title changed; we extended the topic and added some chapters, as an international group of authors from the engineering, urban planning, and architecture fields got together to discuss new trends and paradigms for smart buildings and smart cities.

The original book was about building technologies and sustainable system operations. Nowadays, we integrate information and communication technologies into buildings and call them "smart buildings." However, our discussions resulted in the common understanding that a smart, sustainable building is not just about the building itself. There are things happening in the inside of the building and on the outside. A smart building connects the inside with the outside, provides efficiencies on both sides, synchronizes the outside infrastructure with its inside systems, and integrates nature and its occupants in its design.

The following co-authors extended the discussion on integrated building design from the first edition of this book to a discussion on integrating technologies and nature for smart cities for this new edition: Ingrid Kaltenegger, environmental scientist (Graz, Austria); Helen Santiago Fink, climate urbanist (Washington, D.C.); Peter Mulvaney, water resource manager (Chicago, USA); Dr. Petra Stieninger, urban sustainability advisor (Chicago, USA); Dr. Raj Rajaram, geotechnical and environmental engineer (Chicago, USA); and Stefan Melzer, urban planner (Vienna, Austria).

A smart building doesn't just provide technology solutions. Smart technologies help to facilitate the operation of the building, and they track daily activities, analyze, and evaluate the collected data. But it is not just big data for big data's sake. It is a people-driven design considering aspects such as impacts on human health, the desire for the exposure to nature as well as the connection to the Internet, and the protection of natural resources and the environment. It is about constant exchange between the inside and the outside of the building, the contribution of the building to the quality of the entire neighborhood and the rest of the city, how the smart building can connect people in a sharing community, and how technology can be the key to make it happen.

The book is divided into three parts.

- Part I is all about technologies, how they make building operations more efficient, and how we can take advantage of the outside world of the building by integrating smart technologies.
- Part II talks about the connection to nature and understanding the behavior of people who live, work, or play in the building. It addresses the question on how nature can be integrated in the building, how we can use natural resources in more sustainable ways, and what influence the occupant's behavior has on the functioning of the implemented technology.
- Part III explains why we have to understand the building as a part of a larger urban system and why its connections to its surroundings are crucial for an efficient and sustainable operation of the building.

With this book, we provide you with an international, interdisciplinary perspective on how to integrate nature and technology in order to create smart cities.

Chicago, IL, USA Petra Stieninger

The original version of the preface was revised. An erratum can be found at DOI 10.1007/978-3-319-25715-0_23

Contents

List of Figures

Chapter 1
Smart Trends and Paradigm Shift

Anil Ahuja

The smart trend of new Internet technologies, promoting cloud-based services, the Internet of Things (IoT) and real-world user interfaces through smart phones, smart meters, and network of sensors and RFIDs, will continue to open new ways to collaborative problem solving. The use of ICT (information and communication technologies) to transform life and working environments within the region and territorialization of practices will bring people and nature together to enhance the innovation and knowledge.

Things could be very different. Urban residents could enjoy clean air instead of smog, stress-free mobility instead of congested roads, clean drinking water instead of water sources that carry pathogens, and affordable electricity from renewable sources that is available on demand instead of expensive or dirty energy obtained from fossil fuels.

The path to that future does not lead back into a preindustrial age. Instead, it leads forward to an age of digital technologies in which cities operate like large computers. Many major cities are already on the way to that future—at least in some areas. They are becoming smart in the modern sense: on the basis of data. And they are being helped by many sophisticated technologies.

Due to the breadth of technologies that have been implemented under the smart city label, it is difficult to distil a precise definition of a smart city. A smart city not only possesses ICT technology in particular areas but implements this technology in a manner that impacts the local community, environment, and health. A city can be defined as "smart" when investments in human and social capital and traditional (transport) and modern (ICT) communication infrastructure fuel sustainable economic development. This leads to a high quality of life, with a wise management of natural resources, through participatory action and engagement.

The concept is not static; there is no absolute definition of a smart city, no end point, but rather a process, or series of steps, by which cities become more "livable" and resilient and, hence, able to respond quicker to new challenges. A strong integration of all dimensions of human intelligence, collective intelligence, and also artificial intelligence is the trend going forward. The intelligence of cities "resides

A. Ahuja, *Integration of Nature and Technology for Smart Cities*,
DOI 10.1007/978-3-319-25715-0_1

in the increasingly effective combination of digital telecommunication networks (the nerves), ubiquitously embedded intelligence (the brains), sensors and tags (the sensory organs), and software (the knowledge and cognitive competence)."

One of the paradigm shifts is to make more efficient use of physical infrastructure (roads, built environment, and other physical assets) through artificial intelligence and data analytics to support a strong and healthy economic, social, cultural development.

A cloud-based connected world is promoting online collaborative sensor data management platforms that allow sensor owners to register and connect their devices to feed data into an online database for storage and allow developers to connect to the database and build their own applications based on that data. It can be weather, movement of goods, security, the levels of pollution, noise, traffic, and parking.

The rapid decrease in costs of computing and communication technology has led to the rise of the Internet, "big data," and cloud-based computing. As a result, millions of software- and hardware-based sensors are beginning to be deployed, installed, and connected across our communities—in buildings (e.g., solar power and integrated control systems) and in our infrastructure networks (e.g., water, wastewater, smart-grid energy transmission networks), in our transportation systems (e.g., automated toll tags, signal timing, real-time route data for public transport) and our businesses and workplaces (e.g., virtualization of data and interconnected networks), and even on our personal bodies (e.g., smart phones, watches, and eyeglasses; social software; location-based services; touch and voice interfaces; etc.). In addition, all of these technologies are generating lots and lots of data—data which will provide the bedrock of many smart city technologies still to come. Smart cities will become computer in open air to deliver experience of cloud 9 living.

Electronic cards (known as smart cards) are another common platform in smart city contexts. These cards converge communication to a single platform for efficient integration of services. This will continue to lead innovations in collecting data, communicating data, crunching data, constantly correcting data, and correlating data.

Shifting to Zero Net Life The mega vision of a "Zero Concept" world will shift focus and promote development of products and technologies that "Equate to Zero" in real Hydrocarbon life. We will have driverless cars with zero emissions, zero accidents, and zero fatalities. Cities and buildings will want to be water and carbon neutral, where one could even have a recycled carbon-neutral beverage in a carbon-neutral brewery and spend a night in a carbon-neutral hotel. Organizations are already taking this as a key vision for business as seen by software development companies and are developing a "zero email" or "zero inbox."

Smart is the new green. What used to be called "green" is now called smart, and while green will continue to be important, there will be a shift toward "smart" products, which are intelligent and connected and have the ability to sense, process, report, and take corrective action.

Smart products will be everywhere around us from smart clothing, watches, and phones to smart buildings and smart cities. Smart city market stakeholders will take on one or more of the four main roles in the ecosystem: integrators (the end-to-end service provider), network operators (the M2M and connectivity providers), product vendors (hardware and asset providers), and managed service providers (overseeing management/operation).

1.1 Energy Trends of Future

The energy industry will converge with several related industries to develop efficient and environment friendly solutions. Key trends we will see in this area will be the smart grid, the future "energy Internet." We will see more micro grids and energy storage technologies leveraging electric car batteries second life, and possibly also, for the first time, space-based solar power satellites that beam electricity to the earth.

We will continue transitioning to a carbon-neutral energy supply system (i.e., replacing fossil fuels with more environmentally sustainable wind, solar, and geothermal energy solutions), greater efficiency in both our built environment (e.g., automated climate control systems, life-cycle materials) and work environment (e.g., virtual meetings, telecommuting, etc.), the rise of the sharing energy economy.

1.2 Network Trends and Shifts

All smart city technologies require a robust and reliable broadband network to operate. Today, high-speed broadband networks are as critical as roads, water, sewer, and electricity. News headlines routinely describe how health care, education, public safety, and government institutions are being transformed by new information technologies that rely on high-speed communication networks. People stay connected to the community through social network applications. Video streaming and gaming over the Internet is replacing broadcast television as the main form of home entertainment. Businesses must have access to broadband networks to be competitive. As noted by the Smart City Council, "To build a proper foundation for a Smart City, you must have high-speed broadband and citywide communications."

The growing use of wireless networks will require upgrades to existing network access points and construction of new access points such as cell towers. Even with such investments, however, there are concerns that the existing wireless spectrum will become overcrowded and new spectrum auctions will be required. Although the growth in wireless traffic is impressive, wireless broadband networks still carry far less traffic than fixed wire line networks. One reason is that the growth of mobile data actually increases demand for fixed networks. Mobile connections only travel over the air for a short distance to an access point where Internet traffic is aggregated and carried on high-capacity wired connections such as fiber.

Additionally, wired networks have more capacity and are more reliable than wireless technologies.

A smart city will need both robust wireless networks to provide the mobility and convenience that consumers desire and fiber networks to provide the high capacity, security, and reliability advantages that are desired by businesses, health care, and telecommuters just like electrical grid reliability.

1.3 Trends in Biophilia, Biomimicry, and Green Roofs and Streets

Through the use of technology and innovative design techniques, future biophilia and biomimicry can minimize the amount of polluted gray water runoff from the neighborhood and into our rivers and water sources. To create a naturally smart green infrastructure, planners balance the amount of runoff created and the streetscape's ability to retain and treat runoff, especially applying the following techniques:

- Permeable surfaces—While permeable pavement has been available for over 20 years, it has recently become an acceptable pavement surface in many communities. Even in communities which do not approve of the use of permeable pavement for vehicular travel lanes, it can serve as a mitigation technique for parking areas, bike, and pedestrian facilities.
- Pavement width—Reducing the overall space that is taken up by impervious surface can be the most effective means of limiting the amount of stormwater runoff created by a street.
- Streetscape treatment—Designing the streetscape to retain and treat the diminished amount of stormwater is the other critical piece to creating a green infrastructure. Using landscaping techniques like rain gardens, bioswales, plantings, and trees that thrive in the street environment will allow neighborhoods become more sustainable and a responsible member of the larger community.

1.4 Conclusion

Paradigm shifts will eliminate socioeconomic disparity and broaden broadband infrastructure; big data mines will be active mining; and the blending of technology and sustainability is a powerful combination that has the potential for new levels of efficiency, reshaping our communities, and improving the quality of people's lives.

Tomorrow's cities will be improved not only with ergonomics and technological comforts, but they will be Resilient & disaster proof. Buildings resistant to natural disasters and smart urban planning will force cities to develop new ways of energy sourcing and distribution, to enhance drainage systems, and to create purification systems for contaminated air, water, or soil.

Part I
Smart Buildings and Smart Technologies

Chapter 2
Building Systems Engineering

Anil Ahuja

2.1 An Introduction to Building Systems Engineering

The field of building systems engineering forms a cornerstone in modern building systems. Building systems engineering involves integration of such diverse technologies as civil, architectural, mechanical, electrical, electronic, and computer engineering. Engineering education has traditionally concentrated around single-discipline activities. Mechanical engineering degree courses, for example, have generally included some aspects of electrical and electronic technology, and, similarly, electrical engineering courses have usually incorporated some aspects of mechanical engineering. The reason for this is that both of these engineering schools have been patently aware of other requirements and have at least recognized the need for an appreciation of the other discipline.

Building system applications include heating, ventilation, air conditioning, and cooling (HVAC); lighting; power; security; fire and life safety; building automation; and intelligence, data, video, and audio communications of various kinds. Figure 2.1 shows a small commercial building with mechanical and electrical systems typically found in a building. The majority of industrial enterprises involved with the manufacture of building systems products require engineers who have skills both in electrical and mechanical engineering. They also now require their engineers to be competent in computer technology and software concepts. Computers are mathematical instruments of enormous value in the *science* of building system design. Further, one of the graphics capabilities of computers is drafting; computer-aided drafting is an extremely valuable tool in the *art* of building system design.

Most of our academic engineering study is of the science, and it requires disciplined study and must be learned in sequence. One cannot pick up a little electronics here and some little mechanics there and accumulate either understanding or utility. And using cookbook or code book formulas might be thought of as a black box type of approach, which nonetheless is perfectly viable. There are many other aspects of

A. Ahuja, *Integration of Nature and Technology for Smart Cities*,
DOI 10.1007/978-3-319-25715-0_2

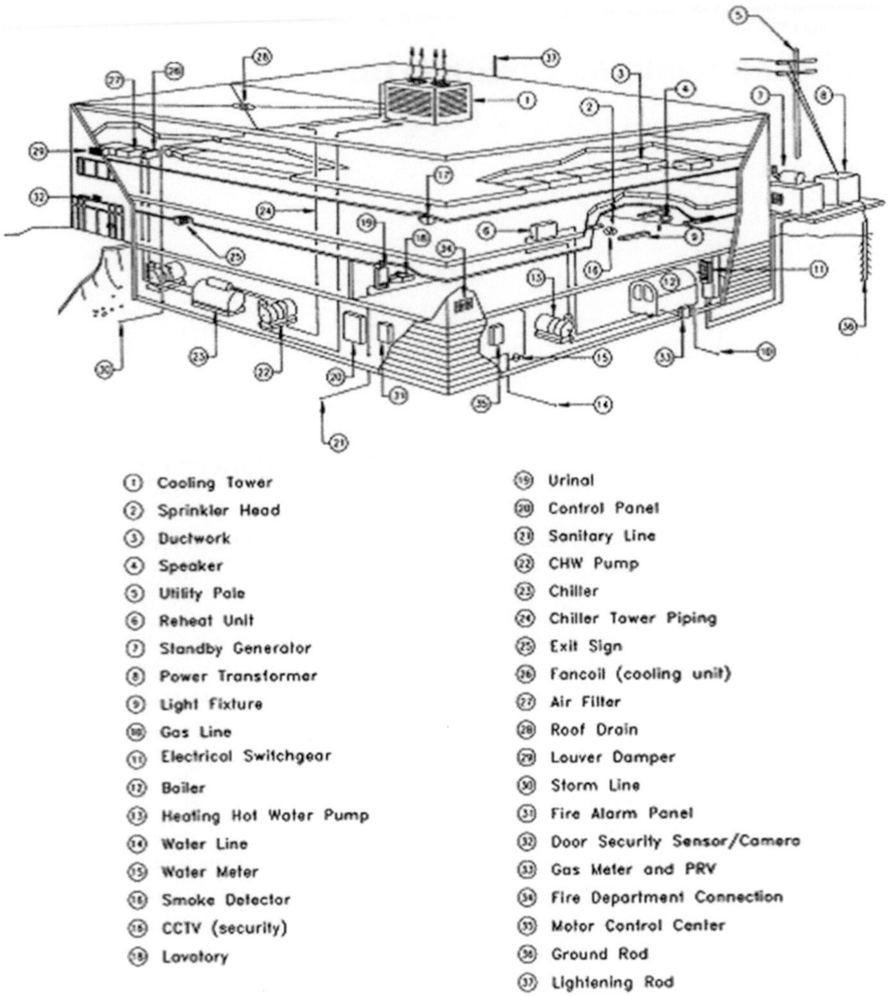

Fig. 2.1 Typical components of building systems in a commercial building

design called follow approaches that are not subject to logic at all. Consider *robustness, aesthetics, and customers' and managers' tastes and prejudices,* which are few of many.

The *science* of building systems engineering design is mathematical analysis of devices and systems. Engineers use the science to *predict performance* and to *size parameters.* The *art* of engineering design is the knowledge of everything else that can be useful in design and the skill to use that knowledge. The art *can* be piecemeal and can go on throughout life. A vital portion of the art of design engineering is a knowledge of building technology and economics.

The building electromechanical engineer needs a working familiarity with the operating characteristics, performance, and practical applications of all building systems. These systems might include complex natural phenomena and mechanisms; diverse hydraulic, thermal, hydronic, and chemical processes; integrated electrical and auxiliary system operations; or any combination of these. A firm grasp of the basic underlying principles is the key to success for the practicing engineer. An intuitive understanding of the behavior of building systems, matter, and energy enables integrated design engineers to imagine and understand the behavior of existing and proposed building systems to a degree that those untutored in the integration of systems cannot match.

It is not particularly surprising that mechanical engineers use an electrical analogy to describe many heat conduction phenomena, while electrical engineers often use heat and fluid flow analogies to enhance an understanding of electrical and electronics principles. Analogies such as electrical voltage to mechanical pressure, electrical resistance to mechanical friction, electric wire to mechanical pipe, and electric switch to mechanical valves are key to better understanding but should be used with caution. It is convenient to establish an analogy between electric and mechanical systems as an aid to comprehension, but it should not be carried so deeply as to lose its parallel relation.

For example, electric current is a measure of power flow and, as such, would correspond to water flow in a hydraulic system. The correspondence is not complete, however, since in the hydraulic system, the velocity of water flow varies, whereas in the electric system, the velocity of propagation is constant. A break in electrical circuit stops the flow of current entirely (voltage is still present but no current can flow), whereas in a hydraulic pipe break, the flow may actually increase until the pump fails or the supply of fluid is exhausted. Similarly, electrical switches are not quite like mechanical valves because they do not leak like a valve, and larger pipes carrying more water flow are not analogous to bigger wire carrying more current, because what limits the allowable current through any wire is how hot the wire gets.

It is clear that every engineer, of whatever discipline, involved in building design will be faced with the need to understand the operation of building equipment and systems in the practice of other disciplines. Further, all building system activities need instrumentation and control equipment that is largely electrical but monitors and controls mechanical equipment. Conversely, mechanical engineers will need to understand motors and motor drives that control and operate their machines. Electrical and mechanical engineers should be particularly concerned with developing system techniques.

This new breed of multidisciplinary engineer is at the heart of current building systems technological developments that continue to exhibit ever higher levels of sophistication and digital-based intelligence within the design of modern building electromechanical systems. This book develops the interrelationship between ambient environment and all individual systems in building systems engineering.

2.2 The Emergence of Building Systems Engineering: Historical Perspective

Throughout history, the development of more sophisticated tools has often been associated with a decrease in dependence on human physical energy as a source of effort. Generally, this is accomplished by control of nonhuman sources of energy in an automated fashion. The building system revolution represented a major thrust in this direction.

A historical path that architecture has followed begins from simple shelters, well adapted to their climate. With the emphasis on energy control, architecture progressed to enclosed, tightly controlled internal climate structures that strive to ignore the conditions outside. The accumulation of energy conservation measures in the buildings resulted in poor air quality in some buildings, created power quality problems, and labeled some buildings as suffering from "sick building syndrome." Present architecture boasts of "intelligent" buildings, equipped with microcomputers and integrated systems—praised for their low annual energy consumption, clean indoor air, and superior power quality.

The basic transistor was invented in 1948, but it was not until the development of the microprocessor that building system design engineers had taken recourse to a much more active involvement in the applications and utilization of microprocessor and integrated circuit devices in intelligent building electromechanical design. The trend in modern construction, except for small or simple structures, is clearly to use integrated system design with centralized monitoring and control of building systems. The subsystems in the 1980s almost always included in a building control system were HVAC, energy management, and lighting control. Inclusion of security, life safety (fire alarm, fire control and suppression, plus emergency aspects of vertical transportation), material handling, maintenance management, data/audio/video communications, and some aspects of office automation is the trend of future. The integration between mechanical, electrical, and computer technology has since become known as mechatronics, and it is now becoming recognized as a curriculum topic in its own right.

A new awareness of the sources, characteristics, and limitation of energy supplies is resulting in new directions in building design, away from many practices of the recent past. This changing pattern in the use of energy sources has been accompanied by a trend in design: integrating building systems so that building's external skin and internal organization work with the surrounding climate for "natural" (passive) heating, cooling, and lighting. In most cases, a new equipment or system integration makes it possible to perform a familiar task in a somewhat new and different way, typically with enhanced efficiency and effectiveness and sometimes with increased understandability as well.

When building systems became so interrelated that it was no longer possible for a single individual to design them, and a design team was then necessary, a host of new problems emerged. This is the situation today. To cope with this, a number of methodologies associated with systems design engineering have evolved. Through these it has been possible to decompose large design issues into smaller component subsystem design issues, design the subsystem, and then build the complete system as a collection of these subsystems. Even so, problems remain. Just simply connecting together the individual subsystems often does not result in a system that performs acceptably, either from an efficiency or from an effectiveness perspective. This has led to the realization that systems integration engineering and systems management throughout an entire system life cycle are necessary.

The purpose of building systems integration is to design, install, and operate the many systems that comprise a facility so that their performances are not counterproductive. The output of an individual system should not jeopardize the performance, protection, and reliability of other systems.

2.3 Basic Vocabulary of Building Systems

Modern building support systems often bring information and signals beside power and gas utilities and process information data and signals at one place similar to producing heating and cooling in one place. The building system distribution structure then distributes them to other building spaces according to their respective needs. A substantial portion of building system design is dealing with the following four Cs to distribute all this flow:

Cessation is achieved with barriers, insulators, breakers, doors, curtains, glass, and switches.

Constraint is achieved through resistors, inductors, capacitors, orifices, pipes, stainers, filters, valves, fittings, insulation, connectors, conductors, and ducts.

Conduction of flow is achieved through the use of conduits/pipes/ducts, conductors, vanes, windows, and doors.

Control of flow is achieved through valves, protectors, controllers, transducers, sensors, and remote signals. What flow? Here is a list:

- Gas/vapor/air
- Heating/cooling/fire protection/domestic-liquid
- Sewage/sanitary/storm
- Electric current/magnetic flux
- Microorganisms/odors/pollens/ions/dirt
- Humans/heat
- Noise/vibration
- Information/light

An integrated system designer has to have a vocabulary of other discipline-specified materials, equipment, accessories, product features, and assemblies required for integration beside distribution. The system designer selects and combines the items in his vocabulary into the building system's design. Expanding your building system design vocabulary is a lifetime program.

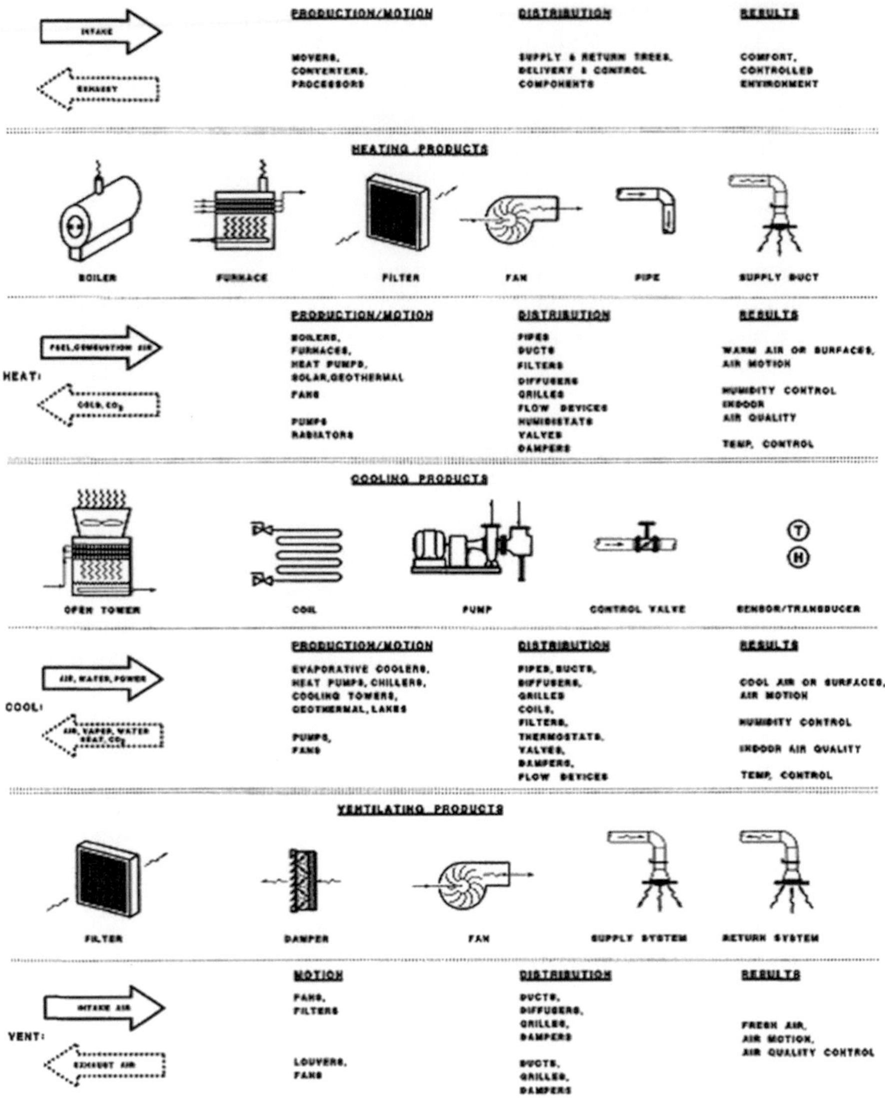

Fig. 2.2 Basic building systems: tasks and components. Heating, ventilating, air conditioning, and cooling (HVAC)

See Figs. 2.2 and 2.3 for integrated building system components: A building system designer's vocabulary should as a minimum include:

Energy and fuel services

Fuel service, storage, handling, piping, and distribution
Electrical service, metering, and distribution equipment
Gas service, metering, and distribution piping

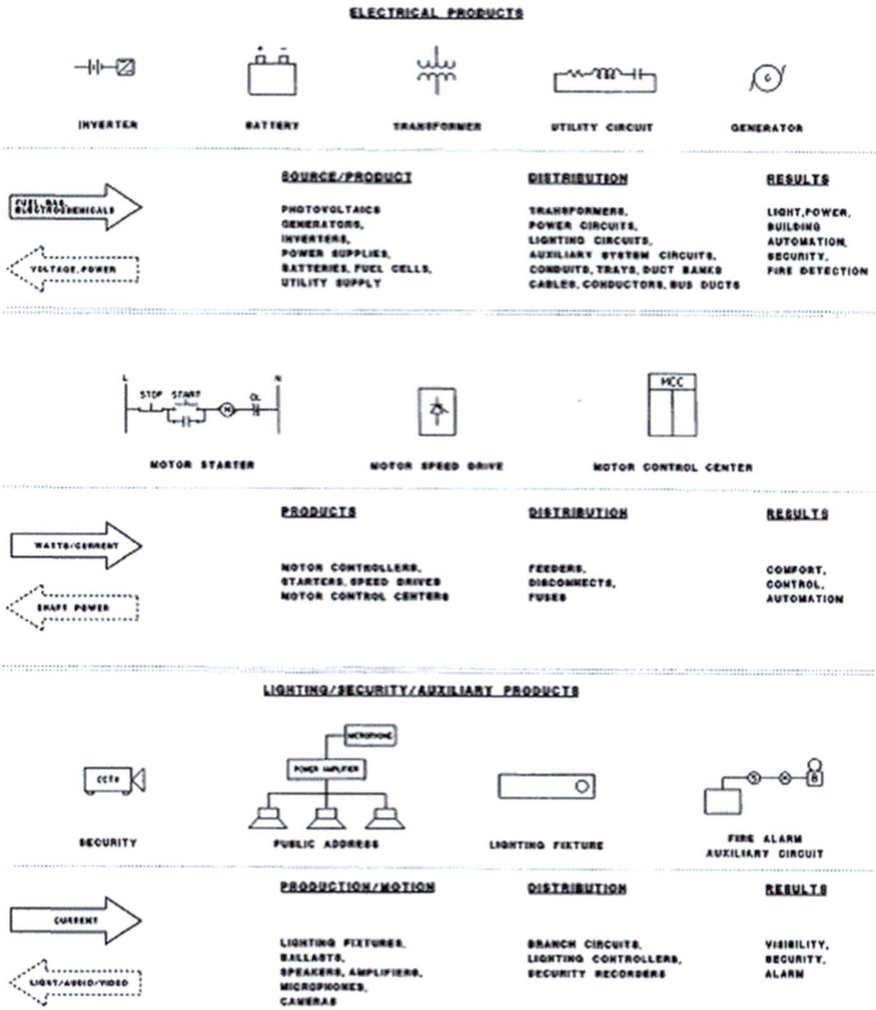

Fig. 2.3 Basic building systems: tasks and components. Electrical

Heat-producing equipment

Boilers and furnaces
Steam–water converters
Heat pumps or resistance heaters
Makeup air heaters
Heat-producing equipment auxiliaries

Power-producing equipment

Generators, batteries, fuel cells
Photovoltaics (solar cells)

Refrigeration equipment

Compressors, chillers, or absorption units
Cooling towers, condensers, evaporators
Refrigeration equipment auxiliaries

Heat distribution equipment

Pumps, valves, piping, piping insulation, etc.
Terminal units or devices, radiators, coils

Cooling distribution equipment

Pumps, piping, valves, coils, condensate drains, etc.
Terminal units, mixing boxes, diffusers, grilles, etc.

Air treatment and distribution equipment

Air heaters, humidifiers, dehumidifiers, filters, etc.
Fans, ducts, duct insulation, dampers, etc.
Exhaust and return systems
Air quality control equipment

Power distribution equipment

Transformers, switchgear, panelboards
Protective devices, switching devices, and PLCs
Feeders, starters, drives, auxiliary power equipment
Power conditioners, power supplies, lightning, and surge protection equipment

Lighting equipment

Lighting fixtures, reflectors, ballasts, daylighting systems
Lighting sources (HID, fluorescent, incandescent)
Lighting controllers, occupancy sensors, dimmers

Fire protection equipment

Standpipe and hose system
Sprinkler system—wet, dry, deluge, or preaction
Dry chemical system
Foam system

Auxiliary system equipment

Audio/video and intelligence communication equipment and media
Wireless network gear, Fiber-optic connectors, receivers, and transmitters
Security system equipment, sound masking, and paging system equipment

System and controls automation

Terminal or zone controls
System program control
System integrators, programmable controllers, network
Protocol, alarms, and indicator system

Building site resources, envelope, and environmental systems

Passive or radiative heating and cooling equipment
Windows, glazing, and daylighting system
Solar and geothermal heating/cooling systems

Building construction and alteration

Mechanical, electric and IT network space
Chimneys and flues
Building insulation
Solar radiation controls
Acoustical and vibration treatment
Distribution shafts, machinery foundations, furring

Some basic nonelectrical engineers need to know: What allows electricity to flow and how can electricity make a magnet? Electricity flashes through copper wire but halts when it bumps up against rubber tubing. Such examples abound in nature—certain substances are good conductors, while others block even the weakest electric current. Whether a material transmits electricity is dictated by the material's atomic structure. How easily a material channels electricity depends on how freely electrons move through it.

Conductors—chiefly metals such as iron, nickel, silver, and copper—have "loose," or free, electrons. Not leashed to any particular atom, these free electrons roam through their atomic neighborhoods, orbiting first one atom and then another. When a conductor is connected to a battery, the electric field organizes the aimless movement of the electrons into a steady flow. For this reason, metals are excellent carriers of electricity. Insulators, however, have few, if any, free electrons. Atoms in materials such as leather, glass, plastics, and rubber keep a tight rein on their electrons. An absence of freewheeling charges keeps insulators from conducting current.

A semiconductor is a crystalline material that conducts electricity but not as well as metals; it also resists electricity but not as well as many insulators. In general, semiconductors' electrons are tightly bound to their nuclei. But if a few atoms of antimony—with a surplus of electrons—are incorporated in a semiconductor such as silicon, the free electrons will give it a negative charge. These properties make semiconductors useful in transistors to amplify current, to block current, or to let current flow in only one direction. In a typical NPN transistor, a layer of positive (P) semiconductor, the base, is sandwiched between two negative (N) layers, the emitter and the collector. When, for example, a small signal from an intercom is channeled through the base, the movement of electrons amplifies the signal.

The discovery that electric currents generate magnetic fields led scientists to develop a magnet using electricity that can be turned on and off. Such electromagnets can consist of a battery attached to a coiled wire—a solenoid—wrapped around a ferromagnetic core (usually iron). The magnetic field produced by the electric current in the wire magnetizes the piece of iron. As long as current flows through the wire, the electromagnet behaves like a permanent magnet.

Magnetic field lines arc from the electromagnet's north to its south pole—usually at right angles to the flow of the current, in keeping with electromagnetic laws. If the current switches direction, the magnet's poles flip and the field lines reverse as well. The overall shape of the magnetic field does not change, however. The pattern of field lines remains the same unless the shape of the wire itself changes. Motors, generators, and other electrical systems operate by electromagnetism.

Chapter 3
Basic Smart Building Systems

Anil Ahuja

The building system of a facility is composed of various types of MIE (mechanical/ electrical) equipment. Systems supporting large installations with closely controlled indoor environments can be very complex. The building system design art has changed in a way that the fine form has not. The architect uses the same basic techniques that have been used for centuries. Meanwhile, the system design engineer has traded in board drafting and hand calculators for a computer and design software. This gives the engineer the capability to analyze libraries of data, perform mathematical calculations, and simulate integrated systems.

Because of the large number of interrelated factors in a system, there can be many solutions to the same building problem, all of which will satisfy the minimum requirements, yet some will be dull and pedestrian while others will display ingenuity and resourcefulness. No single design is the correct one, and for this very reason, it is not entirely desirable to solve a system problem with a step-by-step technique. However, this technique is a good avenue of approach for the uninitiated who lacks the experience necessary to view an entire solution.

The challenge is to involve, but not enslave, the users in the management of their environment. Integration of functions is the answer. In a properly designed system, MIE equipment operates in an integrated manner, so that the output of individual devices will not interfere with or be counter to the performance of the others. Counterproductive performance results in incorrect indoor conditions and wasted energy. Also, maintaining air, power, and water quality is part of internal basic building system design tasks. For example, the protection of sensitive electronic equipment from transients and other interference caused by MIE equipment operation is part of the building system design task.

A. Ahuja, *Integration of Nature and Technology for Smart Cities*,
DOI 10.1007/978-3-319-25715-0_3

3.1 The Lighting System

Almost everything scientists know about the universe comes from observing and analyzing light. Yet little was known about light itself until fairly recently. In the seventeenth century, two compelling theories about the nature of light emerged. The corpuscular theory, championed by Sir Isaac Newton, held that light consisted of tiny particles called corpuscles. Another theory proposed that light was a wave, moving through space in much the same way that ripples move across a pond. Although nearly every discovery about light's behavior during the next 200 years seemed to support the wave theory and discredit the corpuscular theory, the advent of quantum physics in the twentieth century reconciled the two theories: Depending on how it is measured and observed, light may assume the characteristics of either a particle or a wave. See Table 3.1 for all light sources and their light intensity.

Light has five distinct properties: propagation, reflection, refraction, diffraction, and interference. Propagation refers to the transmission of light in straight lines. Reflection causes light to bounce off polished surfaces such as mirrors. Light refracts, or bends, when it travels from one substance to another, for example, from air through a glass lens. Light waves also will bend around an obstacle's edges, a phenomenon known as diffraction. In interference, intersecting light waves alter each other as they meet. Taken together, these properties explain the workings of devices as varied as high-intensity discharge lighting commonly applied in buildings, lasers, and holograms.

How much light does it take for people to see? We know from experience that it takes more light to see small objects and tasks of low contrast than larger, higher-contrast tasks. As our eyes age, they require more light. In fact, a 50-year-old gets about half as much light on the back of the eye (retina) as a 20-year-old. How much light is required to illuminate work spaces or to perform tasks? The Illuminating Engineering Society of North America (IESNA) provides light level (illuminance) recommendations that are consensus standards agreed up on by lighting professionals.

How does a light bulb glow? The electricity that illuminates lamps——and that runs televisions and appliances——consists of flowing electrons, or current electricity. When a free electron is triggered to move, it occasionally will bump against an atom, exciting the atom, which means it gives some of its energy to the atom. The atom then releases this extra energy as electromagnetic radiation and propels other electrons into action. As electrons flow through the metal filament of an incandescent light bulb, the heating of the filament causes it to give off electromagnetic radiation and glow white hot.

In fluorescent lamps, a current flows through a gas instead of a filament. As the current travels through the gas tube, it causes the gas to give off ultraviolet energy, which excites the phosphor coating inside the tube, triggering a chain reaction that releases electromagnetic radiation as visible light.

Lighting quality is complex because it considers visual aspects that are highly subjective and not easily quantified. We can, however, understand certain

Table 3.1 Luminances for various light sources

Light source	Approximate average luminance (cd/m^2)
Natural light sources:	
Sun (at its surface)	2.3×10^9
Sun (as observed from the earth's surface at the meridian)	1.6×10^9
Sun (as observed from the earth's surface at a horizon)	6×10^4
Moon (as observed from the earth's surface at bright spots)	2.5×10^3
Clear sky	8×10^3
Overcast sky	2×10^3
Lightning flash	8×10^{10}
Combustion sources:	
Candle flame (sperm)	1×10^4
Kerosene flame (flat work)	1.2×10^4
Illuminating gas flame	4×10^3
Welsbach mantle	6.2×10^4
Acetylene flame	1.1×10^5
Incandescent Lamps:	
A carbon filament (3.15 lm/w)	5.2×10^5
Tantalum filaments (6.3 lm/w)	7×10^5
Tantalum filaments (vacuum lamp 10 lm/w)	2×10^6
Tantalum filaments (gas filled lamp 20 lm/w)	1.2×10^7
Tantalum filaments (projection lamp 26 lm/w)	2.4×10^7
RF (radiofrequency) lamp	6.2×10^7
Black body at 6500 K	3×10^9
Black body at 4000 K	2.5×10^8
Black body at 2042 K	6×10^5
60-W inside frosted bulb	1.2×10^5
10-W inside frosted bulb	2×10^4
Tungsten–halogen sources:	
3000KCCT	1.3×10^7
3200 KCCT	2.3×10^7
3400KCCT	3.9×10^7
Fluorescent sources:	
T-8 bulb 265 rnA	1.1×10^4
T-12 bulb 430 rnA	8.2×10^3
T12 bulb 800 rnA	1.1×10^4
T-12 bulb 1500 rnA	1.7×10^4
T-17 grooved bulb 1500 rnA	1.5×10^4
Electroluminescent sources:	
Green color at 120 V 60 Hz	27
Green color at 600 V 400 Hz	68
Carbon arc sources:	
Plain carbon arc	1.5×10^8
High-intensity carbon arc	1.0×10^9

(continued)

Table 3.1 (continued)

Light source	Approximate average luminance (cd/m²)
Enclosed electric are sources:	
High-pressure mercury	1.8×10^6
High-intensity short arcs mercury	2.4×10^8
	$(4.3 \times 10^9$ peak)
The xenon short arcs	1.8×10^8
Clear glass neon tube	1.6×10^3
Clear glass blue tube	8×10^2
Fluorescent tubes	
Daylight and white	5×10^9
Green	9.5×10^3
Blue and gold	3×10^3
Pink and coral	2×10^3

Reprinted with permission of the IESNA, 120 Wall Street, 17th floor, New York, NY 10005. Taken from the *IESNA Lighting Handbook, 8th Edition*

quality characteristics. For example, ceiling reflections detract from lighting quality by obscuring task details by reducing contrast. It is sometimes called a reflected glare and is most noticeable from luminaries located in front of and above the viewing task.

Lamp color also affects lighting quality. Recommendations regarding pleasant combinations of lamp color, temperature, and illuminance are changing and are best left to building occupant preferences. However, we do know that when lamps of good color rendering are used, illuminance may be lowered to achieve equivalent brightness and visual clarity. When upgrading from cool white lamps to higher CRI T8 lamps, this effect will be noticeable by employees who may respond that the new lamps are "too bright."

The lamp flicker can also reduce lighting quality. The predominant source of a lamp flicker is from fluorescent lamps operating on magnetic ballasts. The lamps turn on and off 120 times a second, producing distraction, eyestrain, and headaches. The flicker is especially noticeable at high light levels, such as industrial inspection lighting. Electronic ballasts that operate fluorescent lamps at high frequency can reduce the flicker to an imperceptible level. To improve lighting quality, it is important to balance office lighting for visual performance and for visual comfort. Too much contrast will cause workers to be restless. Too little contrast (flat) causes a loss of detail and things appear dull. In retail lighting applications, the lighting lacks the power to attract, and sales will be reduced. Balanced lighting is achieved by avoiding over-diffuse or too strongly directional lighting.

3.1.1 Lighting Terminology and Common Definitions

Lumen Output The lumen is a unit of luminous flux that is a measure of the total light from a source.

Luminous Efficacy

The efficiency of light sources, calculated by dividing the light output (in lumens) by the power input (in watts), commonly called lumens per watt (lm/w). The higher the lm/w, the more efficient the light source.

Average Rated Life
This is the median value of life expectancy assigned to a lamp, in hours, at which half of a large group of lamps have failed. Any particular lamps or group of lamps may vary from the published rated life. For discharge light sources such as fluorescent and HID lamps, the average rated lamp life is affected by the bum cycle or the number of starts and the duration of the operating cycle each time the lamp is started.

Lumen Depreciation
Also known as lumen maintenance. All light sources used for interior lighting lose their ability to produce light as they age. Lamp lumen depreciation (LLD) is a dimension less decimal value representing the percent of initial lumens and is one of several light loss factors used in lighting calculations. LLD can be calculated by dividing the mean (design) lumens by the initial lumen rating. The initial lumens are measured after a burn-in time of 100 h. Values of initial and mean (design) lumens may be found in manufacturers' lamp catalogs.

For example, a 32-W T8 lamp has an initial lumen rating of 3200 lm (255 candle power) and a design lumen rating of 2800 (223 candle power). LLD = 2800/3200 = 0.875. This means that the T8 lamp will retain 87.5 % of its initial light output after 40 % of its average rated life. Lumen depreciation is affected by the ballast used, line-voltage tolerances, and burn cycles.

Color Temperature
The color temperature of a lamp is described in terms of its appearance (when lighted) to the eye, whether it appears "warm" or "cool." It is measured on a Kelvin scale ranging from 1500 K, which appears red–orange, to 9000 K, which appears blue. Light sources lie somewhere between these two, with those of higher color temperature than 4100 K appearing "cool" and those of a lower color temperature than 3100 K being "warm."

Color Rendering Index (CRI)
Color renditions describe the effect a light source has on the appearance of colored objects. The color-rendering capability of a lamp is measured as the CRI. The higher the CRI, the less distortion of the object's color by the lamp's light output. The scale ranges from 0 to 100. A CRI of 100 indicates that there is no color shift as compared to a reference source. The lower the CRI, the more pronounced the shift may be. CRI values should only be compared among lamps of similar color temperature.

Foot Candles
The quantity of light that falls on a work surface is measured in foot candles. One foot candle is equal to one lumen per square foot of area. An illuminance meter is a useful tool to measure the amount of light in work spaces. It is important to understand that the foot candle measure indicates only a level of illuminance. It does not measure the amount of energy to produce that level of light and does not measure the quality of the light produced.

Visual Comfort Probability (VCP)
A rating of lighting systems that is expressed as a percentage of people who, when viewing from a specified location and in a specified direction, will find the lighting system acceptable in terms of discomfort glare. The minimum recommendation for office interiors is 70, and the recommendation for computer applications is 80.

3.1.2 Lamps

The term "lamp" is used to describe light sources commonly called light bulbs and tubes. The total light from a light source is measured in lumens. Lamps are now labeled with measured lumen ratings and efficiency ratings (efficacy). Incandescent, fluorescent, and high-intensity discharge (HID) lamps are the most common lamps used for building lighting systems.

Incandescent Lamps The oldest practical lamp type, the incandescent lamp is what most of us grew up within our homes. They are inexpensive and available in hundreds of sizes, shapes, and wattages and are easily dimmed. However, they are very inefficient——as low as 8 lm/w——have short lamp life, and contribute to building heat.

Incandescent lamps are voltage sensitive, with lamp life, lumen output, and wattage dependent on the applied voltage. Those whose buildings have higher than normal voltage or who want to decrease relampings sometimes use higher voltage rated lamps, usually 130 V. If 130 V lamps are used on 120 V circuits, lumen output will be lower and life will be longer.

Tungsten–halogen lamps are more efficient than standard incandescent lamps. A halogen fill gas combines with the tungsten molecules that boil off the filament. The resulting halogen cycle increases the lm/w, produces whiter light, has longer life, and lowers the LLD.

Fluorescent Lamps

The fluorescent lamps are the most commonly used lamp type in commercial and industrial applications. Fluorescent lamps are an electric discharge source in which light is generated when ultraviolet (UV) energy from a mercury arc strikes a fluorescent phosphor on the inside surface of the tube. The tube contains mercury vapor at low pressure and a small amount of inert argon gas (or krypton in reduced-wattage lamps). Characteristics of fluorescent lamps are long life (12,000–20,000 h), high efficacy (75–90 lm/w), and excellent color rendering, especially with the newer rare-earth (RE) lamps. Fluorescent lamps are, however, temperature sensitive and their rated life depends on the hours per start. A ballast is required to properly start and operate fluorescent lamps.

Compact Fluorescent Lamps (CFLs)

The generic name for a family of single-ended fluorescent lamps of folded or bridged tube design with high CRI and long life (10,000 h). Originally designed as preheat lamps, with the starter built into the base, the lamps are now available in rapid-start models. They are often used as alternatives to incandescent lighting. Configurations now include twin tubes, quad tubes, and triple tubes in both preheat and rapid-start models.

High-Intensity Discharge (HID) Lamps
Lamps are classified as electric arc discharge lamps that operate under high pressure and generate their light directly from the arc. The arc is contained in a small arc tube that is enclosed in a larger outer glass bulb. The outer glass may be clear or coated on the inside with a fluorescent phosphor. Included in this classification are mercury vapor, metal halide, and high-pressure sodium lamps (HID lamps do not operate instantly; they require time to strike and, when power is removed, longer to cool down and restrike). The buildings generally require a backup lighting system for HID lighting for public safety. Advantages include high lumen ratings and long life.

The mercury vapor (MV) lamp is the oldest HID source and is now considered obsolete. The disadvantages of MV lamps include a poor color rendition, lumen depreciation, and high mercury content.

Metal halide (MH) lamps are now considered the replacement lamp for MV lamps, which they closely resemble. In the metal halide lamp, the arc tube contains, in addition to smaller amounts of mercury, additives called metal halides that provide a brighter, whiter light by improving both lumen and color performance. It is important for safety reasons to prevent MH lamps from reaching "nonpassive failure" by turning these systems off at least 15 min every week, group relamping before the end of rated life, and operating them in the correct position and on matching ballasts. Recent advances in MH lamps provide better color consistency. Efficacy ratings are in excess of 100 lm/w, and these lamps have long restrike times, up to 15 min. Applications for MH lamps include commercial lighting interiors, especially high-ceiling applications; sports lighting; and building facades.

High-pressure sodium (HPS) lamps are the primary source for industrial lighting, highway, and street lighting. HPS lamps have a characteristic yellow color, high efficacy——60–140 lm/w depending on wattage——and a long life rating of 24,000 h. Standard HPS lamps cycle at the end of their life, indicating the need for replacement. Recent advances in HPS lamps include higher CRI models and models that do not cycle at an end of life. A double arc-tube HPS lamp is available for safety and security applications.

3.1.3 LEDs

Unlike incandescent lamps, LEDs are not inherently white light sources. Instead, LEDs emit nearly monochromatic light, making them highly efficient for colored light applications such as traffic lights and exit signs. However, to be used as a general light source, white light is needed. White light can be achieved with LEDs in three ways:

- Phosphor conversion, in which a phosphor is used on or near the LED to convert the colored light to white light

- RGB systems, in which light from multiple monochromatic LEDs (e.g., red, green, and blue) is mixed, resulting in white light
- A hybrid method, which uses both phosphor-converted (PC) and monochromatic LEDs

The potential of LED technology to produce high-quality white light with unprecedented energy efficiency is the primary motivation for the intense level of **research and development** currently supported by the US Department of Energy.

There are many white LED products available on the market, and the number continues to grow, with new generations of devices constantly emerging. While many of these products perform quite well, their quality and energy efficiency can vary widely. There are standards, test procedures, and ENERGY STAR® criteria that can enable buyers to make informed decisions when evaluating LED lighting.

3.1.4 Ballasts

Ballast is required to start and operate discharge lamps, fluorescent, and HID. All fluorescent ballasts perform two functions. They provide the right voltage to start the arc discharge, and they regulate the lamp current to stabilize light output. In rapid-start ballasts, a third function is to provide the energy to heat the electrodes. Fluorescent ballasts are provided to operate fluorescent lamps in the following ways:

Preheat
Lamp electrodes are heated prior to the application of a high starting voltage that initiates the arc discharge. Preheat operation is characterized by lamp "flickering" when starting. The starting electrode voltage is applied through a "starter," a thermal switch that, when it opens, applies the high starting voltage across the electrodes. No power is applied to heat the cathodes during operation. Lamps of less than 30 W are usually operated this way.

Rapid Start
Lamp electrodes are heated prior to and during operation. The ballast has two windings to provide the proper low voltage to the electrodes during starting and operation. Rapid-start operation is characterized by smooth starting and long lamp life.

Instant Start
Lamp electrodes are not heated. Instant-start ballasts provide a high open-circuit voltage across the unheated electrodes to initiate the arc discharge. Instant-start operation is more efficient than rapid start, but as in preheat, lamp life is shorter. Eight-foot "slimline" lamps are operated in the instant-start mode.

The ballasting function has traditionally been reliably accomplished with electromagnetic units that consist of a core of magnetic steel laminations surrounded by two copper or aluminum coils. All ballasts have electrical and magnetic losses. These internal losses and the ability of the ballast to operate the lamps efficiently contribute to the overall efficiency of a ballast. The efficiency of magnetic ballasts has been improved by using low-loss magnetic material and copper windings, resulting in lower internal losses.

Ballast efficiency is regulated by the federal appliance standard that prohibits the manufacture of ballasts that do not meet a minimum ballast efficacy factor (BEF). BEF for a specific lamp/ballast combination is calculated by dividing the percent rated light output by the measured input power in watts. The percent rated light output is found by multiplying the ballast factor by 100 %. BEF can be confusing because it is meant to compare the performance of ballasts on a specific lamp and is not of particular value in evaluating efficiency opportunities.

Ballast factor (BF) is the ratio of the lamp lumen output on a commercial ballast to the lamp's rated light output. Fluorescent lamps are rated in lab tests using a loss less, perfect reactor ballast, so ballast factor is needed to derate the catalog rating of lamps.

The efficiency of fluorescent ballasts can be improved beyond the use of energy-efficient magnetic ballasts. Concerns for lighting efficiency have prompted the introduction of electronic ballasts. When fluorescent lamps are operated by an electronic ballast at high frequency, they convert the input power to light output more efficiently. The lm/w of the lamp/electronic ballast combination is increased, which means either producing more light for the same power or producing the same light with lower power. Generally, the lower the ballast factor, the lower the input watts. Proper selections of all electronic ballast parameters are important for successful lighting applications. The total system input watts can actually be less than the total of the lamp wattage. In fact, when a 32-W T8 lamp is operated at high frequency, it consumes only about 28 W, making the system input about 30 W. The actual watt input is controlled by the ballast factor, which can range from 47 to 130 %.

3.1.5 Luminaires

Unlike fixtures, luminaires are complete lighting units consisting of lamps and ballast to convert power to light, lamp holders, an optical system, a means for connecting to power, and a housing. Each luminaire manufacturer provides light distribution data on the integrated assembly called luminaire photometric.

The *photometric* data for a luminaire provides the light distribution in the form of a polar graph and a table whose values represent the variation in candlepower of a luminaire in a given reference plane. This light distribution is the heart of a photometric report generated from laboratory testing by the luminaire manufacturer or an independent testing laboratory. The report also provides luminaire efficiency, zonal lumen output, a coefficient of utilization (CU) tables, and spacing criteria (SC).

Parabolic luminaires use large-cell louvers in a parabolic shape. The resulting light distribution reduces the glare, controls light output, and produces high aesthetic appeal. Paracube louvers are small metallic-coated plastic squares; one-inch squares

Luminaire Efficiency
Defined as the percentage of lamp lumens that leave the luminaire. This rating quantifies the optical and thermal effects that occur within the luminaire under standard test conditions. Judging luminaires on efficiency alone is a faulty ideal, since the most efficient luminaire is a bare lamp.

Coefficient of Utilization (CU)
The percentage of lamp lumens that is received on the work plane and is a function of luminaire efficiency, room geometry, and room surface reflectance. CU values are used to evaluate how effectively a luminaire delivers light to the work plane in a given space. It is inappropriate to judge a luminaire only by its CU, since the CUs are a function not only of efficiency, but also of its application.

Spacing Criteria (SC)
Provides the information regarding how far apart luminaires may be spaced to maintain uniform lighting. To use the spacing criteria, multiply the net mounting height by the spacing criterion's value. The resulting number represents the maximum center-to-center distance that the luminaires may be spaced.

The Luminaire Optical System
Includes the lamp cavity and diffusing media and includes one or more of the following components: reflectors, refractors, lenses, baffles, or louvers. Reflectors redirect light by using the principle of reflection. A refractor is a component that redirects light by refraction. Lenses are made of a transparent or translucent material that has a prismatic configuration on the bottom surface to alter the directional characteristics of light passing through it. Polystyrene lenses are the least expensive but yellow owing to the ultraviolet (UV) radiation from lamps. Lenses made from acrylic are light stabilized and do not discolor. Translucent sheets of milky white plastic, called diffusers, scatter light uniformly in all directions below the ceiling plane, reducing source brightness and shielding the lamps. And baffles are one-way louvers made of metal or plastic.

Louvers
Consist of baffles, arranged in a geometric pattern, that control the light distribution and shield the source from view at certain angles. Products range from small-cell plastic louvers to large-cell aluminum louvers.

are the most common size. Small-cell louvers have high visual comfort probability (VCP) but sacrifice luminaire efficiency and spacing criteria. The highly polished surfaces used in some louvers and reflectors are called specular. Semispecular finishes reflect some of the light directionally, with some amount of diffusion.

Luminaires are also classified according to the manner in which they control or distribute their light output. They can be direct (downward), indirect (upward), or direct–indirect.

Direct luminaires can be open or shielded. Shielded luminaires use lenses, baffles, or louvers. Downlights are direct ceiling luminaires that direct all the light downward. They can be recessed, surface mounted, or suspended. Recessed units include HID downlights and a wide range of fluorescent troffers. Recessed troffers are luminaires mounted within the ceiling so that the bottom surface is in the same plane as the ceiling. Some troffers are designed to lie on the tee-bar flanges of a suspended-grid ceiling system. Surface-mounted troffers are attached directly on the ceiling and project below the ceiling by their depth. Pendant-mounted units, such as direct–indirect, are supported from the ceiling structure by stems or pendants that position the luminaire below the ceiling. Air slot is a term applied to some recessed luminaires that supply or return comfort air. This is accomplished through slots. A static luminaire does not handle comfort air.

Indirect luminaires radiate light up to a reflecting ceiling. Indirect lighting systems usually use luminaires suspended from the ceiling by aircraft cables. Cove lighting, wall-mounted units, and furniture-top units are also considered indirect. Lighting offices with indirect luminaires results in higher, more uniform brightness on both ceilings and walls, and computer screens' glare is reduced significantly. Shadows are reduced from partitions and under-shelf cabinets. The success of indirect systems depends on maintaining a high-ceiling reflectance. Recent designs use T8 lamps and electronic ballasts in a low-profile housing with lenses and special reflectors to achieve high luminaire efficiency and a broad distribution, while allowing mounting close to the ceiling.

Direct–indirect luminaires combine the efficiency and high coefficient of utilization of direct luminaires with the uniformity and glare control of indirect luminaires. Some office and school systems are designed for an equal balance of direct downlight and indirect up light.

Luminaire Light–Loss Factors (LLFs)
The light lost due to reflecting surfaces degrade over time. This degradation of the surface is non-recoverable. In contrast, luminaire dirt depreciation (LDD) is the recoverable light loss factor that describes how light is lost from the initial illuminance provided by clean, new luminaires compared with the reduced illuminance that will be provided after dirt collects on the reflecting surfaces at the time when it is expected that cleaning will be done.

3.1.6 *Lighting and Energy*

To understand the contribution of lighting to building cooling loads, it is important to know that three main energy conversions occur during the process of generating light by a fluorescent lamp. Initially electrical energy is converted into kinetic energy, exciting UV radiation. This UV energy in turn is converted to visible energy by the lamp phosphor. During each of these conversions, some energy is lost so that only small percentage of input is converted into visible radiation. Whereas LED Light use mixing colors to create white light as shown in Fig. 3.1.

Buildings waste lighting energy because of hearsay beliefs like: "Continuously operating fluorescent lights is cheaper than turning them off for brief periods." Or "Lights shouldn't be turned off because it shortens lamp life and increases maintenance costs." Actually, turning off fluorescent lights saves energy, extends overall lamp life, and reduces replacement costs. Although the average rated life of fluorescent lamps is shortened by switching, calendar life is lengthened. Calendar life is

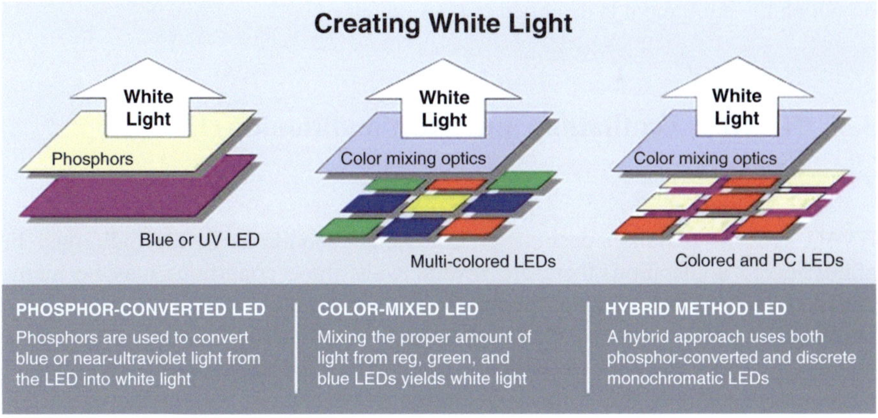

Fig. 3.1 LED technologies

the time between lamp changes and includes the time the lamp is off. For example, standard T32 rapid-start lamps operated continuously result in a rated lamp life of 32,000 h (calendar life of 3.7 years). Turning off T32lamps for 12 h each day decreases the average rated lamp life to 29,000 h, but calendar life is extended to 6.6 years. The belief that there is a high inrush current during the starting is unfounded and is not even of any theoretical concern if an electronic ballast is used.

3.2 Power Systems

Many factors should be considered in the design of electric power distribution system for a modern commercial/industrial building. Factors that influence the system configuration are characteristics of the load, quality of service available and required, the size of the building, and cost. Basic arrangements used for distribution of electric power for commercial/industrial facilities are tabulated in Table 3.2. First investigate power distribution requirements of the building and then select the best system or combination of systems based on the needs of the building function. The commercial and industrial power system classification for voltage classes are as follows: Medium voltage is 2400 V up to and including 69,000 V. High voltage refers to 115,000 up to 230,000 V. Unfortunately, the common usage of the term "high voltage" by the average user can be anything more than 120 V or anything more than 600 V. The discussion on system arrangements described in Table 3.2 covers both medium-voltage and low-voltage distribution.

Table 3.3 lists the US standard nominal voltages in common use in the buildings. The numbers listed in the right-hand column are used in equipment ratings, but these should not be confused with numbers designating the nominal system voltage on which the equipment is designed to operate.

3.3 Heating, Ventilation, and Air-Conditioning (HVAC) Systems

HVAC systems maintain desired environmental conditions inside buildings. In almost every application, there are several ways these conditions may be maintained. Ideally HVAC systems permit people or products to function within the buildings at optimum level. Different systems approach this ideal with varying degree of success. HVAC systems are categorized and segregated by equipment arrangement and how they control environment in the space. Water systems use hot or chilled water to convey heat to or from a controlled space through piping system, connecting a source such as boiler, water heater, or chiller to a suitable terminal heat transfer units. Hot water heating and chilled water cooling systems are frequently called hydronic systems. Water systems are classified by temperature, flow

Table 3.2 Types of power distribution system

Power system type	Features	Discussion
Simple radial system (Fig. 3.2)	Distribution is at the utilization voltage. A single primary service and distribution transformer supply all the feeders. There is no duplication of equipment. System investment is the lowest of all circuit arrangements	This system is satisfactory for small industrial installations where process allows sufficient downtime for adequate maintenance and the plant can be supplied by a single transformer
	Operation and expansion are simple. When quality components and appropriate ratings are used, reliability is high. Loss of a cable, primary supply, or transformer will cut off service. Equipment must be shut down to perform routine maintenance and servicing	
Expanded radial system (Fig. 3.3)	The advantages of the simple radial system may be applied to larger loads by using an expanded radial primary distribution system to supply a number of unit substations located near the load, which in turn supply the load through radial secondary systems	The advantages and disadvantages are the same as those described for the simple radial system. May require shutdown of all systems due to fault on primary main
Primary selective system (Fig. 3.4)	Protection against loss of a primary supply can be gained through use of a primary selective system. Each unit substation is connected to two separate primary feeders through switching equipment to provide a normal and an alternate source. On failure of the normal source, the distribution transformer is switched to the alternate source. Switching can be either manual or automatic, but there will be an interruption until the load is transferred to the alternate source	Cost is somewhat higher than that of a radial system because of duplication of primary cable and switchgear
	If the two sources can be parallel during switching, some maintenance of primary cable and switching equipment, in certain configurations, may be performed with little or no interruption of service	
Primary loop system (Fig. 3.5)	A primary loop system offers improved reliability and service continuity in comparison to a radial system. In typical loop systems, power is supplied continuously from two sources at the ends of the loop. Such a system, if properly designed and operated, can quickly recover from a single cable fault with no continuous loss of power to utilization equipment	To realize optimum service reliability of a primary loop system, the system should be operated (closed-loop mode) A cable fault within the loop may be automatically isolated without loss of transformer capacity. No loss of power will occur, although the system will experience a voltage dip until the circuit breakers clear the fault
	Single electrical power can flow in both directions in a loop system. It is essential that detailed operating instructions be prepared and followed. Additionally, if the two supply points for the loop originate from different buses, the design must consider available short-circuit capacity from both buses, the ability of both buses to supply the total load, and the possibility of a flow of current from one bus to the other bus over the loop	A primary loop system may be operated with one of the series switches in open. A disadvantage of open-loop operation is that a cable failure will result in the temporary loss of service to some portion of the system

(continued)

Table 3.2 (continued)

Power system type	Features	Discussion
Secondary selective system (Fig. 3.6)	If pairs of substations are connected through a secondary tie circuit breaker, the result is a secondary selective system. If the primary feeder or transformer fails, supply is maintained through the secondary tie circuit breaker. The tie circuit breaker can be operated in a normally opened or a normally closed position. In case of the normally opened tie circuit breaker, voltage is maintained to the unaffected transformer's circuits. In the case of the normally closed tie, a voltage depression occurs on the bus until the affected transformer's circuit breaker opens	Normally the systems operate as radial systems. Maintenance of primary feeders, transformer, and main secondary disconnecting means is possible with only momentary power interruption or no interruption if the stations can be operated in parallel during switching, although complete station maintenance will require a shutdown. With the loss of one primary circuit or transformer, the total substation load may be supplied by one transformer The secondary selective system may be combined with the primary selective system to provide a high degree of reliability. This reliability is purchased with additional investment and addition of some operating complexity
Secondary spot network (Fig. 3.7)	In this system two or more distribution transformers are each supplied from a separate primary distribution feeder. The secondaries of the transformers are connected in parallel through a special type of device, called a network protector, to a secondary bus. Radial secondary feeders are tapped from the secondary bus to supply utilization equipment If a primary feeder fails or a fault occurs on a primary feeder or distribution transformer, the other transformers start to feed back through the network protector on the faulted circuit	The secondary spot network is the most reliable power supply for large loads. A power interruption can occur only when there is a simultaneous failure of all primary feeders or when a fault occurs on the secondary bus. There are not momentary interruptions caused by the operation of the transformer switches that occur on primary selective, secondary selective, or loop systems. Voltage sags caused by large transient loads are substantially reduced Networks are expensive because of the extra cost of the network protector and duplication of transformers capacity. In addition, each transformer connected in parallel increases the short-circuit current capacity and may increase the duty ratings of the secondary equipment. This scheme is used only in low-voltage applications with a very high load density
Ring bus (Fig. 3.8)	The ring bus offers the advantage of automatically isolating a fault and restoring service. A fault anywhere in the ring results in two interrupting devices opening to isolate the fault	The ring bus scheme is often considered where there are two or more medium-voltage (i.e., 4.16, 4.8, or 13.2/13.8 kV) distribution services to the facility and the utmost in flexibility and switching options are desired
		Manual isolating switches are installed on each side of the automatic device. This allows maintenance to be performed safely and without interruption of service. This will also allow the system to be expanded without interruption

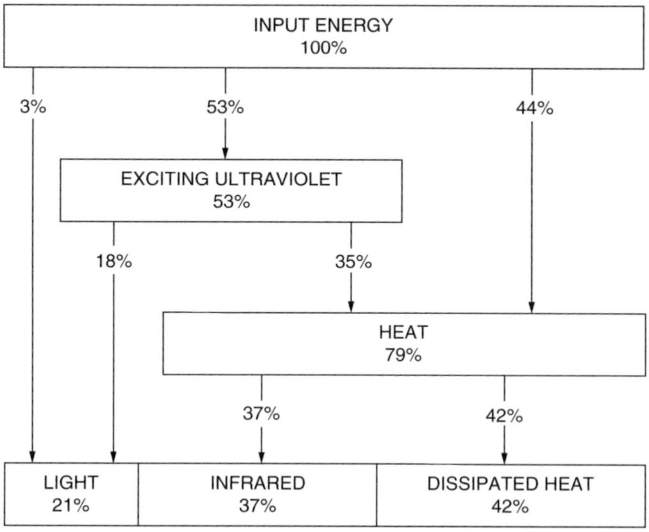

Fig. 3.2 Energy distribution in a typical cool white fluorescent lamp (Reprinted with permission of the IESNA, 120 Wall Street, 17th floor, New York, NY 10005. Taken from the *IESNA Lighting Handbook, 8th Edition*)

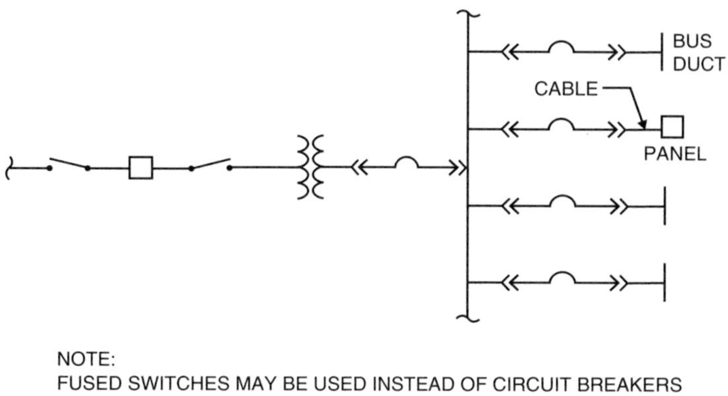

Fig. 3.3 Simple radial system (Copyright © 1994. IEEE. All Rights Reserved)

generation, pressurization, or piping and pumping arrangement. They are either closed (recirculating) or once-through (open) systems. Buildings mostly use recirculating systems.

Dampers and fans are used to control the flow of air into, out of, and within a building. Necessary pressures to move air within ductwork and conditioned spaces are provided by fans, operated either constantly or intermittently. Table 3.4 lists air handling systems commonly used in commercial buildings. Forced-air systems

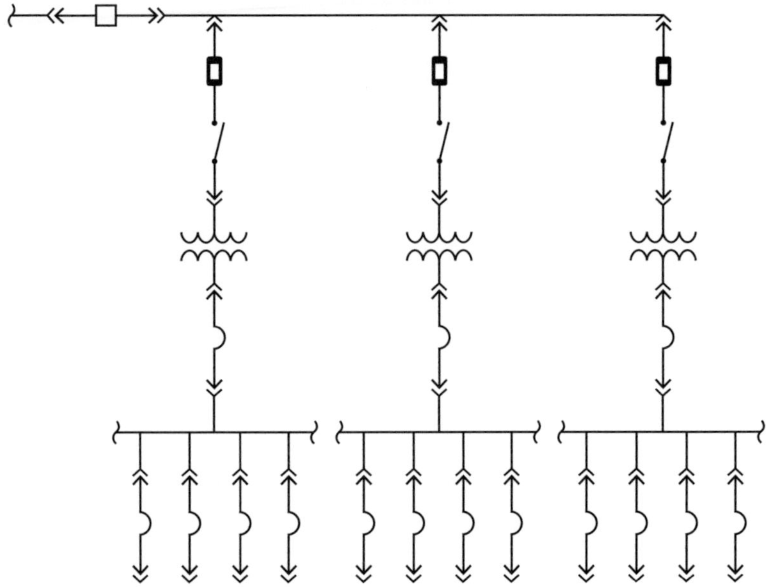

Fig. 3.4 Expanded radial system (Copyright © 1994. IEEE. All Rights Reserved)

range in complexity from single-duct, single-fan networks, found in private residences and small buildings, to sophisticated constant volume (CV) air handling and variable air volume (VAV) systems found in large buildings (Figs. 3.15 and 3.16).

Both CV and VAV systems are important in their own right. Hybrid systems have been developed that combine the advantages of CV and VAV systems and in turn minimize the disadvantages. VAV systems with terminal reheat fan-powered boxes are a good example of a hybrid system. Thermal and transportation savings in these systems are achieved during the period when the system is operating in the VAV mode, that is, cooling only (Table 3.5).

3.4 Variable Refrigerant Flow (VRF) Systems

Currently widely applied in large buildings especially in Japan and Europe, these systems are just starting to be introduced in the United States. The VRF technology/ system was developed and designed by Daikin Industries, Japan, who named and protected the term variable refrigerant volume (VRV) system so other manufacturers use the term VRF "variable refrigerant flow". In essence, both are same.

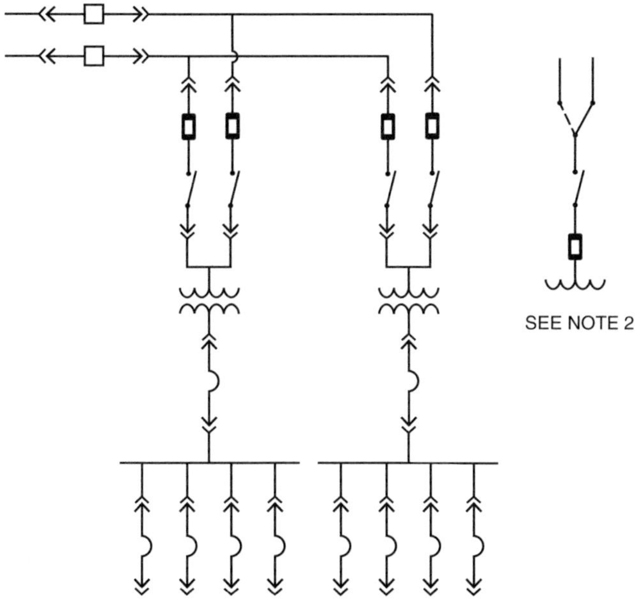

NOTES:
1. IF NON-DRAW-OUT FUSED SWITCHES ARE USED, THE FUSE SHOULD BE ON THE LOAD SIDE OF THE SWITCH,
2. AN ALTERNATE ARRANGEMENT USES A PRIMARY SELECTOR SWITCH WITH A SINGLE FUSED INTERRUPTER SWITCH (WHICH MAY NOT HAVE CERTIFIED CURRENT SWITCHING ABILITY).

Fig. 3.5 Primary selective system (Copyright © 1994. IEEE. All Rights Reserved)

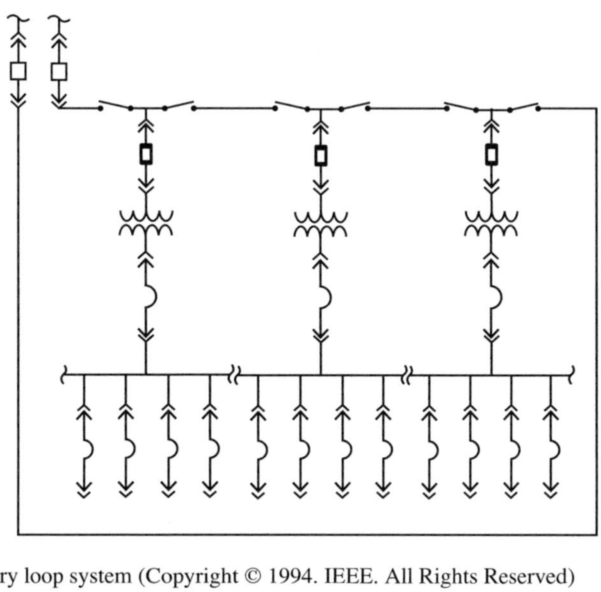

Fig. 3.6 Primary loop system (Copyright © 1994. IEEE. All Rights Reserved)

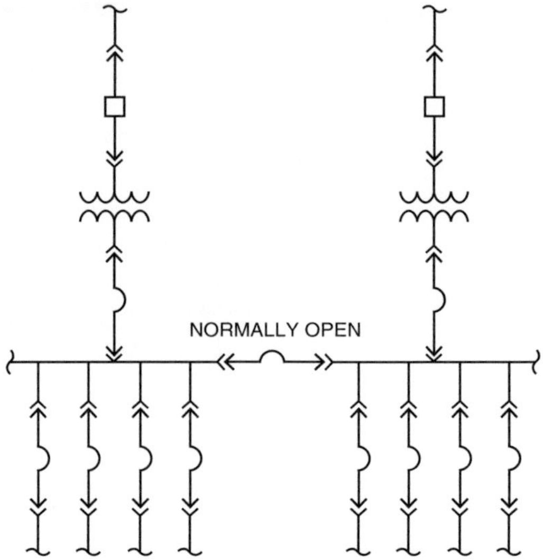

Fig. 3.7 Secondary selective system (Copyright © 1994. IEEE. All Rights Reserved)

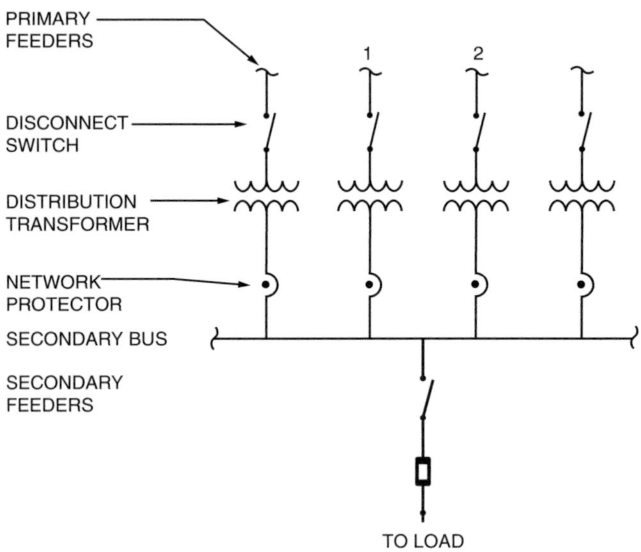

Fig. 3.8 Secondary spot network (Copyright © 1994. IEEE. All Rights Reserved)

Table 3.3 Nominal power system voltages

Standard nominal system voltages	Associated nonstandard nominal system voltages
Low voltages	
120	110, 115, 125
120/240	110/220, 115/230, 125/250
208Y/120	216Y/125
2401120	
240	230, 250
480Y/277	460Y/265
480	440
600	550, 575
Medium voltages	
2400	2200, 2300
4160Y/2400	
4160	4000
4800	4600
6900	6600, 7200
8320Y/4800	11,000, 11,500
12,000Y/6930	
12,470Y/7200	
13,200Y/7620	
13,200	
13,800Y/7970	14,400
13,800	
20,780Y/12,000	
22,860Y/13,200	
23,000	
24,940Y/14,400	
34,500Y/19,920	
34,500	33,000
46,000	44,000
69,000	66,000
High voltages	
115,000	110,000, 120,000
138,000	132,000
161,000	154,000
230,000	220,000

Table 3.4 Brief tabulation of variety of air handling systems used in buildings

System type	Features	Comments
Single-zone system (Fig. 3.9)	AHU serving a single air-conditioned space	Unable to satisfy different needs in different spaces at any given time. Does not heat/cool simultaneously
Constant volume (CV) terminal reheat system	Total supply air is cooled to satisfy worst condition at any given time	Less expensive, wastes both cooling and heating energy and does not offer individual comfort in the conditioned spaces
Mixed air systems (Figs. 3.10 and 3.11) Double-duct (DD) system Multizone (MZ) system	With MZ system, temperature control for all zones is centrally located at AHU, whereas in DD system, the space temperature control is in the mixing terminals located over the occupied spaces	Offers a good temperature control but limited humidity control. Addition of a bypass plenum to hot and cold plenums reduces thermal energy waste but costs more
Induction systems (Fig. 3.12)	In the induction unit primary air from central AHU is discharged through an induction nozzle at high velocity, which induces a certain amount of room air and supplies the mixture into the conditioned space	Uses more air transportation energy (fan HP) and provides limited environment control
VAV systems (Figs. 3.13 and 3.14)	Supply air terminals in a VAV system satisfy space cooling needs by modulating the amount of volume of air rather than varying temperature of the supply air. Since building skin losses are continuous linear functions of the outside air temperature, the use of a CV perimeter heating/cooling system is very common	A VAV fan HP requirement is considerably smaller than a CV system. Variable volume, variable temperature (VVVT) is suitable for smaller buildings in mild climate areas. Using a VVVT system for large buildings or in extreme climates may cause control complexities

With a higher efficiency and increased controllability, the VRF system can help achieve a sustainable design.

3.4.1 Overview of VRF Systems

The primary function of all air-conditioning systems is to provide thermal comfort for building occupants. There are a wide range of air-conditioning systems available, starting from the basic window-fitted units to the small split systems, to the medium-scale package units, to the large chilled water systems, and currently to the variable refrigerant flow (VRF) systems.

The term VRF refers to the ability of the system to control the amount of refrigerant flowing to each of the evaporators, enabling the use of many evaporators of

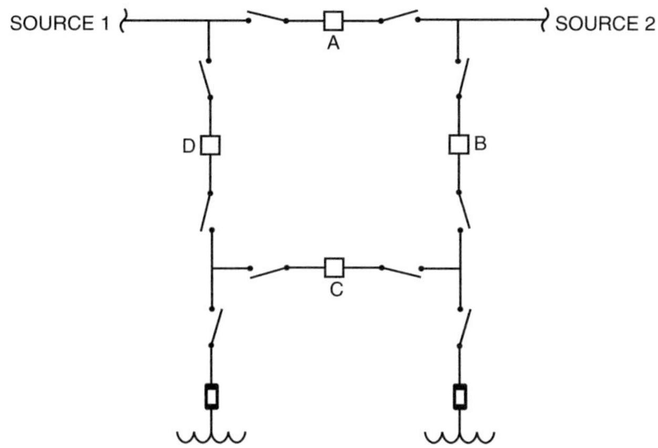

Fig. 3.9 Ring bus system (Copyright © 1994. IEEE. All Rights Reserved)

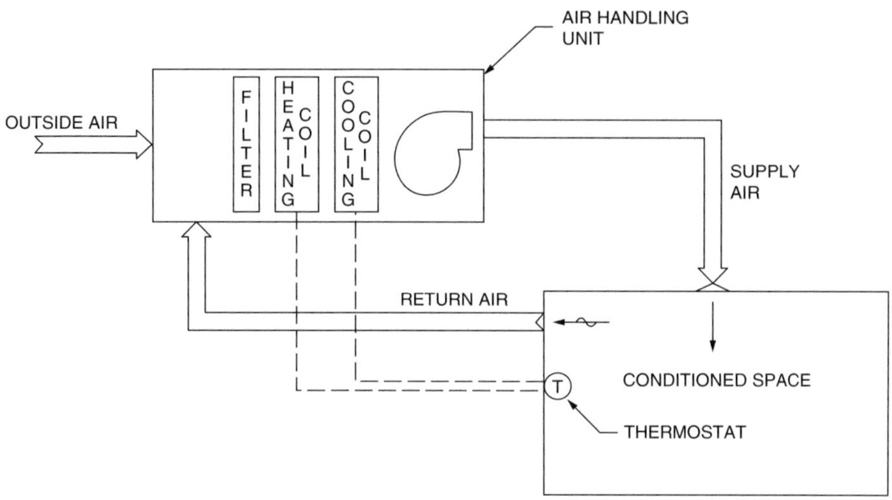

Fig. 3.10 Single-zone air handling system

differing capacities and configurations, individualized comfort control, simultaneous heating and cooling in different zones, and heat recovery from one zone to another. VRF systems operate on the direct expansion (DX) principle meaning that heat is transferred to or from the space directly by circulating refrigerant to evaporators located near or within the conditioned space. Refrigerant flow control is the key to many advantages as well as the major technical challenge of VRF systems.

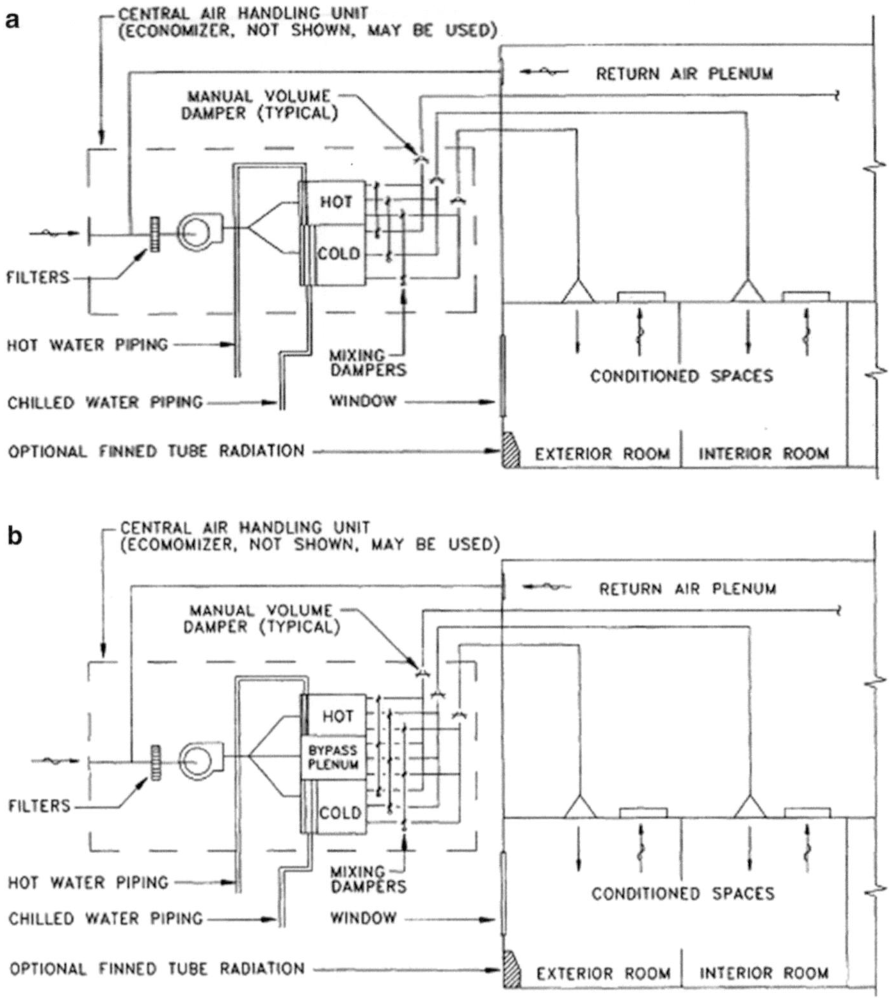

Fig. 3.11 (**a**) Constant volume multizone system with mixing dampers (only three zones shown for clarity). (**b**) Triple deck multizone system with mixing dampers (only three zones shown for clarity)

Note the term VRF systems should not be confused with the centralized VAV (variable air volume) systems, which work by varying the airflow to the conditioned space based on variation in room loads.

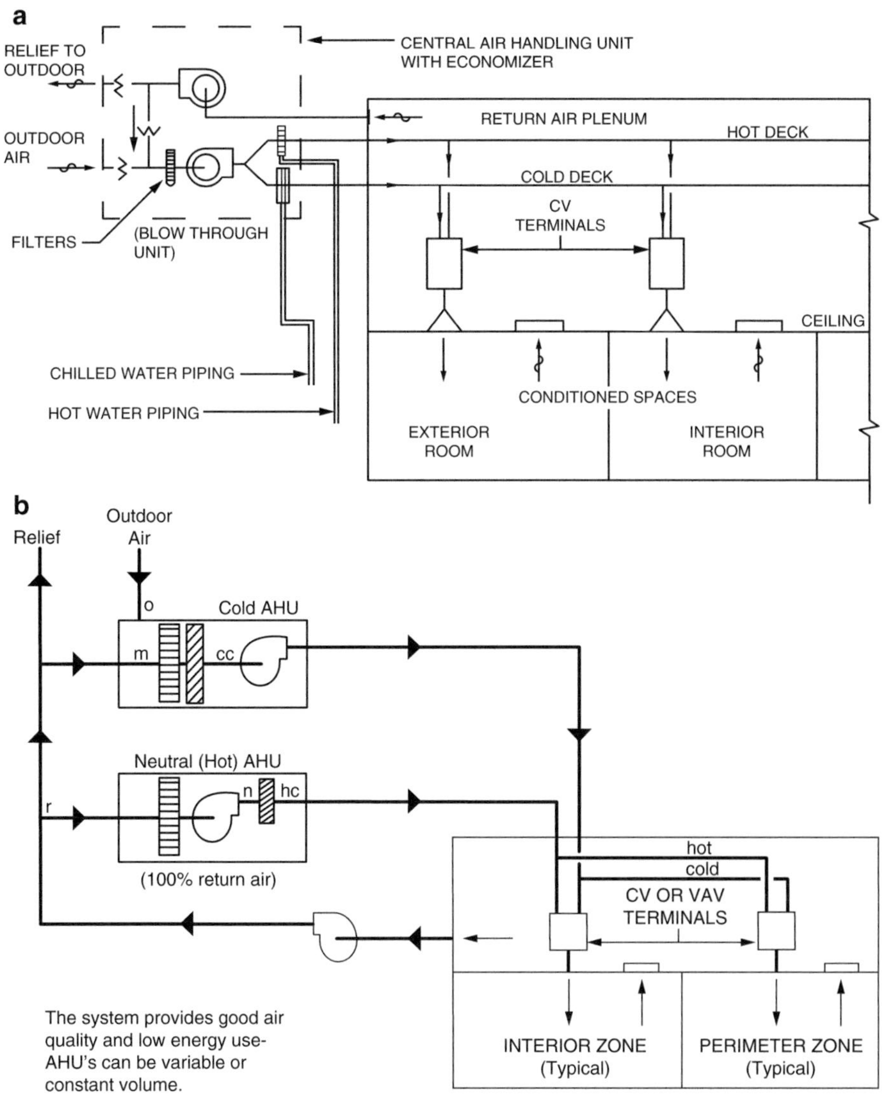

Fig. 3.12 (**a**) Constant volume single-fan double-duct system. (**b**) Dual-fan dual-duct system

3.4.1.1 Variable Refrigerant Flow or VRF Systems

VRF systems are similar to the multi-split systems which connect one outdoor section to several evaporators. However, multi-split systems turn off or on completely in response to one master controller, whereas VRF systems continually adjust the flow of refrigerant to each indoor evaporator. The control is achieved by continually

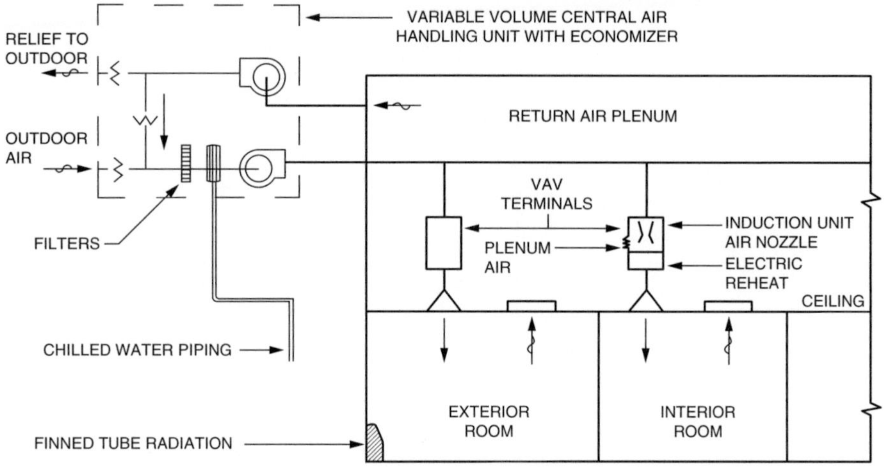

Fig. 3.13 Variable volume system (draw through unit) with independent perimeter heating

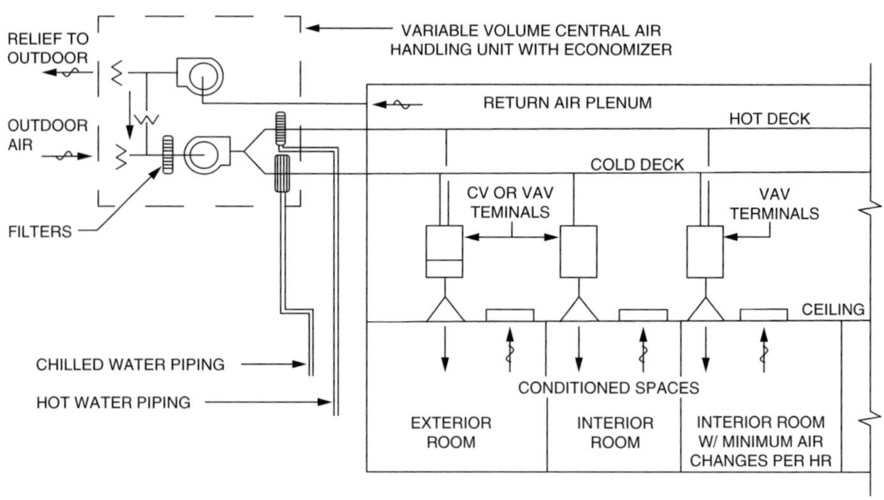

Fig. 3.14 Variable volume double-duct system

varying the flow of refrigerant through a pulse modulating valve (PMV) whose opening is determined by the microprocessor receiving information from the thermistor sensors in each indoor unit. The indoor units are linked by a control wire to the outdoor unit which responds to the demand from the indoor units by varying its compressor speed to match the total cooling and/or heating requirements.

VRF systems promise a more energy-efficient strategy (estimates range from 11 to 17 % less energy compared to conventional units) at a somewhat higher cost.

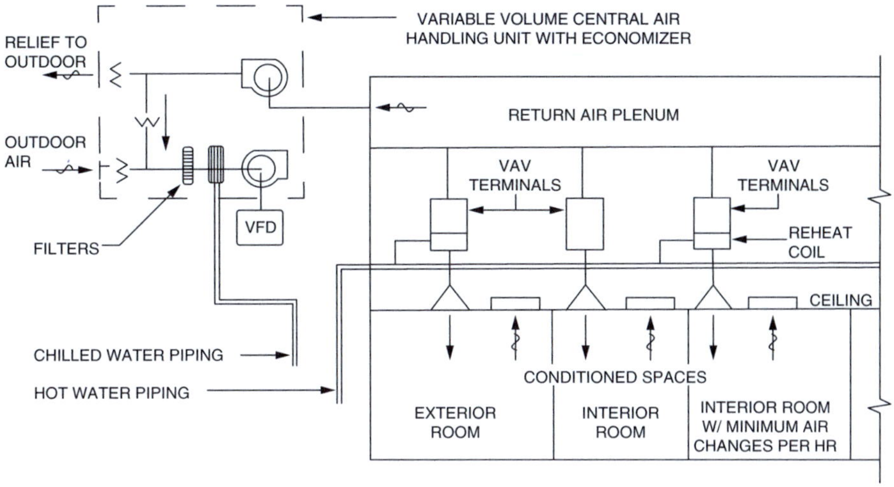

Fig. 3.15 Variable volume terminal reheat system

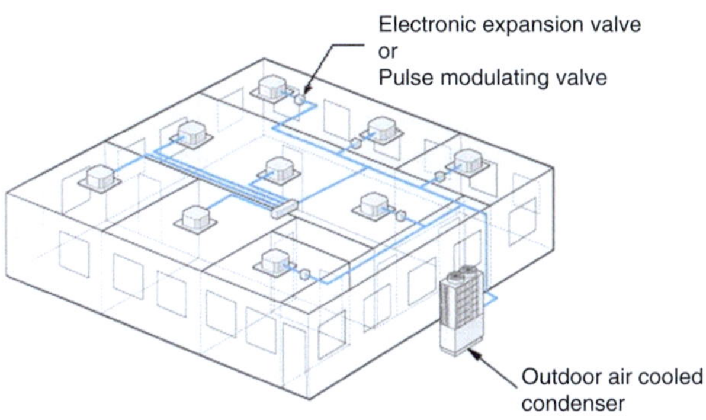

Fig. 3.16 VRF system with multiple indoor evaporator units

The modern VRF technology uses an inverter-driven scroll compressor and permits as many as 48 or more indoor units to operate from one outdoor unit (varies from manufacturer to manufacturer). The inverter scroll compressors are capable of changing the speed to follow the variations in the total cooling/heating load as determined by the suction gas pressure measured on the condensing unit. The capacity control range can be as low as 6–100 %.

Refrigerant piping runs of more than 200 ft are possible, and outdoor units are available in sizes up to 240,000 Btuh (Fig. 3.17).

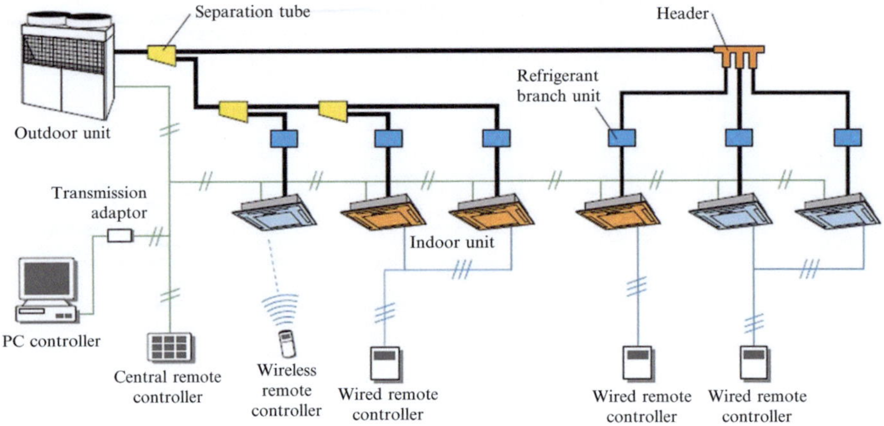

Fig. 3.17 A schematic VRF arrangement (Source: Fujitsu)

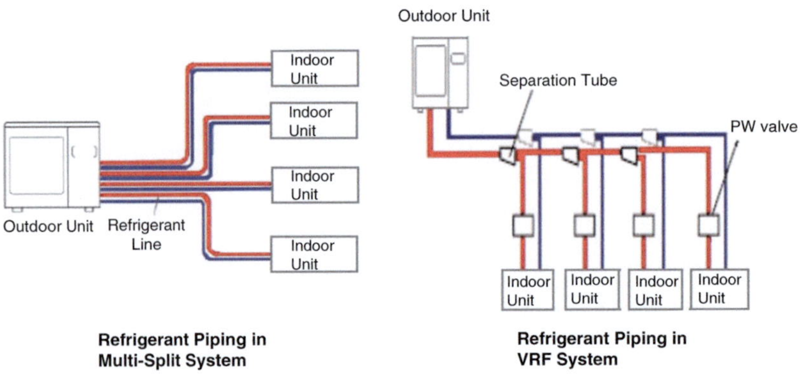

Fig. 3.18 Multi-split and VRF system piping configuration

VRF systems are engineered systems and use complex refrigerant and oil control circuitry. The refrigerant pipework uses a number of separation tubes and/or headers (refer schematic figure above).

A separation tube has two branches whereas a header has more than two branches. Either of the separation tube or header, or both, can be used for branches. However, the separation tube is **never** provided after the header because of balancing issues.

Compared to multi-split systems, VRF systems minimize the refrigerant path and use **less** copper tubing. Minimizing the refrigerant path allows for maximizing the efficiency of refrigerant work (Fig. 3.18).

Table 3.5 Heating, ventilating, and air-conditioning practices for a broad range of applications

Application	Single-zone self-contained		Multizone[b]	Constant volume reheat	W/ separate perimeter heat[d]	Fan-powered parallel[e]	Fan-powered series[f]	Dual-duct constant volume w/reheat[g]	Two-pipe fan coil, separate O.A.[h]	Four-pipe fan-coil separate O.A.[h]	Water-source heat pump	
	Rooftop[a]	Through wall									Central	Modular distribution[h]
Specialty stores	+				+	o	o			+		+
Restaurants	+				+	o	o					
Bowling alleys					o	+	+					
Radio/TV studios	+		o	+								
County clubs	+		+		+							
Funeral homes	+		+		o							
Churches	+		+		+							
Theaters	+				+						+	
Office buildings	o[i]		o[i]		+	+	+			o	+	+
Hotels, dormitories		o							+	+		
Motels/ apartment buildings	o	+							o	+		+
Hospitals				+[j]	+[l]	+[l]	+[l]	+[l]		+[k]		[k]
Schools and colleges	o		o		+	+				+	o	+
Museums			+	+	o	o						
Libraries, standard	+		+		+	+						
Libraries, rare books	o		+		o	o						

(continued)

Table 3.5 (continued)

Application	Single-zone self-contained		Multizone[b]	Constant volume reheat	W/ separate perimeter heat[d]	Fan-powered parallel[e]	Fan-powered series[f]	Dual-duct constant volume w/reheat[g]	Two-pipe fan coil, separate O.A.[h]	Four-pipe fan-coil separate O.A.[h]	Water-source heat pump	
	Rooftop[a]	Through wall									Central	Modular distribution[h]
Department stores	+		+		+	o					o	o
Laboratories			o	+	+	+	+					

+ systems frequently used, o system occasionally used, *O.A.* outside air

[a]In cold climates, often in conjunction with separate perimeter heating system due to high window/wall heat losses and/or to avoid central fan operation during extended unoccupied periods

[b]*Not* recommended for use in hot/burned climates (poor RH control at high cooling load/high outside air humidity level). Limit 12 zones per unit

[c]Pressure-independent boxes recommended for most applications involving long-duration occupancy (fewer expensive pressure-dependent boxes are typically used in short-duration occupancies such as stores and supermarkets, where drafts are tolerable)

[d]Cooling-only VAV box typically used for large interior zones or in perimeter zones provided with separate heating system

[e]Fan-powered parallel VAV boxes (typically used in perimeter zones) are generally more energy efficient than series-type VAV boxes. The small box fans operate only during the heating cycle when fan motor losses contribute to the heating function

[f]Fan-powered series VAV boxes (typically used only for perimeter zones) should be used only where constant airflow to the occupied space is very important design criterion (small, inefficient fan motors operate continuously even during the cooling cycle, increasing annual fan power consumption)

[g]Not recommended in regions where low cooling loads coincide with high outside air humidity (RH levels in space increases quickly). Use in areas with moderately humid climates and less than 78 °F (25.5 °C) outside air dry bulb temperature at design condition

[h]Modular fan-coil units and water-source heat pumps require outside air ventilation in all but residential applications. Separately ducted ventilation systems offer better control, filtering, energy efficiency, and less maintenance than through-wall outside air louvers

[i]Appropriate for small, one- or two-story buildings only

[j]Appropriate only for ORs, ICU, nurseries, delivery rooms, and similar spaces requiring exceptionally high degree or in building applications where recovered heat can *be* used for reheating function

[k]Used in isolation wards to avoid recirculation of air to other areas

[l]Appropriate for general areas and patient rooms

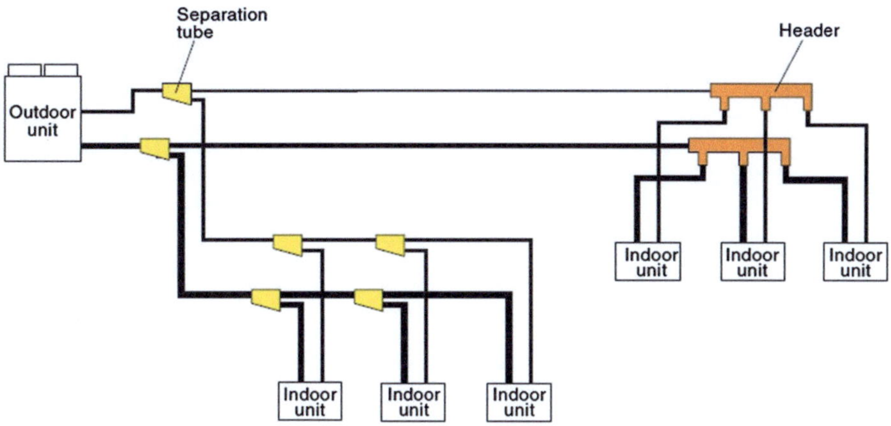

Fig. 3.19 Cooling type VRF system (Source: Fujitsu)

Types of VRF

VRV/VRF systems can be used for cooling only, heat pumping, or heat recovery. On heat pump models, there are two basic types of VRF system: heat pump systems and energy recovery.

VRF Heat Pump Systems

VRF heat pump systems permit heating or cooling in all of the indoor units but **not** simultaneous heating and cooling. When the indoor units are in the cooling mode, they act as evaporators; when they are in the heating mode, they act as condensers. These are also known as two-pipe systems (Fig. 3.19).

VRF heat pump systems are effectively applied in open plan areas, retail stores, cellular offices, and any other area that require cooling or heating during the same operational periods.

Heat Recovery VRF System (VRF-HR)

Variable refrigerant flow systems with heat recovery (VRF-HR) capability can operate simultaneously in heating and/or cooling mode, enabling heat to be used rather than rejected as it would be in traditional heat pump systems. VRF-HR systems are equipped with enhanced features like inverter drives, pulse modulating electronic expansion valves, and distributed controls that allow system to operate in net heating or net cooling mode, as demanded by the space.

Each manufacturer has its own proprietary design (2-pipe or 3-pipe system), but most uses a three-pipe system (liquid line, a hot gas line, and a suction line) and special valving arrangements. Each indoor unit is branched off from the 3 pipes using solenoid valves. An indoor unit requiring cooling will open its liquid line and

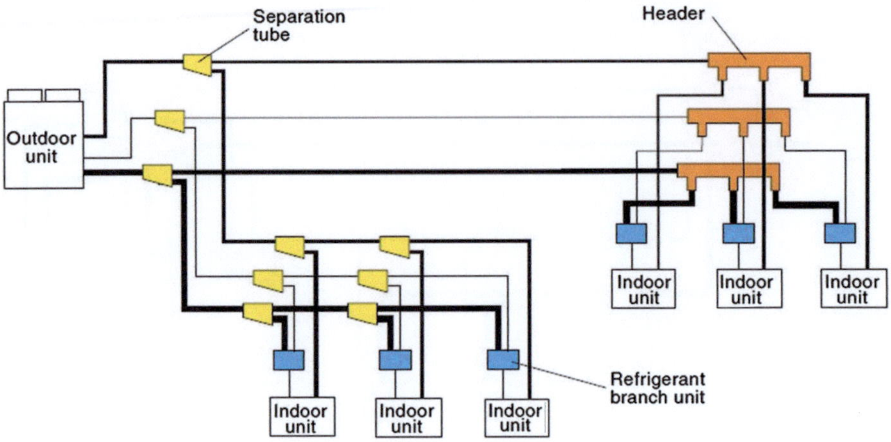

Fig. 3.20 Heat recovery type VRF system (Source: Fujitsu)

suction line valves and act as an evaporator. An indoor unit requiring heating will open its hot gas and liquid line valves and will act as a condenser.

Typically, extra heat exchangers in distribution boxes are used to transfer some reject heat from the superheated refrigerant exiting the zone being cooled to the refrigerant that is going to the zone to be heated. This balancing act has the potential to produce significant energy savings (Fig. 3.20).

VRF-HR mixed mode operation leads to energy savings as both ends of the thermodynamic cycle are delivering useful heat exchange. If a system has a cooling COP (coefficient of performance) of 3 and a heating COP of 4, then heat recovery operation could yield a COP as high as 7. It should be noted that this perfect balance of heating and cooling demand is unlikely to occur for many hours each year, but whenever mixed mode is used, energy is saved. Units are now available to deliver the heat removed from space cooling into hot water for space heating, domestic hot water, or leisure applications, so that mixed mode is utilized for more of the year.

VRF-HR systems work best when there is a need for some of the spaces to be cooled and some of them to be heated during the same period. This often occurs in the winter in medium-sized to large-sized buildings with a substantial core or in the areas on the north and south sides of a building.

The **coefficient of performance (COP)** is the performance rating used primarily in heat pumps. It is defined as the ratio of heat output to the amount of energy input of a heat pump. It compares the heat produced by the heat pump to the heat obtained from resistance heat. COPs vary with the outside temperature; as the temperature falls, the COP falls also since the heat pump is less efficient at lower temperatures. ARI standards compare equipment at two temperatures, 47 and 17 °F, to give you an idea of the COP in both mild and colder temperatures.

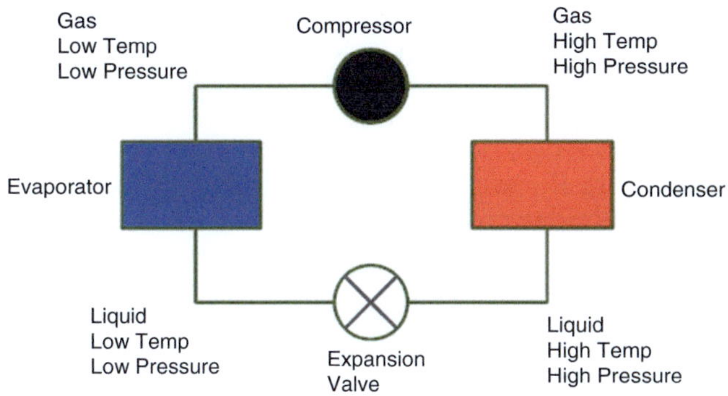

Fig. 3.21 Basic refrigeration cycle

Refrigerant Modulation in a VRF System

VRV/VRF technology is based on the simple vapor compression cycle (same as conventional split air-conditioning systems) but gives you the ability to continuously control and adjust the flow of refrigerant to different internal units, depending on the heating and cooling needs of each area of the building. The refrigerant flow to each evaporator is adjusted precisely through a pulse wave electronic expansion valve in conjunction with an inverter and multiple compressors of varying capacity, in response to changes in the cooling or heating requirement within the air-conditioned space.

We will discuss this further but before that let us refresh the basic refrigeration cycle (Fig. 3.21).

The fundamental of an air-conditioning system is the use of a refrigerant to absorb heat from the indoor environment and transfer it to the external environment. In the cooling mode, indoor units are supplied with liquid refrigerant. The amount of refrigerant flowing through the unit is controlled via an expansion valve located inside the unit. When the refrigerant enters the coil, it undergoes a phase change (evaporation) that extracts heat from the space, thereby cooling the room. The heat extracted from the space is exhausted to the ambient air.

Refrigeration systems can operate on *reverse cycle* mode with an inclusion of special 4-way reversing valve, enabling the absorption of heat from the external environment and using this heat to raise the internal temperature. When in the heating mode, indoor units are supplied with a hot gas refrigerant. Again, the amount of hot gas flowing through the unit is controlled via the same electronic expansion valve. As with the liquid refrigerant, the hot gas undergoes a phase change (condensation), which releases heat energy into the space. These are called *heat pump* systems. Heat pumps provide both heating and cooling from the same unit, and due to added heat of compression, the efficiency of a heat pump in the heating mode is **higher** compared to the cooling cycle.

Expansion valve is the component that controls the rate at which liquid refrigerant can flow into an evaporator coil. The conventional refrigeration cycle uses "thermostatic expansion valve (TXV)" that uses mechanical spring for control. It has its drawbacks.

- TXV operation is totally independent of compressor operation.
- TXV is susceptible to hunting, that is, overfeeding and starvation of refrigerant flow to the evaporator.

As the evaporator load increases, available refrigerant will boil off more rapidly. If it is completely evaporated prior to exiting the evaporator, the vapor will continue to absorb heat (superheat). Although superheating ensures total evaporation of the liquid refrigerant before it goes into the compressor, the density of vapor which quits the evaporator and enters the compressor is reduced leading to reduced refrigeration capacity. The inadequate or high superheat in a system is a concern.

- Too little: liquid refrigerant entering a compressor washes out the oil causing premature failure.
- Too much: valuable evaporator space is wasted and possibly causing compressor overheating problems.

The shortcomings of TXV are offset by the modern electronic expansion valve.

3.4.1.2 Electronic Expansion Valve (EEV)

With an electronic expansion valve (EEV), you can tell the system what superheat you want and it will set it up. The primary characteristic of EEV is its ability to rotate a prescribed small angle (step) in response to each control pulse applied to its windings. EEV consists of a synchronous electronic motor that can divide a full rotation into a large number of steps, 500 steps/rev. With such a wide range, an EEV valve can go from full open to totally closed and closes down when system is satisfied.

EEV in a VRF system functions to maintain the pressure differential and also distribute the precise amount of refrigerant to each indoor unit. It allows for the fine control of the refrigerant to the evaporators and can reduce or stop the flow of refrigerant to the individual evaporator unit while meeting the targeted superheat.

Design Considerations for VRF Systems

Deciding what HVAC system best suits your application will depend on several variables such as building characteristics, cooling and heating load requirements, peak occurrence, simultaneous heating and cooling requirements, fresh air needs, accessibility requirements, minimum and maximum outdoor temperatures, sustainability, and acoustic characteristics.

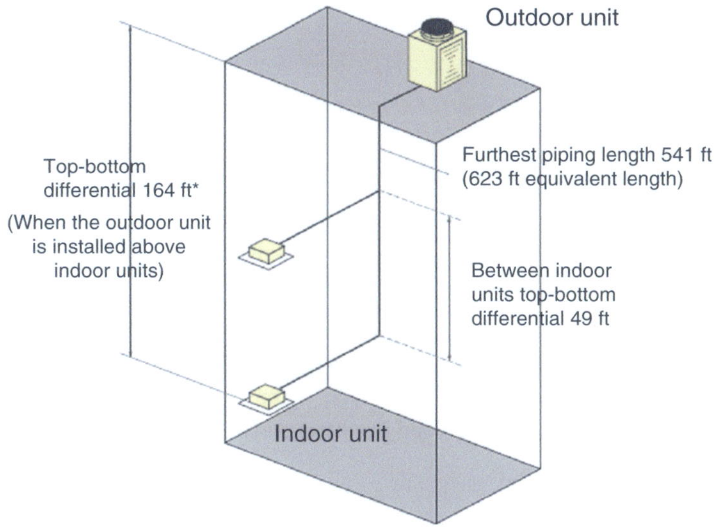

Fig. 3.22 VRF system limitations (Source: ASHRAE Technical Committee on VRF)

Building Characteristics

VRF systems are typically distributed systems—the outdoor unit is kept at a far-off location like the top of the building or remotely at grade level, and all the evaporator units are installed at various locations inside the building. Typically the refrigerant pipework (liquid and suction lines) is very long, running in several hundreds of feet in length for large multistory buildings. Obviously, the long pipe lengths will introduce pressure losses in the suction line and, unless the correct diameter of pipe is selected, the indoor units will be starved of refrigerant resulting in insufficient cooling to the end user. So it is very important to make sure that the pipe sizing is done properly, both for the main header pipe and the feeder pipes that feed each indoor unit.

The maximum allowable length varies among different manufacturers; however, the general guidelines are as follows:

- The maximum allowable vertical distance between an outdoor unit and its farthest indoor unit is 164 ft.
- The maximum permissible vertical distance between two individual indoor units is 49 ft.
- The maximum overall refrigerant piping lengths between outdoor and the farthest indoor unit is up to 541 ft.

Note: The longer the lengths of refrigerant pipes, the more expensive the initial and operating costs (Fig. 3.22).

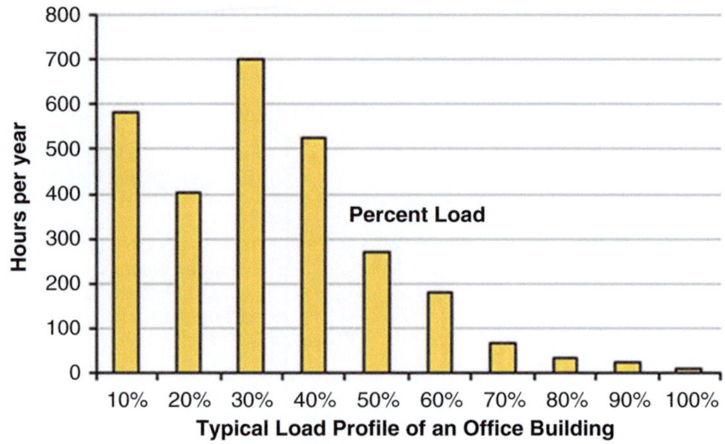

Fig. 3.23 Typical office building load profile

Building Load Profile

When selecting a VRF system for a new or retrofit application, the following assessment tasks should be carried out:

- Determine the functional and operational requirements by assessing the cooling load and load profiles including location, hours of operation, number/type of occupants, equipment being used, etc.
- Determine the required system configuration in terms of the number of indoor units and the outdoor condensing unit capacity by taking into account the total capacity and operational requirements, reliability, and maintenance considerations

Building a load profile helps determine the outdoor condensing unit compressor capacity. For instance, if there are many hours at low load, it is advantageous to install multiple compressors with at least one with inverter (speed adjustment) feature. The figure below shows a typical load profile for an office building (Fig. 3.23).

The combined cooling capacity of the indoor sections can match, exceed, or be lower than the capacity of the outdoor section connected to them. But as a normal practice:

- The indoor units are typically sized and selected based on the **greater** of the heating or cooling loads in the zone it serves, that is, maximum peak load expected in any time of the year.
- The outdoor condensing unit is selected based on the load profile of the facility which is the peak load of all the zones combined at any one given time. The important thing here is that it is unlikely that all zones will peak at a given time so an element of diversity is considered for economic sizing. Adding up the peak load for each indoor unit and using that total number to size the outdoor unit will result in an unnecessarily oversized condensing unit. Although an oversized con-

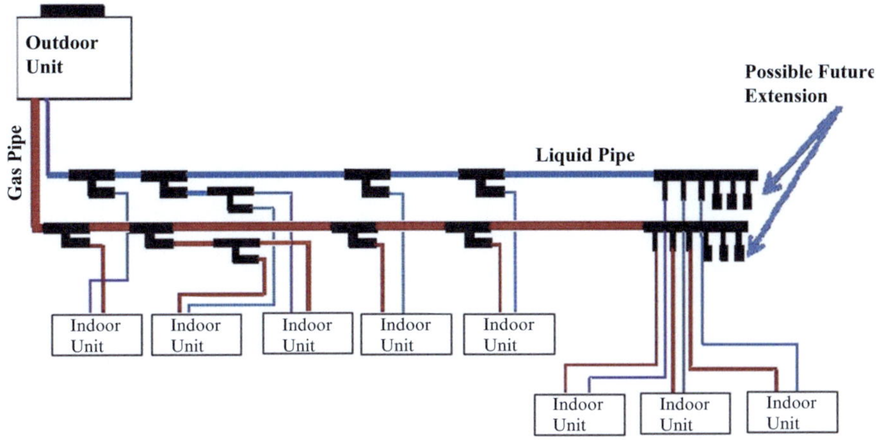

Fig. 3.24 Expandable modular VRF system

densing unit with multiple compressors is capable of operating at lower capacity, too much oversizing sometimes reduces or ceases the modulation function of the expansion valve. As a rule of thumb, an engineer can specify an outdoor unit with a capacity anywhere between 70 and 130 % of the combined capacities of the indoor units.

Sustainability

One attractive feature of the VRF system is its higher efficiency compared to conventional units. Cooling power in a VRF system is regulated by means of adjusting the rotation speed of the compressor which can generate an energy saving around 30 %.

A VRF system permits easy future expansion when the conditions demand. Oversizing, however, should be avoided unless a future expansion is planned (Fig. 3.24).

Other sustainability factors include:

- Use of non-ozone depleting environment-friendly refrigerants such as R-410a.
- Opting for heat pump instead of electrical resistance heating in areas demanding both cooling and heating. Heat pumps offer higher energy efficiency.

Simultaneous Heating and Cooling

Some manufacturers offer a VRF system with heat recovery feature which is capable of providing simultaneous heating and cooling. The cost of a VRF-HR is higher than that of a normal VRF heat pump unit, and therefore its application should be carefully evaluated.

More economical design can sometimes be achieved by combining zones with similar heating or cooling requirements together. For example, the areas that may require simultaneous heating and cooling are the perimetric and interior zones. Perimetric areas with lot of glazing and exposure especially towards west and south will have high load variations. A VRF heat pump type system is capable of providing simultaneous heating and cooling exceeding 6 tons cooling requirement.

Using VRF heat pump units for heating and cooling can increase building energy efficiency. The designer must evaluate the heat output for the units at the outdoor design temperature. Supplemental heating with electric resistors shall be considered only when the heating capacity of the VRF units is below the heating capacity required by the application. Even though supplemental heating is considered, the sequence of operation and commissioning must specify and prevent premature activation of supplemental heating.

Energy Efficiency

- VRF systems benefits from the advantages of linear step control in conjunction with inverter and constant speed compressor combination, which allows more precise control of the necessary refrigerant circulation amount required according to the system load. The inverter technology reacts to indoor and outdoor temperature fluctuations by varying power consumption and adjusting compressor speed to its optimal energy usage. Inverter provides superior energy efficiency performance and also allows for a comfortable environment by use of smooth capacity control. Field testing has indicated that this technology can reduce the energy consumption by as much as 30–40 % a year compared to traditional rotary or reciprocating type compressors.
- VRF technology yields exceptional part-load efficiency. Since most HVAC systems spend most of their operating hours between 30 and 70 % of their maximum capacity, where the coefficient of performance (COP) of the VRF is very high, the seasonal energy efficiency of these systems is excellent.
- A VRF system minimizes or eliminates ductwork completely. This reduces the duct losses often estimated to be 10–20 % of the total airflow in a ducted system.
- Inverter compressor technology is highly responsive and efficient. The modular arrangement permits staged operation, that is, indoor units can easily be turned off in locations requiring no cooling, while the system retains highly efficient operation.
- It is possible to include cooling and heating in a single system which avoids duplicating systems (a reversible heat pump only costs 10 % more than a cooling unit).
- Energy sub-metering with VRF systems is relatively simple and inexpensive by placing an electric meter on one or a few condensing units. This is a very important feature in the multi-tenant buildings if energy costs are charged explicitly to each tenant rather than being hidden in overall leasing costs.

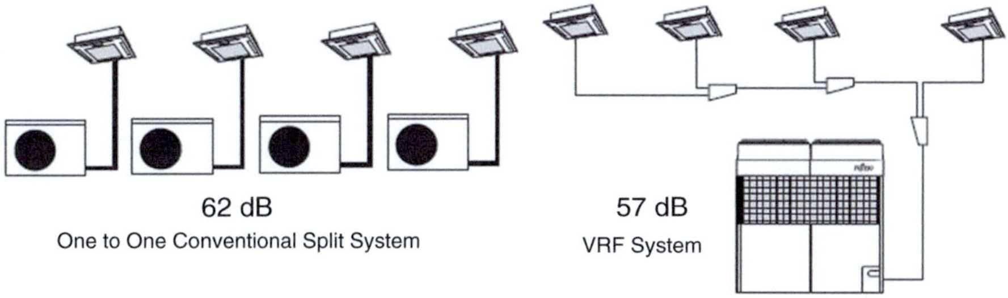

Fig. 3.25 Sound level of VRF system (Source: Fujitsu)

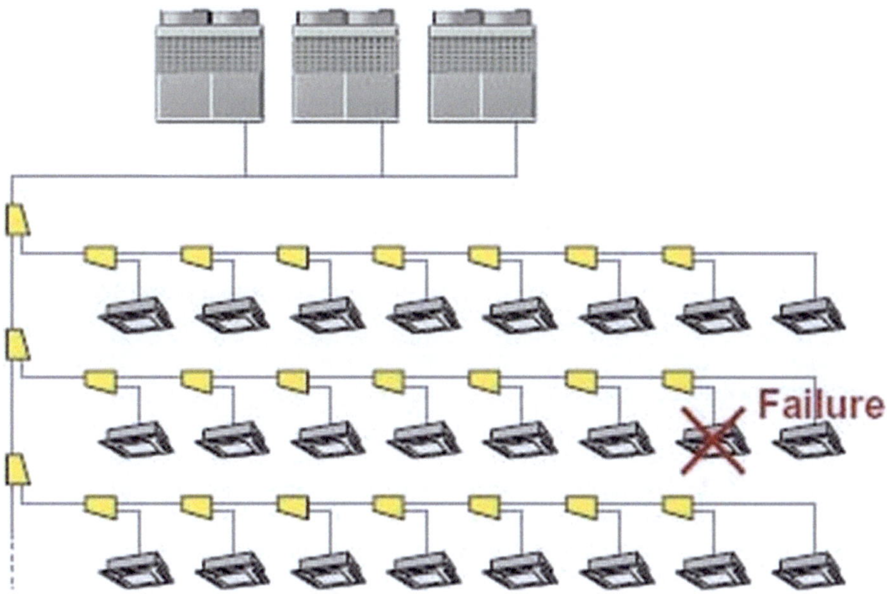

Fig. 3.26 VRF system reliability on compressor failure

Reduced Noise Levels

Indoor and outdoor units are so quiet that they can be placed just about anywhere, providing more flexibility on how to use indoor and outdoor space. Indoor ductless operating sound levels are as low as 27 dB(A) and ducted units sound levels are as low as 29 dB(A).

Outdoor units can even be placed directly under a window, and quiet indoor units are perfect in environments that require minimal disruption like schools, places of worship, libraries, and more. When compared to the single split system, a VRF system reduces outside noise levels by almost 5 dB@1 m (Fig. 3.25).

Reliability

Continuous operation is possible even if trouble occurs at an indoor unit.

Indoor Unit

Each indoor unit is controlled individually on the system network. This allows all indoor units continue to run unaffected even if trouble should occur at any indoor unit in one system (Fig. 3.26).

Outdoor Unit

Continuous operation is possible even in the event of compressor failure. There is no immediate system shutdown if trouble occurs in any compressor. The other compressors can continue to operate on an emergency basis.

Maintenance and Commissioning

- VRF systems with their standardized configurations and sophisticated electronic controls are aiming toward near plug-and-play commissioning. Normal maintenance for a VRF system is similar to that of any DX system and consists mainly of changing filters and cleaning coils.
- Because there are no water pumps to maintain or air ducts to be cleaned, less maintenance is required compared to other technologies.

3.4.1.3 VRF Challenges and Limitations

VRF systems are not suitable for all applications. The key challenges include:

Refrigerant Piping

A VRF system being the split installation is restricted by distance criteria between the condensing unit and the evaporator. The maximum length of refrigerant pipework for a VRF or any other split system is determined by the compressor's ability to overcome the pressure drop and for the system to maintain proper oil return. All "split" systems therefore have a maximum allowable vertical and total refrigeration pipework length. This is a considerable disadvantage compared with hydraulic systems which are pumped, and as the pump may be sized to suit the system, then, theoretically, the hydraulic pipework may be run in almost infinite distances. It is important that the designer/building owner is aware of these limitations.

Each manufacturer specifies both the size of the pipework required for their system and the maximum permissible vertical and total refrigerant pipework runs.

Caution: Although few manufacturers' literature states that the refrigerant lines can be as long as 500 ft, when you read the fine print, after the first "tee" from the condensing unit, you are limited to 135 ft to the furthest unit.

Compliance with ANSI/ASHRAE Standard 15-2001

VRF systems must comply with ASHRAE Standard 15-2011—Safety Standard for Refrigeration Systems (ANSI approved). ASHRAE Standard 15-2001 guides designers on how to apply a refrigeration system in a safe manner and provides information on the type and amount of refrigerant allowed in an *occupied space*.

VRF systems raise the specter of refrigerant leaks which can be difficult to find and repair; particularly in inaccessible spaces. The refrigerant leak, especially if the system serves small rooms, can cause oxygen depletion. So you need to limit the system size within reasonable limits based on the smallest room area served. For example, if the room area is 100 sq-ft, you would need to limit the refrigerant quantity under less than about 30 lbs. Compliance to ASHRAE Standard 15-2011 is sometimes difficult, especially where a long length of refrigerant piping is involved. The total refrigerant charge in the refrigerant loop must be within the limits prescribed by ASHRAE Standard 15-2001. This is to ensure the safety of occupants if the entire charge is released.

Few VRF manufacturers have developed products and protocols to address the concerns of refrigerant leakage. Typically, all joints are brazed joints with no flared fittings. Headers and splitters are specifically designed for the products that do not require flaring or changing wall thicknesses.

Oil Management

As the system has a larger spread, the refrigerant pipes traverse long lengths; hence, their pressure testing and protection becomes critical. Long refrigerant piping loops also raise concerns about oil return. Typically, each compressor has its own oil separator which is optimized for the VRF system. Periodically, the VRF goes into oil retrieval mode during which time the thermostatic expansion valve opens, and the compressor cycles at high pressure to flush oil out of any location where it has accumulated.

Fresh Air Requirements (Compliance to ANSI/ASHRAE Standard 62.1)

One of the most challenging aspects of designing VRF systems is the introduction of outside air to comply with ANSI/ASHRAE Standard 62.1, Ventilation for Acceptable Indoor Air Quality, and building codes, which recommends typically 15–20 CFM of fresh air per person. Like all split systems, VRF systems do not provide ventilation of their own, so a separate ventilation system is necessary.

Ventilation can be integrated with the VRF system in several ways.

A separate ventilation system and conditioning unit could be installed using conventional technology and the VRF system function would be restricted to the recirculation air. Some VRF units have the ability to handle some outside air and could be used accordingly. Bringing the outside air into the room and then conditioning it with the VRF is not recommended except in dry climates where condensation will not create moisture problems. In humid climates, providing preconditioned outside air to each indoor unit ensures good indoor air quality.

Some manufacturers provide a heat recovery unit which provides heat exchange between incoming outside air and the exhaust air from the air-conditioned space independently of the indoor units. With these systems an equal quantity of outside air and exhaust air is supplied and then exhausted from the air-conditioned space. The supply and exhaust air passes over a heat exchanger so heat is recovered from the exhaust air and used to heat or cool the outside air. This solution has the limitation that air is introduced to the space at two different temperatures, i.e., that of the indoor unit and that of the heat recovery unit. If possible, it is always ideal to introduce outside air to the indoor unit.

Particulate Matter Removal

ASHRAE Standard 62.1-2004 (item 5.9) specifically discusses particulate matter removal and how VRF indoor units can or cannot uphold the requirements.

Particulate matter filters or air cleaners have a minimum efficiency reporting value (MERV) of not less than 6 and are rated in accordance with ANSI/ASHRAE 52.2-1999.

The standard filter with 50 % efficiency gravimetric test (which is MERV 1 or 2) is not acceptable.

Note : High MERV rating filters have a higher cost and high pressure drop. These are often suitable for ducted units and some select ductless units such as ceiling-mounted cassette type, double flow; ceiling-mounted cassette type, multi-flow; ceiling-mounted built-in type; ceiling-mounted duct type; slim ceiling-mounted duct type; and console, ceiling suspended type.

Higher rating filters are not available for all types of indoor units such as wall-mounted type, floor standing type/concealed floor standing type, and ceiling suspended cassette type.

Environmental Concerns

Ozone depletion issues have become a global concern, and the issue of a high refrigerant charge associated with long refrigerant lines of VRF systems is a strong negative for the system. But with new refrigerant developments, advances in charge management and controls have transformed the technology to some extent. HFC refrigerants, typically R-410-A and R-407-C, are commonly used.

VRF Systems Are Proprietary Systems

VRF systems are complete proprietary systems from the controls to the condensing units, refrigerant controllers, and all the system components other than the refrigerant piping. That implies users do not have the flexibility to use "anybody's" building control and automation system to run these systems. You will need a BACnet or LonWorks black box to connect from your building DDC system to the VRF system, and you can only monitor what it is doing but you cannot control it.

Reliability and Maintenance

Although suppliers claim that VRF systems are very reliable, contractors and engineers believe that a VRF system with many compressors (e.g., 20 compressors for 100 tons of cooling) is inherently less reliable than a chiller which has a smaller number of compressors (e.g., 1–4 compressors for 100 tons). However, it is also acknowledged as an advantage since, unlike a chiller, a failure of a single compressor would have limited impact on the system's ability to function.

Performance Guarantee

Currently, no approved ARI standard exists for a performance rating of VRF systems. Consequently, manufacturers need to apply for waivers from the Department of Energy to market their products in the United States. Although these waivers have been granted, new applications need to be submitted for new product groups.

3.4.1.4 Concluding

VRF provides an alternative realistic choice to traditional central systems. It captures many of the features of chilled water systems while incorporating the simplicity of DX systems.

Salient Features

- Refrigerant flow rate is constantly adjusted by an electronic expansion valve in response to load variations as rooms require more or less cooling. Also, if reversible heat pumps are used, the heating output can be varied to match the varying heat loss in a room.
- An expansion valve or control valve can reduce or stop the flow of refrigerant to each indoor unit, thus controlling its output to the room.
- This type of system consists of a number of indoor units (up to 48 and varies per the manufacturer) connected to one or more external condensing units.

- The overall refrigerant flow is varied using either an inverter controlled variable speed compressor or multiple compressors of varying capacity in response to changes in the cooling or heating requirement within the air-conditioned space.
- A control system enables switching between the heating and cooling modes if necessary. In more sophisticated versions, the indoor units may operate in heating or cooling modes independently of others.
- A VRF system uses inverters or scroll compressors. They are efficient and quiet and are usually hermetically sealed. Small- to medium-size units may have two compressors.
- Refrigeration pipework up to 500 ft long is feasible.
- Refrigeration pipework level differences between indoor and outdoor units up to 150 ft are possible.
- Ozone-friendly HFC refrigerants; R-410-A and R-407-C are typically used.
- COPs (coefficient of performance) may be as high as 3.8.
- Refrigerant liquid lines tend to be about 3/8″ in diameter and gas lines about 5/8″ to 3/4″ in diameter.
- Central control of a VRV system can be achieved by centralized remote controllers.

VRV/VRF technology is based on the simple vapor compression cycle, but the system capabilities and limitations must be fully understood and evaluated carefully to determine its suitability. Before working with VRV/VRF systems, it is strongly recommended that manufacturer's product training be undertaken (Tables 3.6 and 3.7).

See Table 3.4 for recommended (or frequently used) air handling systems in various types of buildings with varied functions.

The following list provides basic configurations of air, water, and steam HVAC systems typically used in buildings:

All-Air Systems

Single-zone constant air volume system
Single-path systems
Single-duct, variable air volume or constant air volume
Single-duct, variable air volume——induction
Single-duct, variable air volume——fan powered

Dual-Path Systems

Dual-duct constant volume or variable volume system
Multizone constant volume systems

Air and Water Systems

Air–water induction systems
Air–water fan-coil systems
Air–water radiant panel systems

All-Water Systems

Two-pipe hot and chilled water systems
Three-pipe hot and chilled water systems
Four-pipe hot and chilled water systems

Table 3.6 Comparison of VRF and unitary HVAC systems

Item	Description	VRF system	Unitary system
1	*Condensing units components*		
1.1	Single or multiple compressor	Yes	Yes
1.2	Oil separator for each compressor or for all compressors	Yes	Yes
1.3	Oil level control	Yes	Yes
1.4	Active oil return	Yes	In some units
1.5	Option for heating and cooling	Yes	Yes for hot gas defrost
	Simultaneous heating/cooling	Yes	No
1.6	Air-cooled or water-cooled condenser	Yes	Yes
1.7	Liquid receiver	Yes	Yes
1.8	Control of the refrigerant level in the liquid receiver	Yes	Yes
1.9	Condensing temperature control	Yes	It is an option
1.10	Capacity control by the suction pressure	Yes	Yes
1.11	Compressor cooling capacity control by speed (RPM) or steps	Yes	Yes
1.12	Suction accumulator	Depending on the system	Yes
2.0	*Refrigerant lines*		
2.1	Long liquid lines to many evaporators	Yes	Yes
2.2	Refrigerant pipes special design procedure due to pressure drop and oil return	Yes	Yes
3.0	*Internal units*		
3.1	Several units any size	Yes	Yes
3.2	Independent control for each evaporator by an electronic expansion valve	Yes	Yes
3.3	Mechanical subcooling	Provided for pressure drop (if necessary) and to improve performance	Provided to improve performance
3.4	Expansion valve able to handle different cooling capacities and pressure differential	Electronic expansion valve	Thermostatic or electronic expansion valve
3.5	Coil and drain defrost	Only necessary for the external unit heating	Operational and protection
3.6	Air filter	Yes	Not necessary
3.7	Drainage pump	Depends	Depends
4.0	*Controls*		
4.1	Microprocessor control condensing unit	Yes	Yes
4.2	Microprocessor in the evaporator	Yes	Yes
4.3	BMS available	Yes	Yes
4.4	Inverters for power	Yes	Yes
4.5	Alarm codes	Yes	Yes

Source: ASHRAE Handbook

(continued)

Table 3.7 Comparison of VRF and chiller systems

Item	Description	Chilled water system	VRF system	Comments
1.0	Sensible cooling capacity	OK—selection will always meet the thermal load and airflow	There is no option to select equal to the thermal load and airflow	Usually sensible cooling load for VRF is lower than the airflow, you may have to oversize the unit
1.1	Latent cooling capacity	OK—selection will always meet the thermal load. It will be necessary to add a heating device to control humidity	There is no option to select equal to the thermal load	Usually latent cooling load for VRF is a consequence from the sensible load; it will be necessary to add electrical heating for humidity control
1.2	Total cooling capacity	OK—selection will always meet the thermal load	There is no option to select equal to the thermal load and the airflow	Usually total cooling load for VRF is lower than the thermal load or you may have to oversize the unit
1.3	Capacity increase or adjustment for sensible heat factor Air-cooled condenser	Possible—new coil and control valve selection or change in chilled water temperature	There is no option, it should be another equipment or another refrigerant lines	Chilled water is more flexible VRF was design to be compatible with usual Offices and comfort jobs SHF from 0.70 to 0.80
1.4	Capacity increase or	Possible—new coil and control valve	It should be provide room for expansion	VRF was design to be compatible with usual
	adjustment for sensible heat factor Water-cooled condenser	Selection or change in chilled water temperature	Or capacity increase easy to be done Difficult to change the sensible heating factor	Offices and comfort jobs SHF from 0.70 to 0.80
2.0	Airflow in m3/h	Adjustable—it may need a motor change	There is a band for adjustment between a maximum and minimum value	Usually you should oversize the cooling capacity to match airflow
2.1	Airflow pressure drop	Adjustable—it may need a motor change	There is a tap in the motor for adjustment for a higher value	Very narrowband to adjust for VRF. If there is duct, the duct should be calculated according to the external pressure of the internal unit
2.2	Air filter efficiency	Compatible with almost air filter efficiency	It uses a standard low efficiency filter 50 % efficiency gravimetric test not better than MERV 4	VRF there are options up to 85 % efficiency dust spot test MERV 11 but will reduce the external pressure and will have a higher cost

(continued)

Table 3.7 (continued)

Item	Description	Chilled water system	VRF system	Comments
2.3	Electrical motor efficiency internal unit	Higher efficiency, could be better than 90 %	Lower efficiency Depends on the model minimum 60 %	There is no option to change the motor for VRF Not good for ASHRAE Standard 90.1-2004
3.0	Condensate water drainage	Inside the machine room, no problem	It may need pump and needs proper insulation	Very unreliable for VRF Equipment may be over electrical devices
4.0	Long pipes—air-cooled condenser	Increase chilled water pump power but does not change capacity	It reduces the capacity for long lines up to 75 %. It reduces the latent cooling capacity	Very important to verify the real capacity, including the suction pressure drop for VRF. Great issue
4.1	Long condensing water lines	Increase condensing water pump power but does not change capacity	Increase condensing water pump power but does not change capacity	Both are very similar
4.2	Refrigerant lines safety and leakage Air-cooled units	Only in the machine room or in the outside air	It is all over the building, difficult to control and to locate the leakage. High risk for the occupants	Very difficult to certify the VRF system according to ASHRAE 15/1999
4.3	Refrigerant lines safety and leakage Water-cooled units	Only in the machine room or in the outside air	It is all over the same floor, not so difficult to control and to locate the leakage. High risk for the occupants	VRF system may be possible to certify according to ASHRAE 152007
5.0	Coefficient of performance	Easy to calculate; it depends on the % cooling capacity and the outside air	Condensing unit is almost constant regarding the cooling capacity but depends of the outside air	High-efficiency chilled plant could be 0.8 kW/t, and VRF condensing units could be 0.95 kW/t all-year-round average for Sao Paulo, Brazil

(continued)

Table 3.7 (continued)

Item	Description	Chilled water system	VRF system	Comments
5.1	Capacity control	Leaving water temperature keep constant, by the capacity control on the compressor	Suction pressure of the compressor keep constant by the capacity control on the compressor speed or stages	Constant pressure control in the suction line near the compressor keeps the COP constant, but it does not gives the same value for the evaporator due to the pressure drop. Reduces the latent cooling capacity for VRF
5.2	Water-cooled condenser	Shell and tube condensers, standard procedures and easy to clean	Plate heat exchanger or tube in tube, it needs a closed circuit with the use of an intermediate heat exchanger	Higher initial cost but very low maintenance for VRF
6.0	Heating and cooling	Needs four pipes to heat and cool at the same time with heat recovery	Almost standard easy to do and low cost	Advantage for the VRF
6.1	Cooling and heating control	Very sophisticated, not so easy to use for the costumer	Easy to use, is the same as the mini-split	Advantage for VRF, it does not need trained personnel to operate

Source: ASHRAE Handbook

Unitary Refrigerant-Based Systems

Window-mounted air conditioners and heat pumps
Through-the-wall-mounted air conditioners and heat pumps
Air-to-air heat pumps
Outdoor unitary equipment systems
Indoor unitary equipment systems
Water-source heat pump systems

Variable Refrigerant Flow Systems
Heat Recovery Systems

Heat pumps and heat recovery chillers
Energy recovery wheels
Runaround heat recovery loops

Panel Heating and Cooling Systems

Radiant heating panels
Perimeter heating system
Embedded pipe coil system
Ceiling's electric resistance panel
Forced warm air floor panels

Cogeneration Systems

Prime movers
Reciprocating internal combustion engines
Combustion gas turbines
Expansion turbines
Steam boiler turbine combinations
Generators
Waste heat recovery systems
Electrical and thermal distribution system

Steam Systems

Convection-type steam heating systems
Combined steam and water systems
One-pipe steam heating systems
Two-pipe steam heating systems

Central Plant Heating and Cooling System

Central chilling plant
Chilled water and dual-temperature systems
Terminal heating and cooling units

Chilled and Dual-Temperature Water Systems

Two-pipe chilled water systems
Brine systems

Two-pipe dual-temperature water systems
Two-pipe natural cooling systems
Three-pipe systems
Four-pipe systems
Condenser water systems
Once-through systems
Cooling tower systems

Medium- and High-Temperature Water Heating Systems

Direct fired hot water generators
Direct contact heating

Infrared Radiant Heating System

Spot heating system
Total building heating system

Geothermal Heating/Cooling System

Ground-source heat pump system
Ground-coupled heat pump system
Horizontal system
Parallel system

Thermal Storage System

Ice storage system
Chilled or hot water storage system
Building mass thermal storage system
Electrically charged heat storage system

Solar Energy Utilization

Solar heating and cooling systems
Solar collector and storage subsystems
Solar cooling system

Energy Management System
Special Ventilation System
Automatic Control System
Sound and Vibration Control System
Water Treatment System Evaporative Air Cooling System

Indirect/direct evaporative cooling systems

Snow Melting System

Embedded hydronic pipe system
Embedded electric resistance heating
Overhead high-intensity infrared radiant heating

Smoke Management System

Stack effect system
Compartment system
Smoke dilution system
Pressurization system

3.5 Security Systems

Security Systems (SS) Access Control There are numerous methods of providing access control. Among these methods are guard force, mechanical or electronic cipher locks, dial combination locks, key locks, and internally controlled electrical door releases. Recent developments in biometric access control can be used for access when permitted by individual service publications. Biometric access control is the method of identification and verification in which the person seeking access is identified by fingerprints, palm pattern and geometry, retinal pattern, voice analysis, signature dynamics, and other methods.

Closed-Circuit Television (CCTV)
This type of system may be used for physical security inside and outside the facility. This system is particularly effective for monitoring hallways, doors, and perimeters. The CCTV system consists of a monitor station and cameras installed in strategic locations.

Intrusion Detection System (IDS)
An intrusion detection system may be used in facilities to provide additional protection. IDSs typically consist of magnetic switches and motion sensors. All exterior entrances, open areas, and sensitive areas within the facility should be protected. Motion sensors may be installed to provide general coverage of open areas to prevent an intruder from crawling under the area of coverage. The IDS should connect to a remote monitor station through a dedicated line or autodial telephone line. This will aid in providing notification to the appropriate police or security agency).

3.6 Basic Fire Protection Systems (FPS) Used in Buildings

It is necessary to install regularly in buildings automatic fire extinguishing systems that have a fire alarm signaling system. Fire alarm signals where included to initiate extinguishing agent discharge, alarm building personnel, and also provide supervisory service to indicate any off-normal conditions of the system. There are many types of extinguishing systems used with a fire alarm system. Figure 3.15 lists all types of fire protection systems encountered in buildings. At the top are shown regulations and authorities that classify the building and owner's underwriter requirements that determine the type and need.

Sprinkler systems are used in the majority of commercial, institutional, and industrial buildings. There are four types of automatic water sprinkling systems: (1) wet pipe, (2) dry pipe, (3) pre-action, and (4) deluge. Wet-pipe systems employ automatic sprinklers attached to pipes connected to a water supply. The pipes are filled and the sprinklers discharge water immediately when opened by the heat of a fire. The dry-pipe system contains pressurized air or nitrogen. When the sprinklers are opened, as by the heat of a fire, the gas pressure drops water pressure from the supply to open a valve, and water flows into the pipes to be dispensed from the sprinklers. See Fig. 3.16 for the different types of valves. The pre-action system employs automatic sprinklers connected to pipes containing air that may or may not be under pressure. A supplemental fire detection system is installed in the same area as the sprinkling system. When the detection system is activated, a valve opens, permitting water to flow into the pipes and to be discharged through the sprinklers (which have been opened by the fire). The deluge system is similar to the pre-action system but uses open sprinklers instead of closed ones.

Table 3.8 provides a brief description of some important fire protection systems listed in the Fig. 3.15.

3.6.1 Fire Alarm, Detection, and Signaling Systems

Fire alarm systems are classified according to the type of functions they are expected to perform. Regardless of the type of alarm system, national or local codes have very specific requirements for protective system signaling and source of power supply and backup power supply. Fire detection and alarm system circuit can be supervised or nonsupervised. Table 3.9 summarizes basic classification of fire detection and alarm system per NFPA into six types: local, auxiliary, remote, proprietary, central, and emergency.

3.7 MIE Equipment Arrangement

The MIE system arrangement is a combination of pieces of mechanical and electrical and communication equipment and their interconnection with conduits, pipes, wires, channels, ducts, fibers, and fields. Typically gases, vapors, and liquids are conducted by pipes, tubes, and hoses and sometimes open channels; electricity is conducted by solid conductors, ionized liquids, and ionized gases; heat and vibration are conducted by all solids; and people are conducted by paths, ladders and steps, elevators, and escalators.

Table 3.8 Basic fire protection systems and applications

Type	Basic description	Features
1. Wet-pipe sprinkler system	A permanently piped water system, maintained under pressure. When fire occurs, discharges extinguishing medium (water) through heat-activated fusible sprinkler heads exposed to the high heat	Automatically activates and controls or extinguishes the fire. Protects base structure. Causes water damage to building interiors and equipment and any material or device prone to water damage
2. Pre-action automatic sprinkler system	A permanently piped system employing heat-activated sprinklers. Piping system is filled with air that may or may not be under pressure. A fire, heat, or smoke detection system installed in the same area of a sprinkler zone. On detection of smoke, fire, or heat, it activates a "pre-action valve" that permits water to flow into the sprinkler system piping ready to be discharged from any sprinklers that are opened by the heat from the fire	Automatically detects fire. Minimizes the accident discharge of water due to mechanical damage to sprinkler heads or piping and thus is useful for the protection of building interiors and critical equipment
3. On–off automatic pre-action sprinkler system	A pre-action system with fire/heat detector interlock; heat detector actuates "a pre-action valve" to open at a predetermined temperature and close when normal temperature is restored. Should the fire rekindle after its initial actuation and closing, the valve will reopen and water will again flow from the opened heads. The valve will continue to open and close in accordance with the rate of a temperature rise sensed by the heat detectors	In addition to the favorable feature of the automatic pre-action wet-pipe system, this has the ability to further reduce the water damage by automatically stopping the flow of water when no longer needed, thus eliminating unnecessary water damage
4. On–off automatic wet sprinkler system	A wet-pipe system with on–off sprinkler heads. Here, each individual head has incorporated in it a temperature-sensitive device that causes the head to open at a predetermined temperature and close automatically when the temperature at the head is restored to normal	This feature adds the automatic or reclosing control to reduce water damage. It provides the best of wet and pre-action systems

(continued)

Table 3.8 (continued)

Type	Basic description	Features
5. Dry-pipe sprinkler system	It is similar to the pre-action system with no supplemental detection system. It employs "deluge valve," heat-operated sprinklers attached to a piping system containing air under pressure. When a sprinkler operates, the air pressure is reduced, a "deluge valve" is opened by water pressure, and water flows to any opened sprinklers	Provides protection to areas subject to freezing. Water supply and deluge valves must be in a heated area
6. Deluge sprinkler system	It employs open sprinklers attached to a piping system that is empty, with a supplement fire detection system installed in the same area as the sprinklers. A deluge control valve is normally closed to prevent water from entering the system. When fire occurs, the supplement detection system actuates and opens the water control valve, allowing water to enter the system. Water is discharged through all of the sprinklers on the system, thus deluging the protected areas	Used in high hazard areas requiring an immediate application of water over the entire hazard. *Examples:* flammable liquid handling and storage areas, aircraft hangars, oil refineries, chemical plants, hazardous equipment
7. Standpipe and hose system	A permanently installed wet-pipe systems in a building, maintained under pressure, to which a flexible hose is connected for emergency use by building occupants or by the fire department	Features normally provided as a complement to an automatic wet/dry or pre-action sprinkler system. Building personnel need to be familiar with effective hose application
8. Non-CFC agent flooding system	A permanently piped system using a limited stored supply of a non-CFC inert gas under pressure and discharge nozzles arranged to totally flood an enclosed area. Released automatically by a suitable heat/fire or smoke detection system. Extinguishes fires by blanketing the fire zone, thus inhibiting the chemical reaction of fuel and oxygen to extinguish fire	No damage to critical electronic equipment, devices, or other irreplaceable valuable objects. No agent residue. Toxicity level of the agent used governs the limit of agent concentration and duration of extinguishing action. A new breed of non CFC agents may not extinguish deep-seated fires in ordinary solid combustibles, such as paper, fabrics, etc., but are effective on electrical fires or surface fires in these materials. These systems require special airtight areas/enclosures to maintain concentration for a specified time. The high-velocity discharge from nozzles requires extreme care in location of nozzles to avoid damage due to instant release

9. Carbon dioxide flooding system	Uses carbon dioxide gas as flooding agent. Activates through smoke, fire, or heat detection system. Extinguishes fires by reducing oxygen content of air below combustion chemical reaction points	Toxic agent, suitable only for normally unoccupied areas and where water damage is more catastrophic, for example, service and utility areas. Personnel must evacuate before agent discharge to avoid suffocation. May not extinguish deep-seated fires but effective on electrical fires
10. Dry, chemical flooding system	Similar to other flooding systems, except uses a dry chemical powder. Released by mechanical thermal linkage or detection system. Effective for live electrical fires	Can suffocate and should not be used in personnel-occupied areas. Leaves powdery deposit on all exposed surfaces. Requires a cleanup. Excellent for service facilities having kitchen range hoods and ducts. May not extinguish deep-seated fires in ordinary solid combustibles, such as paper, fabrics, etc., but effective on surface fires in these materials
11. High-expansion foam system	A fixed wet foamy extinguishing system that generates a foam agent through mixing for total flooding of confined spaces. It extinguishes fire by cooling and displacing vapor, heat, and smoke. This reduces the oxygen concentration at the fire	High volumetric discharge supplemented by high-expansion foam may inundate personnel in the space, blocking vision, hearing, and creating breathing problems. Leaves the residue and requires a heavy cleanup. Effective on oil storage area fires
12. Water-mist (WM) system	A new type of pre-action sprinkler system that uses high pressure to create a small-droplet water mist to extinguish fire	The small droplets provide cooling, oxygen displacement, and radiant heat attenuation. A WM system could be used in computer room fire suppression

Table 3.9 Basic fire alarm systems and applications

Category	Basic description	Features
1. Local boundary alarm system	Building fire alarm system, operation limited to within building. Alarms on activation of a manual fire alarm box, water flow in a sprinkler system, tamper switches on fire valves, or detection of a fire by a smoke or heat detecting system	The main purpose of this system is to provide an evacuation alarm for the occupants of the building. Requires building personnel to transmit the alarm to local fire authorities
2. Auxiliary fire alarm signaling system	System includes a standard local municipal-coded fire alarm box to transmit a fire alarm from building to local municipal fire station. These alarms are received on the same municipal equipment and are carried over the same transmission lines as are used to connect fire alarm boxes located on streets. Operation is initiated by the building fire detection system, manual pull stations, or water flow in the pipes	Summons immediate help from a local fire department. Some communities are equipped to accept this type of system and others are not
3. Remote station fire alarm signaling system	Utilizes building dedicated telephone lines to transmit alarm to a remote station, such as a fire station or a police station. Includes separate receiver for individual functions being monitored, such as fire alarm signal, manual pull station, or sprinkler water flow alarm	Requires leased telephone lines into each protected building
4. Central station alarming system	Incorporates a privately owned central station whose function is to monitor the connecting lines constantly and record any indication of fire, supervisory, or other trouble signals from the protected premises. When a signal is received, the central station will take such action as is required, such as informing the local fire department of a fire or notifying the police department of the tamper	Flexible system. Can handle many types of alarms, including trouble, security, intrusion, etc., within system at protected premises
5. Proprietary alarming system	An alarm system that serves multiple buildings under one ownership from a central supervising station at the protected building. Similar to a central station system but owned by the protected property	Requires attendant round the clock at central supervising station on the premises
6. Emergency voice/alarm communication system	Includes emergency voice/alarm/or data communications on any of the systems listed above	Provides personal transmission of information to occupants of the building (including fire department personnel)

Heat is transported by moving fluids, including furnace gases, boiled water (steam), boiled refrigerants, air (including refrigerated air to be reheated), and the vapor in heat pipes. Some fluid transportation is driven by convection owing to the different densities of warm and cold fluids, and some is driven by pumps and blowers. Vapor is also driven to the cool end by the pressure drop due to condensation.

Information, signals, video, audio, TV, and data communication is increasingly done via light pulses in optical fibers because of the enormously greater rate of information transport that can be carried over an optical fiber than over a wire or a radio channel. (The common word is fiber optics.) All optical conduits and barriers are mechanical objects that conduct or resist the passage of light (Figs. 3.27 and 3.28).

There is no barrier to the passage of an electric or magnetic field, but iron, nickel, and their alloys provide conduits for it which can bypass a space where the field is unwelcome. A system engineer usually wants an energy flow transportation system to offer least resistance so he can get desired pressure, voltage, or flow distribution in building system networks, and he wants some means to temporarily close or open energy flow either entirely or partially, either by hand, or by remote control, or automatically. The generic word for such a device in a building energy pathway is "valve," but in common building system lingo, a valve is a device for liquids and gases. (In England, vacuum tubes are called valves.) The following is a list of variety of valves used in building systems including both fluid and other valves so you can see the relationships among them.

1. Electrical valves include relays, switches, circuit breakers, fuses, vacuum tubes, diodes, SCRs, MOVs, and transistors.
2. Fluid valves include on–off valves, transfer valves, check valves (fluid diodes), and throttling valves. For on–off, full or no flow requirements, ball, cock, or plug, or gate valves are favored; where tight shutoff is not required, butterfly or slide valves are used; for throttling purpose globe or for finer adjustment of flow, needle valves are used. They may be operated or adjusted by hand; by electrical, electronic, pneumatic, or hydraulic signals; or by mechanical signals (such as the float-operated valve in a cooling tower). They may be operated by the direction, pressure, or speed of the fluid itself, such as check valves, safety valves, or flow control valves. See Fig. 3.17 for the different types of valves used in building systems.
3. Most heat valves are fluid valves for the fluids that carry the heat. Radiant heat valves are made as radiation absorbers or reflectors.
4. Doors with locks are valves for people. Doors with latches are valves for animals and babies.

3.8 Building Systems Distribution Tree

The distribution tree is the means for delivering information, power, heating, and cooling: the "roots" are the machines and processors that provide information, power, heat, and cold; the "trunk" is the main duct or pipe that carries air, water,

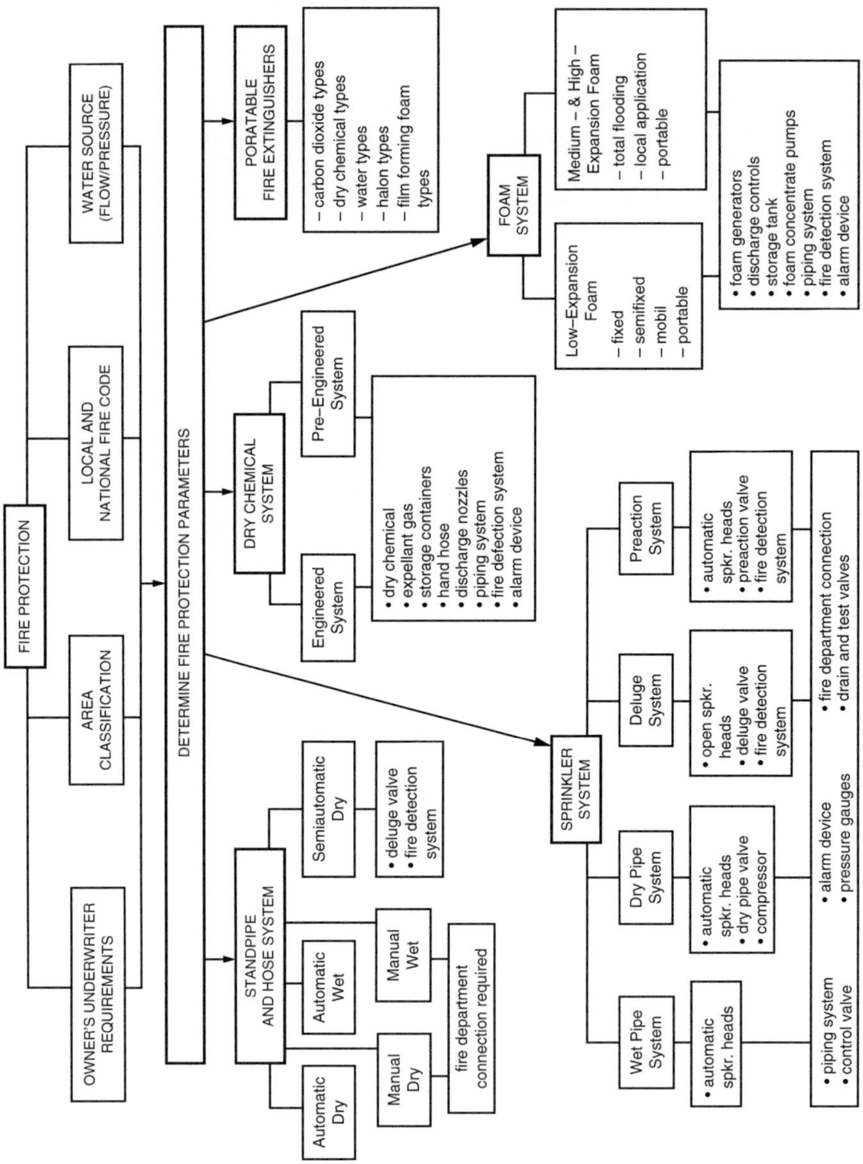

Fig. 3.27 Fire protection system design criteria and types

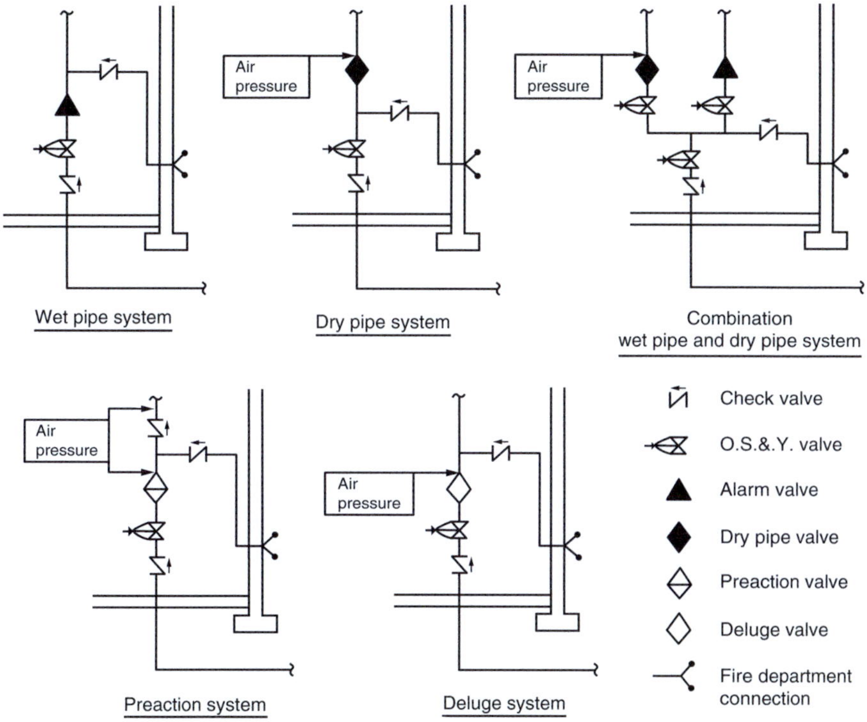

Fig. 3.28 Basic fire protection system valve arrangements

current conductors, and optical fibers from the information centers and mechanical/ electrical equipment to the zone to be served; and the "branches" are the many smaller ducts or pipes or fiber nerves that lead to individual spaces.

For now, the questions to be answered about distribution trees for buildings are: How many? What kind? Where? A building can have one giant distribution tree, several medium-sized trees, or an orchard of much smaller trees. At one extreme, a large building system room is the scene of all power, heating, cooling, and information production; leading from this room is a very large trunk of duct and pipes with perhaps hundreds of branches. Power and signal pipes are like tree creepers attached to the trunk and branches of the tree. At the other extreme, each zone is self-contained and has its own electrical/mechanical equipment (such as a rooftop heat pump, transformer, panel board, and processor), with short trunks and relatively few branches on each tree (Fig. 3.29).

What kind of distribution tree? Most simply, it can be an air (ducts), water (pipes), power (pipes and ladders), or information (fiber conduits and wire pairs) tree. Air distribution trees are bulky and therefore likely to have major visual impacts unless they are concealed above ceilings or within vertical chases. Water and power distribution trees consume much less space (a given volume of water carries vastly more heat that than does the same volume of air at the same temperature) and can

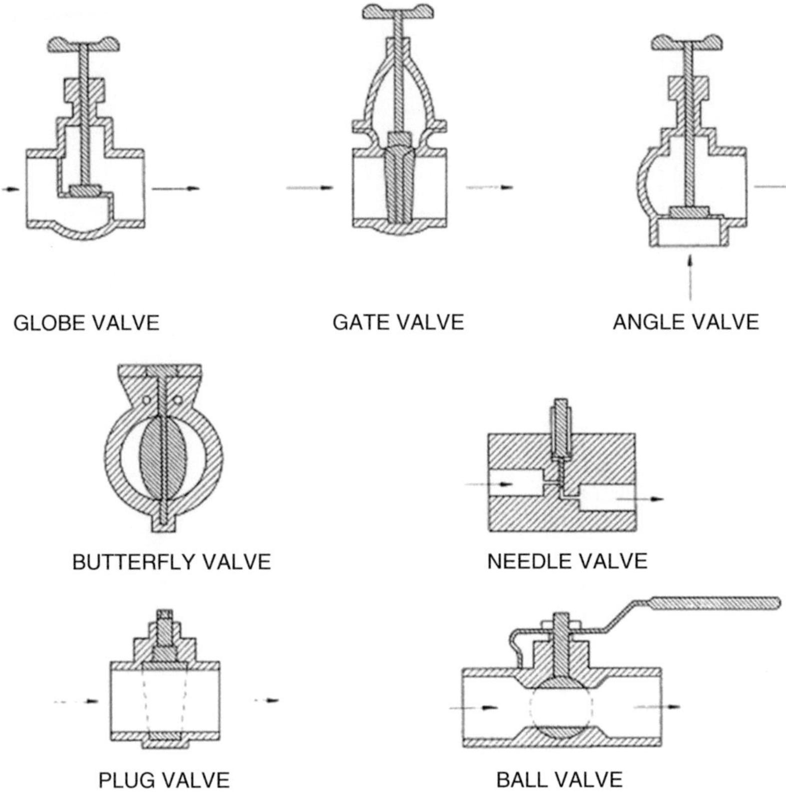

Fig. 3.29 Valves commonly used in building systems

be easily integrated within structural members, such as columns. Further, information distribution trees consume less space and can be supported off any main tree provided they maintain distance to power and water pipes. Both air and water trees can be sources of noise.

Distribution trees are rich in form possibilities; their variations are enormous. For now, assume that the choices of how many, what kind, and where have tentatively been made and that the details and variations are to be considered later. The designer who has chosen the basic form, type of envelope components, and a support system distribution tree with different branches can now consider questions of size.

Another important factor in system choice is the amount of space the system requires. In some cases, it is easy to provide small building system equipment rooms at regular intervals throughout a building, such that little or nothing in the way of a distribution tree will be required. In other cases, a network of distribution trees and central, large equipment spaces are easier to accommodate.

The relocation of risers at the perimeter (where they are most needed) provides a reduction in the length——and consequently in the size——of pipe and duct branches. This avoids the need to move large assemblies of building "arteries and veins" to the surfaces they serve and eliminates bulky ceiling crossovers. It is logical to place at the perimeter the parts of the system that deal with the effects of sun, shade, and temperature change in the different perimeter zones, leaving at the core a separate network to handle the more stable interior areas. The disadvantages of perimeter distribution include (usually) higher construction costs and an environment that is more thermally hostile, owing to the extremes of outdoor temperature.

The integration of distribution tree with building structures is required owing to increases in the complexity and size of the building distribution systems with technological development (typically, in modern buildings, more air and power is required to power and cool a space owing to high internal loads). The size of structural systems is reduced owing to the increased strength of materials. The "uncluttered" floor areas between the more widely spaced columns are now desirable for flexibility in spatial layouts. With the building system branches at or within these columns, floor areas remain clear, thus giving building system structural integration further impetus. With the high-efficiency expectations for cooling, the refrigeration cycle's cooling tower often moved to the roof, taking the bulky air handling machinery with it. This further encourages the merging of systems, for one system is growing wider as the other is diminishing.

Thus, while it is possible to wrap the building systems in a structural envelope, it is of questionable long-term value, given the differing life spans and characteristics of these systems. The functions of these systems differ widely; compared to the dynamic on–off air, water, and electrical distribution systems, the structural system is static. Changes in occupancy can mean enormous changes in systems, requiring entirely different equipment; structural changes of such magnitude usually occur only at demolition. Building systems can invite user adjustment; structural systems rarely do.

To overcome the above flexibility issues, *horizontal distribution* above corridors is very common, since reduced headroom here is more acceptable than in the main activity areas. Further, corridors tend to be away from windows, so their lower ceilings do not interfere with daylight penetration. Since corridors connect nearly all spaces, horizontal service distribution to such spaces is also provided. Furthermore, exposure of these services above corridors can heighten the contrast between such serving spaces and the uncluttered, higher ceiling offices that are served.

Rooftop equipment and distribution tree offer the *economizer cycle* which uses cool outdoor air, as available, to ease the burden on a refrigeration cycle as it cools the recirculated indoor air. The economizer cycle can thus be thought of as a central mechanical substitute for the open window. Relative to open windows, this cycle has several advantages: energy-optimizing automatic thermal control, filtering of fresh air, tempering of the cool outdoor air to avoid unpleasant drafts, and an orderly diffusion of fresh air throughout the building. Its disadvantages are the loss of personal control that windows offer and little awareness of exterior–interior interaction.

3.8.1 Tree Care Cost and Conservation

One big advantage of central equipment rooms is the opportunity they present for energy conservation. Regular maintenance is simplified when all the equipment lives in a generous space kept at optimum conditions; with regular maintenance comes increased efficiency of operation. Another conservation opportunity is that of heat transfer between various machines, or between distribution trees, where one's waste meets another's need.

Within equipment, heat transfer can occur in a *boiler flue economizer,* through which the hot gases in a boiler's stack are passed for use in preheating of the incoming boiler water. For cooling equipment, *dual-condenser chillers* can reject their heat either to a cooling tower (via the heat rejection condenser) or to the building (via the heat recovery condenser).

There are numerous methods for heat transfer between distribution trees—especially between building exhaust air and fresh air. When these two air streams are rather far apart, a set of *runaround coils* can be used. *Energy transfer wheels* go further, in that they transfer latent as well as sensible heat. In winter, they recover both sensible and latent heat from exhaust air; in summer, they can both cool and dehumidify the incoming fresh air. Seals and laminar flow of air through the wheels prevent mixing of exhaust air and incoming air. We commonly experience daily changes from warmer to colder conditions, even in summer. Thermal storage at the base of equipment tree for large buildings can take advantage of this cycle to increase operating efficiency and save energy.

To carry the tree analogy to its logical conclusion, in the spaces served, consider the "leaves," the points of interchange between the piped and ducted heating or cooling; cabled or conducted power and lighting; wired or fibered information and signals; and piped or hosed fire suppression and drinking water. In return leaves shade, comfort, nourish, and protect the spaces served.

Chapter 4
Air, Water, Power, and Internet Quality

Anil Ahuja and Afi M. Hassan

Consider a traditional building. Yet in it, you provide security locks to the building entrance as filters to pass certain humans and not others. You provide water filters to avoid microorganisms and bacteria in drinking water. You provide electrical filters as barriers to unwanted frequency and power distortion. You provide air filters; thus supplying air is dirt filtered, heated, or cooled. You provide vapor filters against leakage moisture transmission. All these basic filters basically provide clean air, power, and water to building habitats. In a high-tech building, this requirement is expanded to microlevels and expectation of building systems are at that level.

4.1 Indoor Air Quality (IAQ) and Modem Buildings

Airborne materials include gases and particulates that may be generated by occupants and their activities in a space, from outgassing and/or shedding of building materials and systems, originate in outside air, and/or created from building operating and maintenance programs and procedures. Airborne particles include bioaerosols, asbestos, man-made mineral fibers, and silica, and gaseous contaminants include radon and soil gases as well as volatile organic compounds.

4.1.1 Contaminants and Their Sources

Bioaerosols These are airborne microbiological particulate matter derived from viruses, bacteria, fungi, protozoa, mites, pollen, and their cellular or cell mass components. Bioaerosols are everywhere in the indoor and outdoor environments.

A. Ahuja, *Integration of Nature and Technology for Smart Cities*,
DOI 10.1007/978-3-319-25715-0_4

Some common bioaerosol sources are saprophytic bacteria and fungi in the soil and in the atmosphere, for example, *Cladosporium*, a fungus commonly found on dead vegetation and almost always found in outdoor air. *Cladosporium* spores are also found in indoor air, depending on the amount of outdoor air that infiltrates into interior spaces or is brought into the HVAC system. Bacteria that are saprophytic on human skin (*Staphylococcus*) and viruses (e.g., influenza A) that are parasites in the human respiratory tract are shed from humans and are thus commonly present in indoor environments.

Although microorganisms are normally present in indoor environments, the presence of abundant moisture and nutrients in interior niches amplifies the growth of some microbial agents to the extent that the interior environment is microbiologically rich. Thus, certain types of humidifiers, water spray systems, and wet porous surfaces can be reservoirs and sites for growth of fungi, bacteria, protozoa, or even parasites. Turbulence associated with the start-up of air-handling unit plenums may also elevate concentrations of bacteria and fungi in occupied spaces.

The presence of bioaerosols or microorganisms in indoor environments may cause infective and/or allergic building-related illnesses. Some microorganisms under certain conditions may produce bad odors and volatile chemicals that are irritative, thus contributing to the development of what is called the *sick building syndrome*. Outbreaks of infectious illness in the indoor air may be caused by other types of microorganisms, such as viruses from ill persons.

Asbestos The term asbestos refers to a group of naturally occurring silicates that occur in fiber bundles with unusual tensile strength and fire resistance. In older buildings they are found in insulation products. This material is banned in building materials. In old buildings, exposure can occur owing to natural degradation, renovation, and removal.

A variety of diseases have been associated with asbestos exposure, such as asbestosis (a chronic disease of the lung tissue leading to respiratory failure because of inadequate ability to transport oxygen across the lung tissue) and cancers of the lungs, larynx, and gastrointestinal tract (esophagus, stomach, colon, and rectum).

Asbestos use in the United States is strictly regulated through federal agencies. The Occupational Safety and Health Administration (OSHA) has jurisdiction in the workplace, with an overall permissible exposure limit set the same for both general industry and construction. The Environmental Protection Agency (EPA) regulates asbestos use and exposure in a variety of other settings, including schools and residential environments. Strict guidelines are provided, including a document on safe handling and abatement/remediation.

Man-Made Mineral Fibers (MMMFs) Also known as man-made vitreous fibers (MMVF), these are synthetic, amorphous, noncrystalline vitreous structures. They are used in buildings for insulation materials. Since they are amorphous, they do not split lengthwise, as do crystalline fibers such as asbestos. They break across their diameter, leading to progressively shorter pieces. Under the influence of high temperatures, refractory fibers may change their chemical/physical characteristics to crystalline silica.

Possible health effects from man-made mineral fibers are still very controversial. Broad categories of possible health effect are as follows:

Dermatitis: itching and erythema (reddening of skin) occur after dermal exposure.
Microbial contamination: exposure to specific microbial agents may sensitize some individuals.
Lung disease: several surveys have examined the relationship between MMMF exposure and interstitial lung disease, the kind of disease associated with asbestos exposure. At present, there is no evidence that MMMF causes interstitial lung disease.
Cancer and mesothelioma: man-made mineral fibers have been suspected of leading to elevated cancer rates. Still, at present there is no clear evidence that any forms of MMMF are carcinogenic in humans. However, the International Agency for Research on Cancer and the EPA have classified mineral wool and rock wool as possible human carcinogens.

OSHA regulates man-made mineral fibers as nuisance dust, at 15 mg/m^3 for the construction industry. The National Institute for Occupational Safety and Health (NIOSH) proposes a recommended exposure level of 3 f/mL (fibers/mL) for respirable fibers and a level of 5 mg/m^3 for total fibrous glass. The American Conference of Governmental Industrial Hygienists (ACGIH) recommends 10 mg/m^3. Sweden has set a level of 1 f/mL for respirable fibers, and Denmark has set a level of 2 f/mL.

Silica Silica consists of silicon dioxide, which is abundant in the earth's crust and is used in the manufacture of glass, refractories, abrasives, buffing and scouring compounds, and lubricants. Its biological effects depend on the grade of the silica. Silica is associated with three forms of silicosis: acute, accelerated, and chronic (nodular). They are characterized by an immunologic action of silica in the lungs, leading to progressive tissue destruction and immunological abnormalities. Standards have been established separately by OSHA and Mine Safety and Health Administration (MSHA). The environmental criterion is five (5) fibers per milliliter. Sampling may occur on particulate filters. Quantification is through low-temperature ashing or X-ray diffraction spectrometry.

Radon and Other Soil Gases Radon is a naturally occurring, chemically inert, odorless, tasteless radioactive gas. It is produced from the radioactive decay of radium, which is formed through several intermediate steps from the decay of uranium. Additional, but secondary, sources of indoor radon include groundwater and radium-containing building materials.

Radon gas enters buildings through cracks or openings such as sewer pipe and sump pump openings, cracks in concrete, and wall–floor joints. The amount of radon entering and subsequent indoor concentration distribution depends on several factors, including the concentration of radium in the surrounding soil or rock, the soil porosity and permeability, and the air pressure differential between the building and the soil or between various indoor spaces which may result from the stack effect, operation of exhaust fans, or operation (or lack of) HVAC equipment. While several sources of radon may contribute to average radon levels in buildings, pressure-driven flow of soil gas constitutes the principal source.

The health effects associated with radon are mainly due to exposures to radon decay products, and the amount of risk is assumed to be directly related to the total exposure. Indoor concentrations of radon can vary hourly, daily, and seasonally, in some cases by as much as a factor of 10 to 20 on a daily basis. For example, measurements made during a mild spring may underestimate the annual average level because of ventilation from open windows or the operation of HVAC equipment (e.g., economizer operation, which increases outdoor air ventilation rates). Similarly, indoor levels during a cold winter may be higher than average because the building is sealed and the outdoor air ventilation rates are minimized. Thus, longer-term measurements (6 months to 1 year) made during normal use generally provide more reliable estimates of the average indoor concentration.

Exposure to indoor radon may be reduced by preventing radon entry, or removing or diluting radon or radon decay products after entry. To prevent entry, sealing various parts of the building substructure has been attempted, but it is often not completely effective by itself, since some openings in the building shell may not be accessible or new openings can develop with time. Because furnaces or HVAC systems can contribute to the depressurization of buildings, the location and tightness of supply and return ducts are important, especially in new construction.

Any volatile organic compounds present in soil gases in the vicinity of buildings may enter those buildings under certain, as yet undefined, soil and building conditions. Therefore, their presence, or possible presence, should be considered in the design, construction, and operation of buildings—particularly in the vicinity of inactive landfills or hazardous waste sites. Buildings in these areas should be made to minimize substructure leakage, to prevent substructures from being depressurized by the operation of HVAC systems, and to seal duct systems in crawl spaces or beneath slab foundations against the possible intrusion of gases. Techniques installed to prevent entry of radon gas will also prevent entry of other soil gases.

Volatile Organic Compounds (VOCs) These are air pollutants found in all nonindustrial environments. After ventilation, VOCs are probably the first concern when diagnosing an IAQ problem. VOCs are organic compounds with vapor pressures greater than about 10^{-3} to 10^{-4} mmHg (torr). While building and furniture materials are known to emit VOCs, ventilation may also transport outdoor pollutants to the indoor environment; so the ventilation system itself may be a source of VOCs (refer back to Table 4.1 for indoor air quality safe limits of some gases and chemicals naturally present in ambient air).

Adverse health effects potentially caused by VOCs in buildings indoor environments fall into three categories: (1) irritant effects, including perception of unpleasant odors and mucous membrane irritation; (2) systemic effects, such as fatigue and difficulty concentrating; and (3) toxic chronic effects, such as carcinogenicity. Approaches to reducing indoor exposures to VOCs include the following:

- Use low-emitting products indoors.
- Increase general ventilation, although this may not be energy-efficient and may not be effective for some sources, such as building materials, where increasing the ventilation six times decreases the VOC only by 50 %.
- Install local ventilation, that is, local exhaust ventilation, near photocopiers, printers, and other point sources.

Table 4.1 EPA data on drinking water

Pollutant	Maximum contaminant level (MCL) mg/l or ppm
Arsenic	0.05
Bacteria	4/100 mL
Barium	1.00
Benzene (organic)	0.005
Cadmium	0.01
Carbon tetrachloride (organic)	0.005
Chloride	250.0
Coliform	<1/100 mL
Color (platinum–cobalt scale)	15 units
Copper.	1.0
Chromium (hexavalent)	0.05
Cyanide	0.01
1,1 Dichloroethylene (organic)	0.007
1,2 Dichloroethylene (organic)	0.005
Endrin (organic)	0.0002
Fluoride	4.0
Foaming agents	0.5
Iron (>0.3 makes red water)	0.3
Lead	0.05
Lindate (organic)	0.004
Manganese (>0.1 forms brown-black stain)	0.05
Mercury	0.002
Methoxychlor (organic)	0.1
Nitrate	10.0
Odor (threshold odor)	3
p-Dichlorobenzene (organic)	0.075
pH	6.5–8.5
Selenium	0.01
Silver	0.05
Sulfate (SO_4) (>500 has a laxative effect)	250.0
Total Dissolved Solids	500.0
Toxaphene (organic)	0.005
1,1,1 Trichloroethane (organic)	0.2
Trichloroethylene (organic)	0.005
Trihalomethanes (organic)	0.1
Turbidity (silica scale)	1–5 TU
Vinyl chloride (organic)	0.002
Zinc	5.0
2,4–D (organic)	0.1
2,4,5–TP Silvex (organic)	0.01
Radionuclides:	
Gross Alpha particle activity	15 pCi/L
Beta particle and photon radioactivity	4 mrem
Radium–226 and Radium–228	5 pCi/L

Odors Good or bad odors are caused by molecules released from the surface of various substances. A fresh fish will have hardly any smell at all, but bacteria on a decomposing fish will release strong odor molecules. The olfactory receptors in the nose will immediately detect even a few of these molecules. Odors can be eliminated physically, chemically, or biologically. Removing odors physically can be accomplished with activated carbon filters, as in the air handling units. A tiny crystal of activated carbon contains countless holes, or pores, that give the crystal a large surface area relative to its volume. Odor molecules floating on air currents collide with the carbon particles and become trapped in these pores. Other agents that work in a similar way include silica gel and activated bauxite.

Odors can be chemically removed by using acids to neutralize alkaline odors and alkalies to neutralize acid odors. But only a limited number of substances can be treated chemically.

For biological odor removal, microorganisms are used to break down odor molecules, rendering them odorless. Biological agents are limited, however, by conditions such as temperature. Because of such limitations, activated carbon is the most effective means of odor removal.

Pollens These are allergens for a substantial fraction of the population. It may not be such a good idea for agriculture and horticulture to eliminate them, but removal in indoor spaces in buildings is desired. *Ventilation* involves the provision of "fresh air" to interiors to replenish the oxygen used by people and to help carry away their by-products of carbon dioxide and bodily odors. Ventilation is desirable all year round; local and national codes of advanced countries recommend minimum circulation rates of fresh air required based on building occupancy and functions.

Air Handling Systems These are not only potential conduits for the spread of disease, but they are sometimes the cause of the problem. *Legionella* microorganisms, which enter and grow in HVAC systems, are the most notorious example. Less dangerous but far more typical is the growth of mold, the most common form of allergen. In the dark, moist environment of an HVAC system, mold spores can proliferate year-round. In allergic individuals, these spores initiate a chain of reactions starting with the release of histamines and inflammation of mucus membranes. These symptoms may lead to congestion, breathing difficulties, or even asthma and other complications. There are several categories of organisms that can grow and/or spread in modem air handling systems:

- Pathogens—viruses, bacteria, and fungi that cause a range of infectious diseases
- Allergens—bacteria and mold that cause allergic rhinitis, asthma, humidifier fever, and hypersensitivity pneumonitis
- Toxins—endotoxins and mycotoxins that cause a variety of toxic effects, irritation, and odors

As HVAC systems move large amounts of outdoor and recirculated air through occupied buildings, they become the conduits by which these unhealthful organisms are spread throughout the spaces they serve.

In the ongoing quest for better indoor air quality (IAQ), experts have come to recognize that these biological contaminants in indoor air are major contributors to *sick building syndrome* and building-related illness. In fact, according to the World Health Organization, biological contaminants in buildings are believed to account for a substantial portion of absences from work and school as well as days where activity is impaired or restricted. As a comparison, in most cases, the cost of losses in productivity far exceeds the cost of operating and maintaining the HVAC system.

4.1.2 Clean Indoor Air Strategies

Biological contaminants are also among the most difficult to control. Though high-efficiency ASHRAE grade or HEPA filters are helpful, many systems do not lend themselves to filter upgrades without major changes. And since many microorganisms are typically less than 1 μm in size (with some viruses as small as 0.003–0.004 μm), even high-efficiency filtration may be inadequate. Another very important but overlooked issue is a condition that may occur when time-clock systems are turned off. Natural temperature differentials between the system and the space create a convection flow or "back-draft" effect that returns space contaminants back through the ductwork to the downstream side of the filters. When combined with system leaks, these conditions often compromise the filter's role.

Where biological contamination is known to exist, a common control strategy is duct cleaning, sometimes followed by a biocidal treatment. In subsequent swab sample testing, however, a biological activity of concern has often been demonstrated to return in as little as 3 months after cleaning and treatment. In cases where *Legionella* microorganisms are present or suspected, acid washing (or other treatment) of fan coils, drain pans, cooling towers, etc., is required—an expensive and often destructive procedure that shortens equipment life. Other methods for micro-organism control tend to be impractical, potentially toxic, detrimental to equipment operation and efficiency, or simply too costly.

A *UVC light* system can be a significant control strategy to help reduce discomfort caused by microbiological reactions, *Legionella* microorganism growth and airborne dissemination, circulation of tuberculosis in air handling systems, and spread of cold and flu viruses. They kill harmful microorganisms without posing a risk to building occupants, maintenance personnel, mechanical equipment, and interior furnishings. If UVC emitters are to be used for general microbial control, *E. coli* is accepted as the standard target. However, if a specific organism is to be targeted, one must determine the right UVC dosage for that organism.

A kill rate of 90 % (based on single-pass testing with conventional measurements) is accepted as the standard for general microbial control in IAQ situations. This rate will deliver nearly a log reduction in microbial contamination, achieving indoor air quality that is roughly equivalent to that of outdoor air. The light irradiance value—flux in microwatts per square centimeter—determines the number of photons available to strike a target microorganism. Reflectance within the cavity creates a global ricochet effect of generated photons, making a higher number of them available and bouncing them in all directions to penetrate every nook and cranny—in effect, potentially increasing the dosage received by any organism. Reflectance works literally to "unhide" these organisms for proper destruction.

The greater the irradiance, the higher the kill rate for that microorganism. For example, achieving a 99.99 % kill versus a 90 % kill mathematically requires four times the irradiance or number of UV sources or high reflectance. What is considered a good reflectant for visible light, however, is not always a good reflectant for the invisible light energy of UVC. For example, common glass totally attenuates UVC; therefore, a typical rear-surfaced glass mirror does not reflect UVC at all. Since a typical duct liner has little or no reflectance, practical solutions include coating surfaces with aluminum paint or lining them with aluminum foil or sheeting.

Air velocity, expressed in fpm, is a key factor in determining the time an organism spends within the physical cavity (sometimes referred to as "dwell time") and the amount of heat removed from the UVC lamp, which directly affects its output. Thus, the higher the velocity or lower the air temperature, the more UVC energy required to achieve desired performance. Humidity is another attenuator to UVC energy. If higher (greater than 60 %) relative humidity exists, more UVC energy is needed to compensate for the absorption effect. Owing to variations in temperature, velocity, and other factors (as above), location has an impact on performance. For example, a UVC sterilization system upstream of the evaporator coil has different energy requirements than one located downstream of the coil.

Industrial Buildings and IAQ Control In the industrial building environment where the quality of the air is a critical parameter, this refers only to the strict temperature and the moisture content. Moisture content is measured in terms of the mass of water vapor contained in a unit mass of dry air. The numerical values are quite small since at any particular temperature the air can only contain a finite amount of water vapor. The following are the main processes carried out in an industrial building air-conditioning plant that need to be controlled to maintain these air quality parameters in the plant:

1. *Heating*: Heating causes the temperature to rise. The moisture content is not affected, however, and the relative humidity decreases with heating.
2. *Cooling*: Cooling causes the temperature to fall. Provided the temperature does not fall below the "dew point" of the mixture, there is no change in the moisture content and the relative humidity increases.
3. *Cooling below the dew point*: When the temperature of the mixture falls below the dew point, then the mixture can no longer hold the same mass of water vapor. The water vapor then condenses out of the mixture and appears on any cold surface which, in general, is the source of the cooling. This degree of cooling is often used in air-conditioning systems to remove water vapor from the mixture. Subsequent reheating if the mixture can result in the same original mixture temperature, but with a reduced relative humidity. Another way to achieve it conservatively is through desiccant cooling as shown in Fig. 4.1.

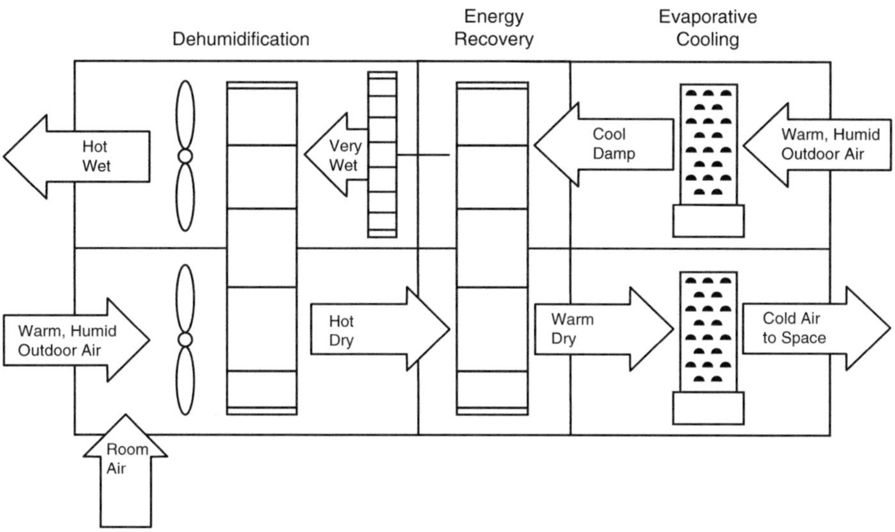

Fig. 4.1 Desiccant cooling

4. *Humidifying:* This process refers to the injection of water into the air to give an increased moisture content. Humidifying is usually associated with an attendant decrease in the mixture temperature. If the mixture is reheated, then it can be brought back to the same original temperature, but the relative humidity will be increased.

4.1.3 IAQ and Energy Conservation

In modern buildings, reduction of the fan-driving power is an indispensable part of any energy conservation program. Underventilation can create poor IAQ, while overventilation wastes energy if outside air must be heated or cooled. One solution is Demand-Controlled Ventilation levels on the basis of measured quantities of CO_2 or other air quality parameters. There is sometimes valid overventilation—particularly in spaces with unpredictable swings in occupancy, such as auditoriums, cafeterias, theaters, retail stores, class rooms, and conference rooms.

Decentralized types of air conditioners are sometimes the best choice. These air-cooling/reheating conditioners, which are concealed in the plenum, minimize the transport resistance. The use of a variable air volume (VAV) further enhances the reduction of the fan-driving power. This allows separate outdoor and indoor air conditioners concurrently tempering the condition of outdoor air through energy recovery units (dehumidifying or heating and humidifying) and the indoor air conditioners cooling the sensible heat generated in a room.

The reason for this "division of labor" is because, as long as the two cooling systems are of a completely different nature, they should not be mixed. To be more specific, one system supplies chilled water of a relatively low temperature to the outdoor air conditioners, which then perform dehumidifying functions. The other system supplies chilled water of a relatively high temperature to the decentralized air conditioners, which then cool the sensible heat. The system as a whole improves the energy performance of the chiller, thereby contributing to energy conservation.

4.2 Water Resources, Quality, and Building Systems

One basic principle of the earth's hydrologic cycle is that water is a closed system; there is no new water input of any significance in the environment. (Some scientists may disagree with this principle, but on the whole, unless the Arctic ice caps melt, no significant increase in the earth's volume of water is likely to occur.) Two thirds of the earth's surface is covered by water. Of that, 96.8 % is salt water in the oceans, and 3.2 % is freshwater. Of the freshwater, 75 % (2.4 % of the earth's total supply) is frozen in ice and glaciers, another 24.4 % (0.78 % of the total) is in the form of ground water, and 0.6 % (0.02 % of the total) is in rivers, lakes, the atmosphere, and soil. The amount of water used by facilities and associated building systems is a very small portion of all the water on earth—less than 1 %. To maintain the quality of this water economically, it is important to reclaim, recycle, and reuse our most precious natural resource surrounding our habitats.

4.2.1 Water Reclamation

Reclamation is defined as the conversion of naturally occurring nonusable water supplies into usable sources. Water reclamation generally falls into three broad categories: desalination, waste water reclamation, and surface water and groundwater reclamation. These reclamation activities generally take place on a municipal or regional level. However, in some remote industrial settings and owing to government regulations such as the Clean Water Act, many of these techniques and concepts are necessary to ensure a high-quality water supply. Building systems must be coordinated with reclamation systems so that there is an integrated and efficient approach to keep system cost down and provide an overall planned water system.

Given the vast quantities of salt water available, desalination provides rich opportunities to increase our supply of usable water. Common techniques for treatment include:

- *Multiple-effect distillation.* Heat transfer tubes are immersed in a tank filled with salt water, which is heated to the boiling point. It vaporizes and passes into a lower tank, where new incoming salt water cools it through a heat-exchange process, yielding a distillate of practically pure water.

- *Multiple-state flash.* Salt water is heated in a shell-and-tube heat exchanger then passed into a second heat exchanger at lower pressure. The pressure change causes the salt water to "flash" or change back into a liquid form. The vapor is condensed, yielding a distillate of pure water.
- *Vapor-compression distillation.* In this process, salt water is heated in a heat exchanger and vaporized. Then energy is applied by a compressor, which takes the vapor and compresses it to a higher pressure and temperature. This, in turn, furnishes heat for vaporization of more salt water. The vapor is then condensed to yield distilled water.
- *Electrodialysis.* This desalination method is effective on waters with salinity below 28 ppt. The process pulls ions of dissolved salts out of the water by electrical forces, and these pass from the salt water compartment into adjacent compartments though permeable plastic membranes.
- *Reverse osmosis.* This method is effective for brackish water. Water is passed through membranes under high pressure, leaving the salts behind. On some locations, reverse osmosis desalination systems are used to reclaim high-quality freshwater from wells located below ground. Although these systems are expensive to install and operate, the fact that brackish well water is readily accessible within 50 to 75 feet below ground makes desalination more economical than drilling hundreds of feet to find freshwater, which may have to be treated in any event.
- *Other methods.* These include freezing, solar evaporation, and ion exchange.

Additional treatment may be necessary depending on the water's intended use. For example, some industrial processes may require deionizing systems or other means to polish out specific contaminants that might be harmful to that process.

The second category for reclamation is waste water. Proper treatment of waste water is one way to help keep freshwater supplies from becoming contaminated, since the discharge of most of this water is returned back to the environment through irrigation systems, ponds, lakes, rivers, oceans, and infiltration and percolation through the soil.

The third area for reclamation relates to surface and groundwaters. Surface water is directly related to the amount of precipitation that can be expected annually. This water is not distributed evenly; much of it falls on mountainous areas and areas located near the tropics or equator.

Groundwater is water stored at or above atmospheric pressure. (Water stored in the soil is under tension.) The amount of groundwater stored is approximately 0.78 % of the total freshwater on earth. Groundwater is extremely important. In the United States alone, it supplies approximately 25 % of all the freshwater used for public supply, livestock, irrigation, industrial thermoelectric power, and other manufacturing uses; out of this 10 % is used for building systems.

Treatment required for both surface water and groundwater varies, depending on the quality of the water and its contaminants. These contaminants include heavy metals, trichloroethylenes, nitrates, radionuclides, herbicides, pesticides, trihalomethanes, and other organic materials such as dissolved solids, sodium, and chloride.

The Safe Drinking Water Act also requires stricter regulations as it relates to contaminant levels (refer to following EPA data on safe limit of contaminants in drinking water).

Exposures over safe limits can result in a variety of serious health problems ranging from liver and kidney damage, high cancer risk, nervous system disorders, skin discoloration, and hypertension.

Treatment options are extremely varied. The following is a summary of some of the most common methods:

- *Chlorination and ozonation*: Control pathogenic or nonpathogenic organisms.
- *Activated carbon*: Removes dissolved organics, color, and taste- and odor-causing compounds.
- Flocculants: Make particles cling together for more efficient water filtration.
- *Mechanical filtration*: Removes suspended solids. These devices include gravity sand filters, slow-rate sand filters and high-rate sand filters, etc.
- *IR stripping/aeration*: Removes carbon dioxide, hydrogen sulfide, and other taste- and odor-causing compounds, as well as volatile organic compounds.
- *Softening (ion exchange):* Removes hardness.
- *Chemical treatment:* Controls acidity.

In addition to basic treatment, many municipalities offer a reverse osmosis treatment system to remove trichlorethylene pesticides, nitrates, and radionuclide contaminants.

4.2.2 Water Recycling

Recycling refers to recovering water that would otherwise go into the waste stream and cleaning it for reuse. Ideally, the goal of the building systems engineer working in the commercial, industrial, and institutional arena should be to develop a closed-loop system by which waste water that is produced from human habitation could be recycled continuously with minimum introduction of freshwater into the cycle. Of course, to achieve this naturally, the population would have to be kept low enough relative to land density so that the natural biological processes can develop to allow a closed or semi-closed system to work in densely populated areas, which is not possible. Therefore, other ways to recycle water must be found. This includes recycling semigray, gray, and black water.

4.2.3 Water Reuse

Reuse refers to the reapplication of recycled water in other building systems or for landscaping, etc. Water reuse has been practiced since ancient times. Athenians reused waste water for crop irrigation. In the Far East, human waste has been used for thousands of years to fertilize ponds and produce aquatic animal and plant life. One large potential source of water for reuse in buildings is storm water. Many building sites have storm water retention, either in on-site ponds or underground storage tanks. This allows for a controlled discharge of storm water into sewers. It also allows storm sewers to be sized for average rainfall, thus reducing their installation cost. The incremental cost of recycling storm water is generally a fraction of the total cost of building a retention pond or storage tank. Thus, if such a storage facility is required, the cost of the additional pumps and piping for recycling should be evaluated. These storage basins also can be used for recycled gray and black water after on-site treatment. Commercial building water supply and drainage systems are shown in Figs. 4.2 and 4.3. Conservation on these systems can be done as follows:

- *Flushing toilets and urinals.* This is an acceptable use for recycled waste water. Piping must be clearly marked "nonpotable" and a dye is added to the water for safety. This could potentially save 15–20 % of building total water use per day. These types of systems should be considered for use in drought-prone areas.
- *Makeup water for cooling towers and mechanical cooling equipment.* Approximately 38 % of the freshwater used in the United States is for these purposes. Therefore, saving 5 % through recycling could result in a substantial savings in water bills per year.

Water conservation is not a plumbing issue alone. For example, reused water may be used for fire sprinklers, and conservation certainly will have an impact on electrical requirements for pumps, etc.

This type of integrated initiative offers the potential not only to save water but also to save energy costs associated with pumping water.

4.3 Power Quality and Intelligent Buildings

The concept of load and source compatibility is not new. The need to provide power with steady voltage and frequency has been recognized since the inception of the electric utility industry. However, the definition of "steady" has changed over the years, reflecting the greater susceptibility of increasingly sophisticated electronic equipment to the departure from "steady" conditions. Some of the early concerns were flicker of light bulbs due to voltage variations and overheating of electromagnetic loads or interference of communication loads due to voltage waveform distortion. Recognition of these problems led to the development of voluntary standards that contributed significantly to reducing occurrences.

The advent of electronic power conversion has been widely applauded by users, but the drawbacks from the point of view of power quality have not always been recognized. The very advantages of solid-state devices that made possible modern switching power supplies, electronic ballast, invertor rectifiers, high-frequency

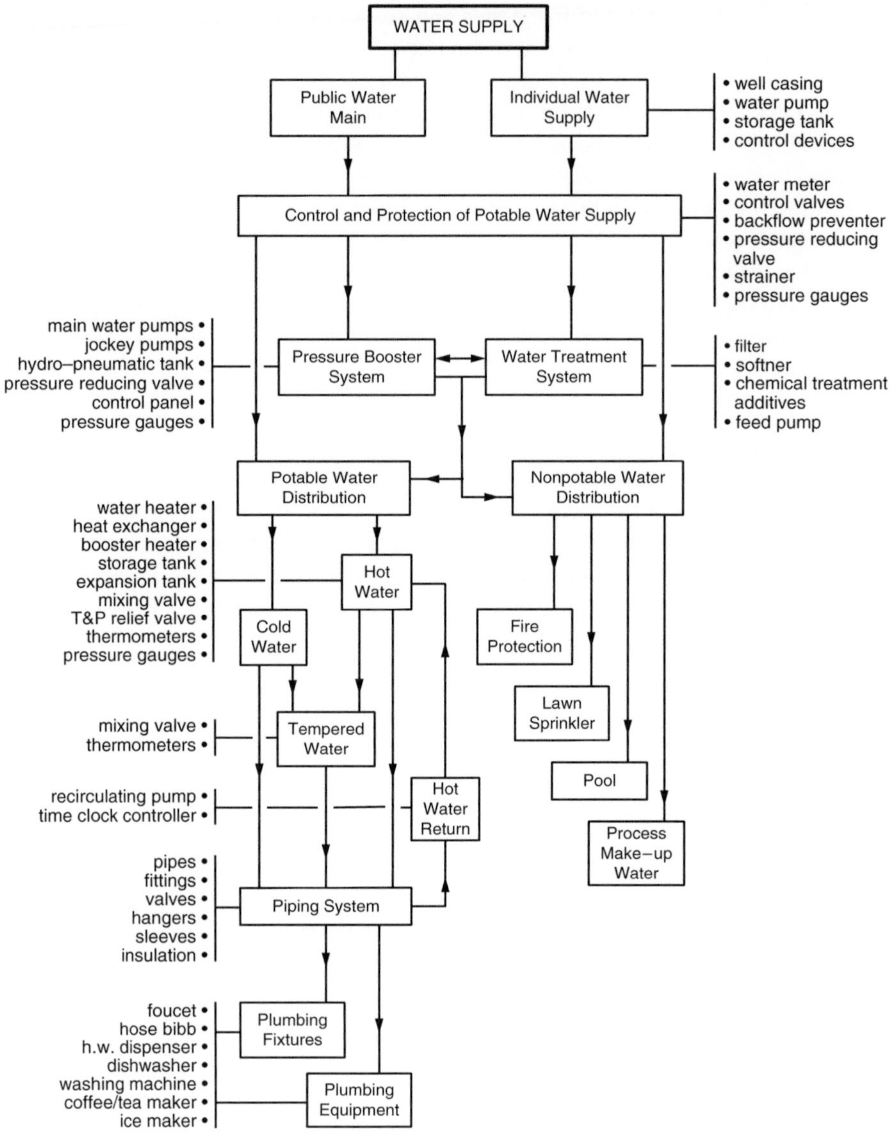

Fig. 4.2 Building water supply and distribution system

induction heating, and adjustable-speed drives also make these power converters into generators of harmonic currents and additional sources of line voltage drops. Thus, in addition to the disturbances generated by the normal operation of the familiar power delivery and load equipment, the disturbances resulting from the new electronic loads must be taken into consideration.

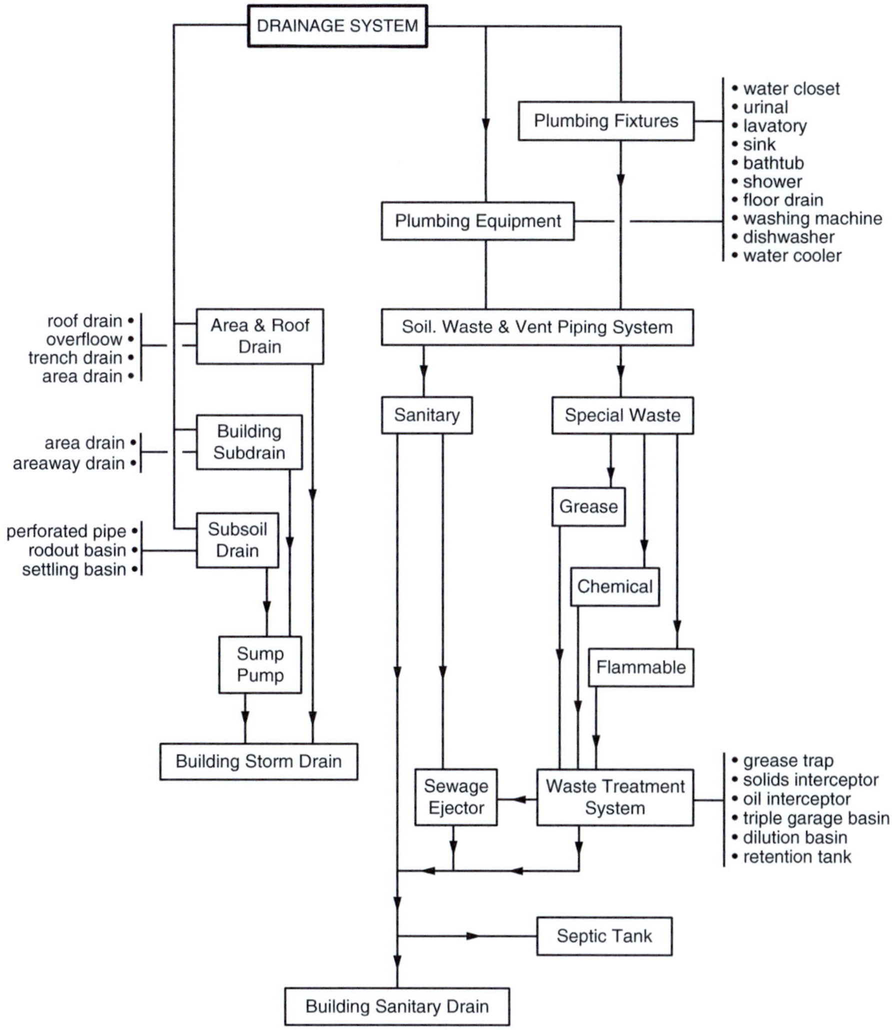

Fig. 4.3 Building sanitary and storm drainage system

Power semiconductor devices that constitute the heart of modern power electronics have been undergoing dynamic evolution in recent years. Never before in the history of power semiconductor devices have we seen the emergence of so many exotic devices in such a short span of time. A power semiconductor device is indeed the most complex, delicate, and "fragile" element in a converter. The building system engineer needs to understand the device thoroughly for efficient, reliable, and cost-effective specification of converter. In addition, the cost of silicon-based power and control devices is continuously falling along with the improvement of performance, whereas the cost of passive system components, such as inductor, capacitor, transformer, etc., is essentially constant and in some cases is gradually increasing, motivating integrated engineers to search for a "silicon solution" of passive components. Today's electronics industry is built on the integrated circuit—a tiny chip of silicon engraved with as many as a million microscopic electronic switches for controlling electric current. Because silicon's crystalline structure can be made to carry a nearly infinite array of electric pathways, it has become a vital electronic resource.

Only the purest silicon, however, can serve these purposes. Natural silicon consists of differently oriented crystals grouped into polycrystalline grains. Where these grains meet, they form irregular boundaries that can disrupt electricity flowing across them. To get around this problem, scientists devised a way to produce single-crystal silicon, a substance with a crystalline structure so uniform that it will accept any electrical pattern imposed on it.

The availability of high-power high-frequency devices at economical price will eventually permit mass application of motor drives, UPS systems, and active power line conditioners in building systems. HVAC engineers have always dreamed of using ideal switching devices in drives. Such devices should have large voltage and current ratings, zero conduction drop, zero leakage current in blocking condition, high temperature and radiation withstand capability, high mean time between failures (MTBF), and instant turn-on and turn-off characteristics. Of course, even with all these ideal features, the device should be available at an economical price. This dream will never materialize, but historically we have moved step by step in that direction.

It is important to understand what pure power looks like. Perfectly clean power will have a perfectly sinusoidal voltage of constant amplitude and frequency. Voltage amplitude will be adequate for the application, the voltage source will have no impedance, and the frequency will be 60Hz (or 50 Hz in some foreign countries). The wave shape will be perfectly free of harmonics, noise, and transients. Of course, such a perfect source of power does not exist, even in a laboratory.

"Power quality" (strictly speaking, lack of quality) is a term often used today in describing an aspect of the electricity supply. A power quality problem is any occurrence manifested in failure or misoperation of electronic equipment. The newness of the term reflects the newness of the concern. Decades ago, when the traditional building system workhorse—the induction motor—was hit by a sag, it did not shut itself off but produced a lower output until the sag was ended. Probably the most noticeable effect of a voltage reduction would be the dimming of the lights inside a building. But today, with the worldwide proliferation of sensitive power electronic equipment in building automation, these shortfalls in power quality can be very expensive in terms of building system shutdowns and equipment malfunctions.

4.3.1 Power Impurities

Let us have a closer look at some power impurities. The kinds of system events that contribute to the problem of power quality warrant closer examination, if only to better appreciate the remedies being proposed. So first, we will take a quick survey of everything from short interruptions to unbalance and flicker.

Voltage Variations Voltage swells, which are brief increases in system voltage, can upset system electric/electronic controls and electric motor drives, including common adjustable-speed drives, which trip because of their built-in protective circuitry. Swells may also stress delicate computer components to the point of premature failure. Besides sags and swells, longer-lasting increases and decreases of voltage occur on occasion. An overvoltage has a less immediate effect than a swell, but it may shorten the life of building power system equipment and motors.

Undervoltages are sometimes due to the deliberate reduction of voltage by the utility in order to lessen the load during periods of peak demand. These planned undervoltages, often called brownouts, are the blight of those hot summer days when air conditioners are at full blast and utilities lack enough generation to keep them going.

Voltage disturbances even shorter than sags and swells are classified as *transients.* They fall into two basic classes: impulsive transients, attributable in many cases to lighting and building load switching, and oscillatory transients, usually due to utility capacitor bank switching. Utility capacitor banks are customarily switched into service early in the morning in anticipation of a higher power demand. With the exception of lightning, almost all transients are generated as the result of interaction between stored electrical energy in building power system inductances and capacitances.

When current is interrupted at peak current flow (for example), the building system inductive load (L) is left with considerable stored energy. When the flux collapses, this energy interacts with system capacitance (C), causing an LC circuit oscillation with a theoretical peak voltage of perhaps ten times the normal peak voltage. This type of interruption is known as "current chopping" and is a form of "current suppression." Current chopping refers to a vertical cut in the current wave and is caused by circuit breakers and switches interrupting light inductive loads such as the excitation current of an unloaded transformer. Current chopping is also common with high-speed circuit breakers. Current-limiting fuses, thyristor, or SCR switches used in building power system configurations commonly cause current suppression and generate large spikes or transients. The effect of transients on the computer can be errors due to the *dv/dt* (change in voltage/delta time) coupling through the stray and interwinding capacitances of the power supply.

Resonance voltage spikes are traveling waves on electrical circuits and follow all of the laws of transmission line theory. For this reason, it is extremely difficult to predict accurately their rise time, amplitude, or frequency of occurrence. At each change of impedance (such as a wire size change or splice) and at each transformer, a portion of the spike is reflected and a portion is refracted or passed through.

Another disturbance—and a very common one—is the *powerfail.* The powerfail is defined as the total removal of the input voltage for at least half a cycle per utility and manufacturer standards as shown in Fig. 4.5. Powerfails can cause the floating heads of disk drives of building system computers to "crash down" on the disk, causing memory loss, unscheduled shutdown, or equipment damage. Some disk drives have heads that automatically retract upon loss of power, but the designer cannot assume that this is the case.

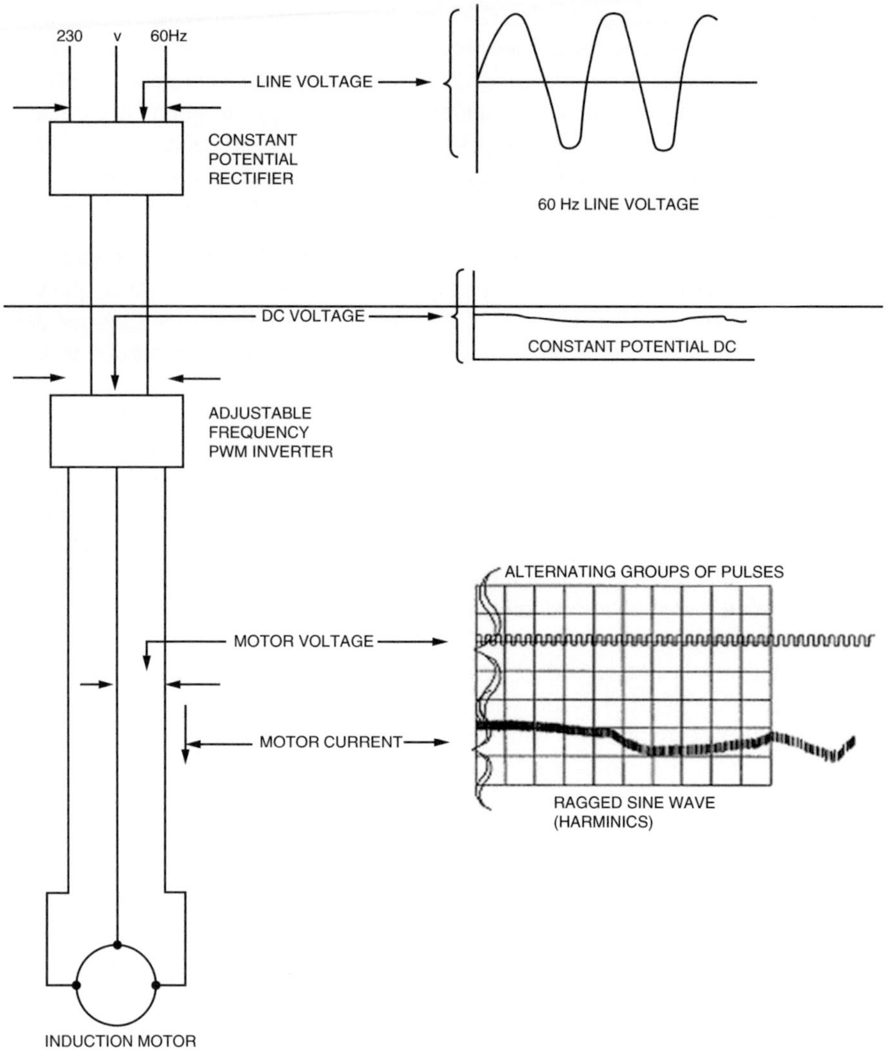

Fig. 4.4 Simplified circuit, voltage, and current characteristics for a PWM adjustable frequency drive with fixed-pulse-rate system operating at reduced speed and generating harmonics

Harmonic Distortion Harmonic currents are a result of building system equipment that require currents other than a sinusoid (see Fig. 4.4 and word harmonic on Table 4.2 refer to components of building system that cause harmonic distortion). The amount of harmonic voltage distortion occurring on any building power distribution system will depend on the impedance versus frequency characteristic seen by the equipment and by the magnitude of the currents. The distortion factor can refer to either voltage or current. A more common term that has come into use is *total harmonic distortion* (THD). To calculate THD, the periodic wave is broken into its sinusoidal components then a quantitative analysis of its parts is made. IEEE or IEC

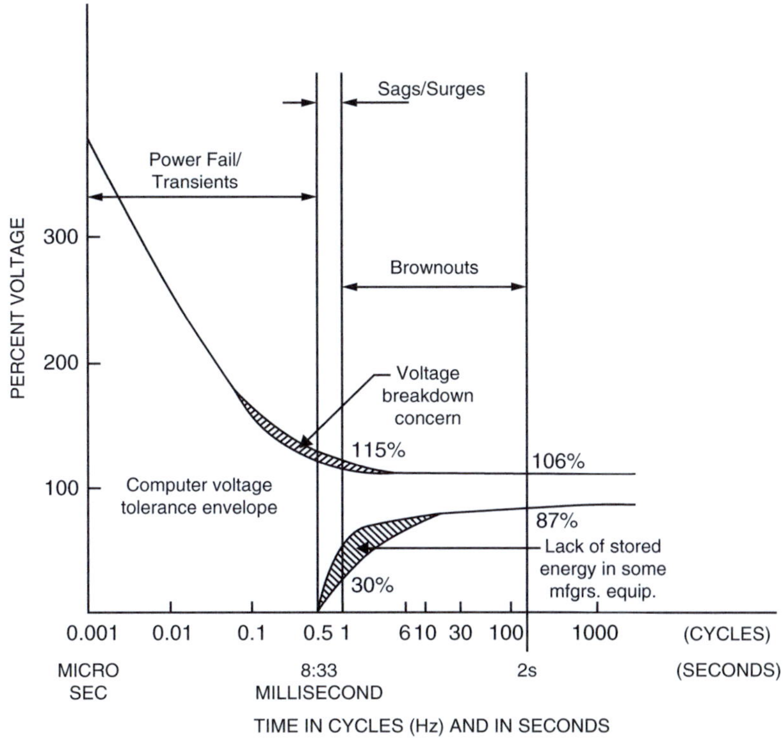

Fig. 4.5 Typical power quality design goals. Note: this graph addresses only the magnitude of the voltage and its duration. It lacks information on rate of change in voltage (power) disturbance

Standards make recommendations for limits within which current and voltage harmonics distortion should be kept. These standards are a system standard and not an equipment standard and contain application information (Fig. 4.5).

4.3.2 Effects on Building Systems

The major effect on building systems that have a power quality problem is increased heating due to iron and copper losses. It can give rise to high audible noise emission and can cause a phenomenon called cogging (the refusal to start smoothly) or crawling (very high slip) in building system induction motors. Table 4.2 provides a summary of power problem along with possible causes and their effects.

A system resonance condition is the most important factor affecting system harmonic levels. Parallel resonance is a high impedance to the flow of harmonic current, while series resonance is a low impedance to the flow of harmonic current. In building systems, capacitor banks are used for voltage control and/or power factor improvement and the manner in which capacitors are connected can cause resonance conditions (both series and parallel that can magnify harmonic current levels.

Table 4.2 Summary of power problems: cause and effects

Power problems	Possible causes	Effects
High-voltage spikes and surges	Lightning	Equipment failures
	Utility grid switching	System lock-up
	Heavy building system equipment	Data loss
Low-voltage electrical noise	Electronic equipment	Building system data corruption
	Switching devices	Erroneous command functions
	Motorized equipment	Variations in system timing signals
	Improper building power system grounding	Changes in building system processing states
	Power system protective devices, contractors, and relays	Loss of building systems synchronization
	Copiers	Control instability
		Mal-activation of protective devices
Harmonics	Building system controllers	High power system neutral currents
	Switch-mode power supplies	Overhead distribution and power equipments, e.g., transformers, panel-boards, neutral conductors.
	Uninterruptible power supplies	
	Building system motor drives	
	Electronic lighting ballasts	
Voltage fluctuations	Overburdened power distribution networks	System lockup
	Power system faults	Motor overheating
	Planned and unplanned brownouts	System shutdown
	Unstable generators	Bulbs bum-out
		Data corruption and loss
		Reduced performance
		Loss of control
Power outages and interruptions	Blackouts	System crash
	Backup generator start-up	System lockup
	Backup battery system discharges	Lost data
		Loss of control
		Lost communication and system link
		Complete shutdown

Parallel resonance occurs when the system current oscillates between the energy storage in the system inductance and the energy storage in the capacitance. The frequency at which parallel resonance occurs can be estimated by the following simple equation:

$$H_{\text{resonance}} = \frac{\text{short circuit MVA}}{\text{capacitor bank size in MVA}} = \frac{X_{\text{C}}}{X_{\text{L}}}$$

where H is the harmonic order and X_{C} and X_{L} are capacitive and inductive reactances of building power system at the fundamental frequency.

Series resonance occurs as a result of the series combination of building power system capacitor banks and line or transformer inductances. Series resonance presents a low-impedance path to harmonic currents and tends to draw in, or "trap," any harmonic current to which it is tuned.

Building cables involved in system resonances may be subjected to voltage stress and corona which can lead to dielectric (insulation) failure. Cables that are subjected to "ordinary" levels of harmonic current are prone to parasitic heating. The flow of nonsinusoidal current in a conductor will cause additional heating over and above that expected for the RMS value of the waveform. The harmonic currents flowing through the resistance of the building power system represent heat as

$$P_{\text{loss}} = \sum_{h=1}^{h_{\max}} I^2_{\text{harmonic}} \cdot R_{\text{harmonic}}$$

R_{harmonic} for a given building power system can vary with applied harmonics because of skin effect, proximity effect, stray currents, eddy currents, etc. These vary as a function of frequency as well as conductor size and spacing. As a result of these, the effective alternating-current resistance (R_{AC}) is raised even higher, increasing the I^2R_{AC} loss.

4.3.3 *Pure Power Solutions*

The pure power need of intelligent buildings is analogous to its pure water need. Let us look into how water is purified. The purification of drinking water proceeds in several stages. The first step is sedimentation, in which large particles suspended in the water settle to the bottom. The second is filtration, in which suspended solids and harmful bacteria are strained out. In the third stage, chlorine, a powerful disinfectant, is added to the water to kill the remaining microorganisms. Unfortunately, chlorine can give water a bad taste and in large doses can even cause serious health problems.

Similarly impure power is purified in steps based on level and type of impurity and large doses of certain treatment type can cause serious power stability problems. If one provides a "cure for which there was no disease," the remedy can sometime cause disease. The most effective and least expensive way to remove building power system transient overvoltages is through surgery—simply cutting them off or clipping them. The problem is that of all the power conditioning techniques in use, this is probably the least understood and the most misapplied. Table 4.3 summarizes various types of power conditioners and their applications.

The most common lightning surge suppressor is the lightning arrester. Since the lightning arrestor responds slowly, the fast transients often pass through. Probably the most common low-voltage fast transient suppressor is the metal oxide varistor, which is fast enough to clamp or clip off most system transients. One of the disadvantages of system transient suppressors is that they pump transients to ground, thus

Table 4.3 Summary of performance features for various types of power conditioning equipment

Power quality condition		Power conditioning technology								
		Transient voltage surge suppressor	EMI/RFI filter	Isolation transformer	Voltage regulator (electronic)	Voltage regulator (ferroresonant)	Motor generator	Standby power system	Uninterruptible power supply	Standby engine generator
Transient	Common mode	D		Y	D	Y	Y	D	D	
Voltage surge	Normal mode	D			D	Y	Y	D	Y	
Noise	Common mode		D	Y	D	Y	Y	D	D	
	Normal mode		D	D	D	Y	Y	D	Y	
Notches				D		Y	Y		Y	
Voltage distortion						D	Y		D	
Sag					D	D	D	D	Y	
Swell					D	Y	Y	D	Y	
Undervoltage					Y	Y	Y	D	D	
Overvoltage					Y	Y	Y	D	D	
Momentary interruption							D	Y	Y	
Long-term interruption										Y
Frequency variation								D	Y	D

Y (Yes)—It is reasonable to expect that the indicated condition will be corrected by the indicated power conditioning technology
D (Doubtful)—There is a significant variation in power conditioning product performance. The indicated condition may or may not be fully correctable by the indicated technology

shifting a line power quality problem to a building ground potential problem. Because of this, surge suppressors and lightning arrestors should always be separated from the building critical equipment ground by an isolation transformer. The shielded isolation transformer is often used in tandem with voltage regulators, transient suppressors, and other devices and power conditioners because it offers excellent power dirt rejection and a clean ground.

Most computers used in building systems do not need external voltage regulation. Those that do are generally served best by ferroresonant regulators which are extremely fast and do not "fight" with the computer regulator or by very slow regulators that likewise avoid fighting with the computer. Where a high level of power quality and total isolation from sags, dips, surges, and transients is required, motor generator (MG) sets and uninterruptible power systems (UPS) are used. Since MG sets and UPS actually reconstruct the power sine wave by converting the system ac input voltage to another form of energy and then regenerate an ac voltage, these units are not pure power conditioners. UPS often require power conditioners themselves to prevent damage to their solid-state components.

4.3.4 Power Quality and Load Synchronization

Right now, providing power quality to specific data on utility systems is still in its infancy. Most probably, its course over the next few years will be directed by the economics of problems related to building system power quality and their solutions.

The regulated computer power supply has an input window that will vary from one unit to the next. The power supply will operate properly if the input voltage remains within the window. If power supply data for building computers are not available, the building system design engineer must insist on receiving data.

Table 4.4 gives indices of source power quality requirement in relation to load.

4.4 Quality of Internet

Quality of the Internet Versus the Sustainability of the Internet. This truly amplifies the argument of global parity relative to internet access to "all humans around the world" level. Many subject matter experts have authored brilliant opinions regarding the GD2 and there is not a right or wrong. We have a tendency for the most part to reference the "future" of the internet.

The future of the Internet is a widely debated public policy theme all over the world. Questions are raised on how to preserve the public "best-effort" Internet as an "open" platform for innovation and competition and how to combine economic and societal value creation and sustainable returns on investment.

Table 4.4 Matching sensitive load and power source requirements with expected environments

A. Voltage				
Voltage parameter affecting loads	**Typical range of power sources**	**Typical immunity of electronic loads**		
		Normal range	**Pure power range**	**Units affected and comments**
Over- and undervoltage	+6 %, −13.3 %	+10 %, −15 %	±5 %	Power supplies, capacitors, motors; components overheating and data upset
Swells/sags	+10 %, −15 %	+20 %, −30 %	±5 %	Same as above
Transients, impulsive and oscillatory, power lines	Varies: 100–6000 V	Varies: 500–1500 V	Varies: 200–500 V	Dielectric breakdown, voltage over stress; component failure and data upset
Transients. impulsive and oscillatory, signal lines	Varies: 100–6000 V	Varies: 50–300 V	Varies: 15–50 V	Same as above
ESD	<45 kV 1000–1500 V	Varies widely: 200–500 V	Varies widely: 15–50 V	Signal circuits; dielectric breakdown, voltage over stress, component, failure, data upset; rapid changes in signal reference voltage
RFI/EMI (conducted normal and common mode)	10 V up to 200 kHz less at higher frequencies	Varies widely 3 V typical	Varies widely 0.3 V typical	Signal circuits; data upset, rapid changes in signal reference voltage
RFI/EMI (radiated)	<50 kV/m <200 kHz <1.5kVm >200 kHz	Varies widely w/shielding	Varies widely w/shielding	Same as above
Voltage distortion (from sine wave)	5–50 % THD	5–10 %	3–5 %	Voltage regulators, signal circuits, capacitor filters. capacitor banks; overheating; undercharging
Phase imbalance	2–10 %	5 %	3 %	Polyphase rectifiers, motors; overheating

B. Current				
Current parameter affecting sources	**Typical range of load current**	**Typical susceptibility of power sources**		
		Normal range	**Pure power range**	**Units affected and comments**
Power factor	0.85–0.6 lagging	0.8 lagging	<0.6 lagging or 0.9 leading	Power source derating or greater capacity source with reduced overall efficiency
Crest factor	1.4–2.5	1.0–2.5	>2.5	1.414 normal; impact function of impedances at third and higher harmonics (3–6 % Z); voltage shape distortion
Current distortion	0–10 % total RMS	5–10 % total — 5 % largest	5 % max total 3 % largest	Regulators, power circuits; overheating
dc current	Negligible to 5 % or more	<1 %	As low as 0.5 %	Half-wave rectifier loads can saturate some power sources; can trip circuit breakers
Ground current	0–10 A RMS + noise and surge currents	>0.5 A	<0.1 A	Can trip GFI devices, violate code, cause rapid signal reference voltage changes
C. Frequency				
Frequency parameter affecting loads	**Typical range of power sources**	**Typical immunity of electronic loads**		
		Normal range	**Pure power range**	**Units affected and comments**
Line frequency	±1 %	±1 %	±0.5 %	Zero-crossing counters
Rate of frequency change	1.5 Hz/s	1.5 Hz/s	0.3 Hz/s	Phase synchronization circuits

Although so far the history of the Internet has been an incredible success in organically developing a self-adapting complex of network business relations, concerns are raised about how the Internet will be able to sustain an adequate quality of experience for the end user in the future.

This ability may come under pressure by a spectacular boom in Internet traffic volumes in the coming years, resulting in unprecedented demand for reliable, ubiquitous Internet access and mass uptake of bandwidth-intensive services and applications. To illustrate this point, by 2020, more than 50 % of the world's population will be online. This means an increase from 2.7 billion users in 2014 to 5.0 billion users by 2020. By 2025, the "GD2" will comprise around 50 billion connected devices.

By 2030, machine-to-machine ("M2M") communication is expected to constitute more than 50 % of IP traffic. A global, immersive, invisible, ambient networked computing environment built through the continued proliferation of smart sensors, cameras, software, databases, and massive data centers in a world-spanning information fabric known as the GD2.

In the last decade, a range of digital technologies and services have hit the market and moved quickly from niche use to the mainstream. Consider that just seven years after being founded, Facebook is used by more than 50 % of the online population worldwide. Flat-panel TVs, e-readers, smartphones, and even augmented-reality apps—all largely missing from the consumer landscape just a few years ago—continue to be eagerly adopted even in the face of economic uncertainty.

As we look toward the next decade, it is clear that we are in for even more dramatic changes in the roles that technology will play in daily life. But what technologies are poised to move from niche toward the mainstream in the next years? And how will these technologies change everyday activities?

To bring this into sharper focus, let us revisit some of the key themes that we feel will help define the tech experience in the coming decade. These "green technology paths" will give people a powerful new "toolkit"—new devices, services, and capabilities—that will forever alter the way that we go about everyday activities, from dating and shopping to learning and working.

4.4.1 Technology Paths

Adaptive Environments Advances in materials will make the home and work environment "smart." Everyday objects, surfaces, and coatings will gain the ability to adapt to changing conditions or people's needs—e.g., becoming self-cleaning, self-insulating, or protective. The built environment will no longer be simply structural and passive; it will become adaptive, functional, and smart.

Cloud Intelligence The cloud will evolve from being a static repository of data into an active resource that people rely on throughout their daily lives. With new capabilities for accessing online expert systems and applications, we will tap into information, analysis, and contextual advice in more integrated ways. Virtual agents will migrate from being an automated form of phone-based customer service to a personalized form.

Collaboration Economy The power of collective intelligence will enable us to accomplish cognitive tasks not easily handled by virtual agents and machines in the cloud. We will get advice and recommendations and solve problems by tapping into the social graph, and this cognitive outsourcing will be applied to both business issues and personal and lifestyle questions (e.g., “Which diet will work best for me?”).

Circumstantial–Contextual Reality People will navigate through their daily activities thanks to multiple layers of real-time and location-specific information. This contextual overlay for everyday life will give us a new way to see our surroundings and provide new forms of decision support. We will move from a world where information and connections are hidden to one where real-time, contextual information generates ambient awareness.

Information Fusion It will become possible for people to generate useful insights about their own habits and behaviors by fusing personal data (e.g., social media profiles, tweets, location data, purchasing histories, health sensor data). But these insights will only be as good as a user’s ability to understand and act on them. Personal data will become comprehensible through visualization and other services.

Interface Anywhere, Any Way. Intuitive interfaces will become the dominant form of interaction with personal electronics and computing devices. We will be freed from the rigidity of conventional input devices (e.g., keyboard, mouse, screen, remotes) and able to interact with the digital world anywhere—and any way—using a combination of gesture, touch, verbal commands, and targeted use of traditional interfaces.

Manufacturing Manufacturing will be reconceived—from a far-flung, global activity to more of a human-scale and re-localized endeavor. As consumers continue to call for both personalization and attention to environmental pressures, demand will grow for a more local manufacturing infrastructure where product schematics in certain categories are digitized and distributed to commercial “prefabs/OEMS” services (or in-home 3D printers) for final fabrication.

Personal Analytics Data analytics will increasingly become a consumer tool as much as a business tool. This will open up analytics to a wide variety of personal and lifestyle applications. We will collect, store, interpret, and apply the vast amounts of data being created by and about ourselves during our everyday activities.

Socially Networked Stuff Many of our possessions will interact with each other and with the broader digital infrastructure. This will create a world of socially networked stuff, where things can actively sense, communicate, and share data.

Fig. 4.6 Sustainable digital integration

Rather than owning a fragmented set of possessions and devices, passively sitting next to each other, we will manage a dynamic ecosystem of belongings that interact and work in concert for our benefit.

4.4.1.1 Societal Drivers Influence Technological Advancements

So, how will the new capabilities described in the Technology Paths change home and family life? What will our homes look and feel like? How will they support our activities and lifestyles?

Technology is not the only driver at play here, and the technology paths are not emerging in a vacuum. There are numerous social, generational, and value drivers at play as well. We especially noted the impact of digital natives on adoption of technology in the home, shifting demographics, and economic considerations (Fig. 4.6):

"Augmented reality" enhancements to the real-world input that people perceive through the use of portable/wearable/implantable technologies

Disruption of business models established in the twentieth century (most notably impacting finance, entertainment, publishers of all sorts, and education)

Tagging, knowledge reuse, and intelligent analytical mapping of the physical and social realms

These technologists, including the majority of the expert respondents here, predict the GD2 will be evident in:

Bodies: Many people will wear devices that let them connect to the Internet and will give them feedback on their activities, health, and fitness. They will also monitor others (e.g., their children or employees) who are also wearing sensors or moving in and out of places that have sensors.

Homes: People will be able to control nearly everything remotely, from how their residences are heated and cooled to how often their gardens are watered. Homes will also have sensors that warn about everything from prowlers to broken water pipes.

Communities: Embedded devices and smartphone apps will enable more efficient transportation and give readouts on pollution levels. "Smart systems" might deliver electricity and water and warn about infrastructure problems.

Goods and services: Factories and supply chains will have sensors and readers that more precisely track materials to speed up and smooth out the manufacture and distribution of goods.

Environment: There will be real-time readings from fields, forests, oceans, and cities about pollution levels, soil moisture, and resource extraction that allow for closer monitoring of problems.

RFID, egress, ingress, video analytics and sensored roadways, buildings, bridges, dams, and other parts of infrastructure that give regular readings on their state of wear and tear and provide alerts when repairs or upgrades are needed;

Vastly improved productivity in manufacturing at every stage, as supply chain logistics are coordinated.

Paper towel dispensers in restrooms that signal when they need to be refilled. Municipal trash cans that signal when they need to be emptied. Alarm clocks that start the coffee maker.

Smart appliances working with smart electric grids that run themselves or perform their chores after peak loads subside/

More than anything, this speaks to the momentum of the Universal Integrated Telecomm (UIT) Box (the Internet, Smart TV, VOIP, home security/video analytics, etc.) and one of its primary subcategories, the connected home.

With UIT, connected machines and devices talk to each other, allowing for heightened levels of control, monitoring, and automation. This extends to environments both public (think city infrastructure, utilities, and transportation) and private (like your house and the appliances in it). It is an exciting thought (Fig. 4.7).

A Global Nerve Center (GNC) is the physical location where an organization comes together during an emergency to coordinate response and recovery actions and resources. These centers may alternatively be called command centers, situation rooms, war rooms, crisis management centers, or other similar terms. Regardless of the term, this is where the coordination of information and resources takes place. The Global Nerve Center (GNC) is not an incident command post; rather, it is the operations center where coordination and management decisions are facilitated.

Fig. 4.7 Nerve center of smart city

In the domain and in conclusion, the quality of the green digital divide–Internet must be paretoized, taking a deeper dive and establishing a spectrum and an inventory of global needs. The term IRP represents Internet requirements planning (we know what they need and who needs it). This is a rudimentary approach to ascertain that accessibility, availability, adaptability, and accuracy relative to GD2 and the sustainable internet are readily available to all humankind.

Suggested Reading

Figure 4.6 via Epidemiology and Energy Environment-Not a google photo
GD2-Green Digital Divide: Afi Hasan Dissertation page 168 EACO (Economically Advanced Cities & Organizations)
M2M-Goldas-Author and via M2M Now I.T. news article
Nerve Center: Design and Diagram via Afi Hasan and www.eacopregenta.com
U.I.T. via the Teleric.net technology: Developer Mr. Khaliq Rehman. Naperville, Illinois

Chapter 5
Acoustics, Vibrations, and Integrated Building Systems

Anil Ahuja

5.1 Introduction

Sound is transmitted by waves. These waves travel through gases, liquids, and solids alike. Wave action is mainly a transfer of energy. In the case of sound, this transfer takes the form of tiny motions at the molecular level. In gases and liquids, a sound wave shifts molecules slightly in a direction parallel to itself, that is, in a lengthwise direction. In solids, motion may also occur perpendicular to the wave. Sound waves spread from their source in all directions. The speed of sound is independent of loudness of tone. The sounds from a radio in a room, whether they are loud or soft, of high pitch or low, all reach a listener simultaneously.

Mechanical oscillation occurs in a broad range of frequencies. Since an adult can hear oscillations between approximately 100 cycles per second and 15,000 cycles per second, we call that range audible sound. Below the audible sound range, we call it oscillation vibration, and above the sound range we call it oscillation ultrasound or ultrasonics. Vibration, audible sound, and ultrasound are exactly the same thing except for frequency.

Ultrasound is conducted by the air, although humans do not hear it. The burners of a hot air balloon generate enough ultrasound to drive dogs on the ground into a frenzy. And, of course, there are ultrasonic dog whistles. Ultrasound is severely damped by most materials, so there is not much of a problem in providing barriers. In fact, the principal mechanical design problems involving ultrasound are to provide conduits for it.

The ideal building acoustical environment would permit any occupant to talk easily or talk on a telephone without distracting or being understood by other occupants. The best way to design such an environment is to look at building systems as a man–machine system. This system may be considered in three principle parts:

Noise Sources Building systems electrical and mechanical equipment; flow of liquid in pipes, air in ducts or pipes, current in conductors, or flux in magnetics; people

A. Ahuja, *Integration of Nature and Technology for Smart Cities*,
DOI 10.1007/978-3-319-25715-0_5

who talk and move about; office equipment; and anything else that is likely to make annoying noises.

Noise Transmission Elements Constitute the bulk of the architectural elements of building system, such as ceilings, walls, floors, screens, furniture layout, and anything that influences the amount of noise reaching the receiver.

Noise Receiver Predominantly people with a common, but not completely singular, sensitivity to these noises. Their sensitivity may be modified by adding masking sound, which does not annoy them, to their environment.

5.2 Acoustical Environment of Buildings

It is the responsibility of the system designer to provide the proper environment by confining noise sources under his or her control, achieving the best attenuation along noise transmission paths, and understanding the sensitivity of the people-receivers. The bulk of quantitative knowledge developed to date centers on the second item—attenuating, absorbing, canceling, and masking subsystems. Not much has been quantified on receiver sensitivity in this context.

Interfering Sounds Common office equipment beside building systems can be equally annoying. Sometimes the most irritating noise in an open area is the clamoring telephone bell. Other machines, such as copiers, tend to be less numerous and may be given special placement to minimize their noise. Other sources of mechanical noise, such as ventilation systems, should be kept below noise criteria curve 35 (NC35) (or approximately 45 dB A) throughout the office. This is not a stringent requirement and is usually achievable. Noise control procedures are generally most effective when applied directly to the source.

Building System Noise Mechanical devices obviously make noise, and, generally, the more power they consume, the more noise they make. In many of today's buildings, 40 % of the total cost is spent on mechanical systems. These systems are located throughout a building.

In most buildings, the primary sources of mechanical noise are the components of the air-conditioning and air-handling systems such as fans, compressors, cooling towers, condensers, ductwork, dampers, mixing boxes, induction units, and diffusers. Figure 5.1 shows noise and vibration sources of a rooftop unit commonly used in commercial buildings. The increased use of variable air volume (VAV) systems has introduced some noise problems that should not be neglected. VAV system noise can be minimized by maintaining minimum system static pressure since fan noise increases exponentially with static pressure. Ceiling diffuser acoustic characteristics must be integrated to meet requirement of design flow and masking noise. Normally piping is not a major noise source, but where flow velocities are high and the pipe is in contact with the structure, it can create noise problems. Pumps are another source of mechanical noise. Pump noise is frequently transmitted along pipes to remote points.

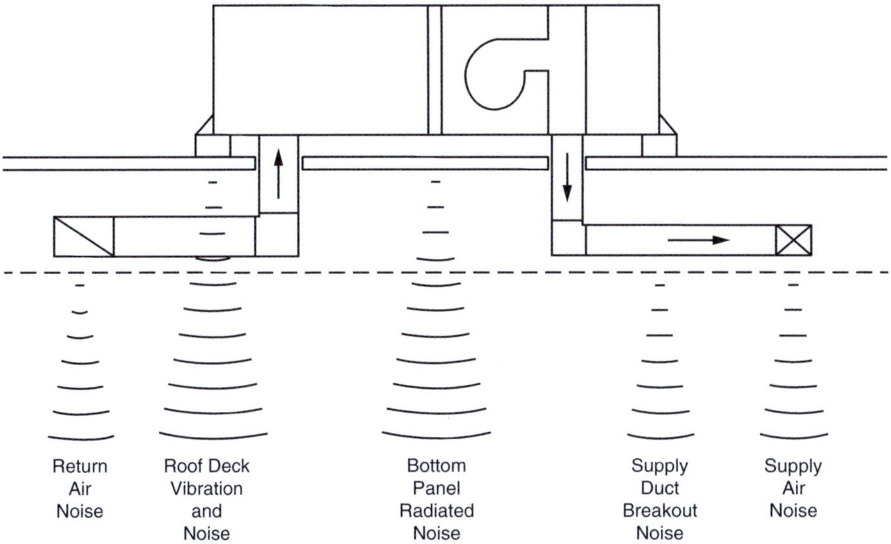

Fig. 5.1 Rooftop unit noise sources

5.3 Noise and Acoustic Planning Issues

Too often, building systems are designed with little concern for controlling noise and vibration. Design team members may ask equipment vendors for their opinions during the design process, or an acoustical consultant may be retained just before the drawings are released for bids to check things over and make last-minute recommendations. In such instances, the site plan, space plan, equipment selections, duct and pipe routing, and structural design are all complete and are unlikely to be changed.

On the other hand, if the project's acoustical aspects are given adequate priority during the early design phases, the design team has the necessary flexibility to select and locate HVAC equipment for proper operation and effective noise and vibration control. In the early design phases, this is done by effective site planning, space planning, equipment selection, and equipment room sizing. Proper consideration of acoustics early in the design may allow solutions to potential problems at little or no cost.

This planning issue is important in the placement of ground-based, outdoor HVAC equipment such as cooling towers, air-cooled condensers, and air-cooled chillers. Many cities and counties have ordinances that place limits on the permitted level of building support system noise at adjacent property lines. Complying involves keeping noisy equipment away from the property line, thereby bringing the equipment closer to the building that it serves. This planning option must consider whether or not the equipment's proximity to the building may cause interior noise problems from sound penetration through the building's windows, doors, or ventilation openings.

The mechanical engineer should make tentative equipment selections as early as possible to allow for a preliminary noise analysis and to determine the probable sizes of the mechanical rooms. The primary acoustical design guideline is to select the quietest equipment possible. This usually means selecting equipment that will perform at its most efficient operating point. For fan-based equipment, select a fan with the lowest practical rpm, while avoiding proximity to the fan's stall region for all expected operating conditions.

After the equipment locations have been determined and the tentative selections have been made, the equipment locations and operating weights should be given to the structural engineer. The structural engineer will use the information to help select beam sizes and column spacing, aspects of the design that are very difficult to change later.

Economic pressure to maximize a building's rentable floor area has resulted in less space being available for the HVAC system and other building services. This trend often forces the mechanical engineer to specify small, inefficient equipment or to shoehorn properly sized equipment into a restricted space. Both options can lead to excessive noise in neighboring spaces, which may then become nonrentable.

Electrical Equipment is generally overlooked as a noise source and this is unwise and a drawback of not integrating acoustic design concerns. Most electrical equipment noise is a 120-Hz hum on a 60 Hz system or 100 Hz on a 50 Hz system. This can be very disturbing because it is so low a frequency, and low-frequency noise is difficult to attenuate.

A typical motor/blower exhibits mechanical, aerodynamic, and electromagnetic noise. Its level and frequency are directly proportional to motor speed. The noise spectrum typically will include both discrete frequency components (tones) and random broadband sounds. The bearings are the primary sources of mechanical noise in electric motors together with any unbalances of the rotor. A well-balanced rotor operating below its critical bending speed (operating frequency coincident with the bending frequency of the shaft) is essential for quiet high-speed motor operation.

However, all rotating elements have some degree of unbalance, characterized by a sinusoidal vibration that may be transformed into acoustical energy, possibly at a frequency of once per revolution. The aerodynamic noise of concern is produced by the impeller (rotor), which generates a frequency spectrum containing the two major components, discrete tonal noise and its broadband counterpart:

- Tonal noise is produced when the air leaving the equally spaced blades hits a stationary obstacle near the impeller. The tonal frequency is commonly called the blade passing frequency (BPF).
- Broadband noise is generated by turbulence and other fluid instabilities of the airflow over the blades, plus interaction with local structures. It is characterized by equal amounts of energy across the audible frequency range which usually dominates the radiated sound power spectrum of the motor/blower.

In addition, the electrical noise such as that created as the rotor slot passes a stator air gap is produced by the electric motor. However, these levels are small in comparison to those created by the motor/blower. Another major source of 120-Hz hum is conventional, core-and-coil discharge light fixture ballasts. (Electronic ballasts are practically noiseless.) This includes fluorescent plus all the HID sources.

5.4 Masking Sound

This is commonly used or applied in open office areas where personnel are separated by partitions. An individual speaker in an average office or work area will normally modulate his voice so that his listener will enjoy a nearly 20 dB signal-to-noise ratio at his listening distance, typically 1.2 m (4 ft).

In addition to noise transmission through or diffraction about building partitions, intruding sounds may reach the listener by bouncing from nearby flat surfaces. This process is termed "specular reflection" and is analogous to the bounce of a billiard ball from a cushion or the reflection of light by a mirror. The amplitude of this reflected sound depends on the reflective properties of the surface. Reflectivity is not always predictable by the noise reduction coefficient (NRC) of the surface, primarily because NRC averages absorptivity at all angles of incident sound.

Privacy Criteria At the onset of the design process, it is important to determine the need for sound level, since measures necessary to provide acoustic privacy later can be expensive. The noise criterion for a given function depends on that function. Executive and sensitive areas require confidential privacy, defined as zero phase intelligibility. Other areas may require minimal privacy, such as secretarial pools. To date it has been customary to relate privacy requirements to speech communication as discussed in the following sections:

Speech Sounds in the Open Office Average male conversational speech at a distance of 1 m (3 ft) has a broadband level of 60–65 dB. The dynamic range of component levels is 30 dB across the frequency range from 200 to 5000 Hz. These limits are not firm. It is usually accepted that the middle-frequency components are more important for communication than the low and high frequencies. For 100 % intelligibility, the full 30 dB dynamic range of the complex frequency spectrum should lie above the prevailing background noise. Intelligibility of speech is a function of the signal-to-noise ratio. Apart from the dynamic range of 30 dB, the mean value of a normal voice sound has a range of 20 dB from "quiet" through "normal" to "raised," depending on the amount of speech effort applied. Merely raising one's voice may destroy the state of acoustical privacy at nearby work stations. This is perhaps the most irritating human disturbance in the open-plan office. Fortunately the higher absorptive surroundings in the open-plan office appear to induce lower voice levels.

It is common, therefore, to use electronic random noise generators feeding loudspeakers hidden in the ceiling plenum to raise the level of the background. The spectrum of this sound is shaped to effectively mask speech with the least

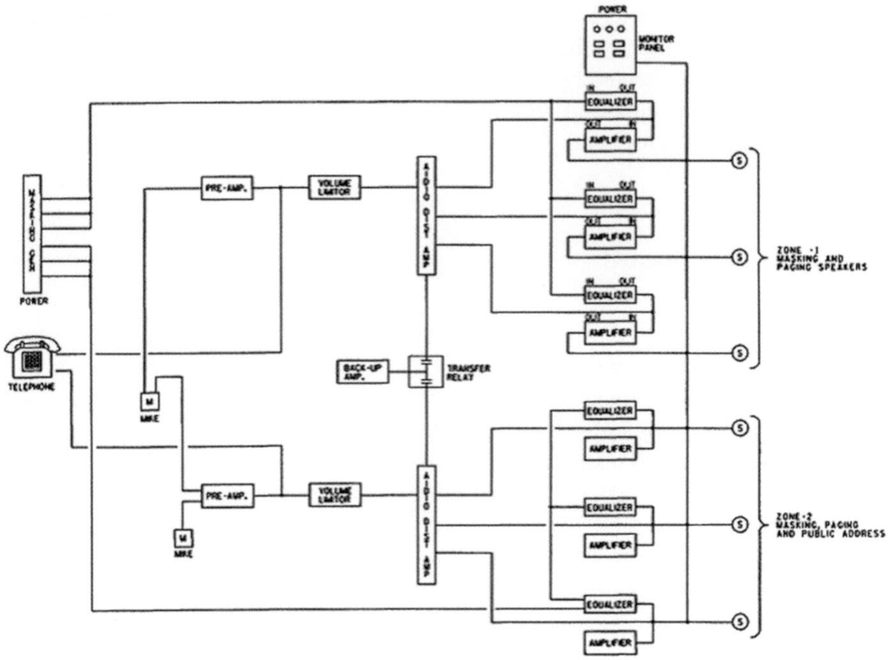

Fig. 5.2 Commercial building public address paging and masking system. Two zones are shown for clarity

distraction. Figure 5.2 is a block diagram of an electronic masking sound and paging system. Signals from the sound generator are shaped with narrowband filters (e.g., one third or one half octave) into the spectrum (as heard by the listener) that is considered optimum.

Continuous background noises provided by air-conditioning systems, though an attractive expedient, have not been generally successful because of the difficulty in generating the right level with the preferred spectrum. Electronic generators are more flexible. Likewise, music alone is unacceptable because it has pure tone components and transient level characteristics; however, music has been successfully introduced as a supplement to a properly adjusted masking system.

The masking sound field produced in the office must be uniform in space and time to avoid occupant complaints or detection of the masking system. The uniformity of the sound field is governed by the spacing and orientation of loudspeakers, transmission characteristics of ceiling systems, reflective properties of the plenum surfaces, and openings in the ceiling such as plenum air return slots and light fixtures.

A common arrangement is to distribute individual loudspeakers in a square or triangular array with approximately 3–6 m (10–20 ft) spacing between units. The exact distribution will be influenced by plenum depth, obstructions, ceiling tile properties, and speaker radiation patterns. Ceiling materials that are relatively transparent to sound sometimes use an impervious layer on the back side to diffuse the sound and produce a uniform sound field in the listener areas. Another approach is to mount the speakers over fairly large baffles, about 1.25 m (4 ft) square, so that the sound must reflect and diffract around them, producing a more uniform field.

The monitoring and adjustments should be carried out in the fully furnished office area minus staff. The level chosen for this background sound will depend on the degree of privacy required and on the physical properties of the screens and ceiling, although there is an upper limit. Clearly, the introduction of masking sound must not arouse hostility or adverse reactions in the office. Although various occupants have different reactions, most people seem to accept broadband, steady-state sounds having A-weighted levels up to 52 dB A.

It is best not to have other areas in the building quieter than the open-plan office, lest occupants become aware that their work area is not as quiet as it might be. This suggests the use of masking sound throughout the building. Some work areas require more masking sound level than others, however, and a compromise must be found. One successful technique is a gradual transition zone between levels. A side benefit of the electronic masking system is that it amplifies and loudspeakers can be used for public address announcements, life safety programs, or emergency paging.

5.5 Acoustic Analysis

All sound generation mechanisms and sound transmission paths are potential candidates for analysis. Adding to the computational workload is the necessity of extending the analysis over, at a minimum, octave bands 1 through 8 (63 Hz through 8 kHz). A computer can save a great amount of time and difficulty in the analysis of any noise situation, but the system designer should be wary of using unfamiliar software.

Caution and critical acceptance of analytical results are mandatory at all frequencies, particularly at low frequencies. Not all manufacturers of equipment and sound control devices provide data below 125 Hz. Thus, the system designer conducting the analysis and the programmer developing the software must make assumptions based on experience for these critical low-frequency ranges.

The designer/analyst should be well satisfied if predictions are within 5 dB of field-measured results. In the low-frequency "rumble" regions, results within 10 dB are often as accurate as can be expected, particularly in areas of fan discharge. Conservative analysis and application of the results is necessary, especially if the acoustic environment of the space being served is critical.

Noise computations for systems should account not only for obvious noise generators such as fans and transformers but also for the potential for flow generation in duct elements (e.g., trees and branches) and downstream elements (e.g., VAV boxes). Whenever flow velocities exceed 1500 fpm, flow-generated noise is a significant possibility. Sound in ducts propagates upstream as well as downstream, and sound from the fan of a single system travels down at least four paths.

The sound pressure level in a space is the composite of all sound paths to the space. For example, one diffuser in a room may meet noise criteria curve NC-30, but the combined sound pressure level of ten diffusers along with transformer hum in a room may be NC-40 or greater.

Several currently available acoustics programs are generally easy to use but are often less detailed than the custom programs developed by acoustic consultants for their own use. Acoustics programs are designed for comparative sound studies and allow the design of a comparatively quiet system. Acoustic analysis should address the following key areas of the building system:

- Sound generation by HVAC equipment
- Sound generated by passive equipment
- Sound attenuation and regeneration in duct elements
- Wall and floor sound attenuation
- Ceiling sound attenuation
- Sound breakout or break-in ducts or casings
- Room absorption effect (relation of sound power criteria to sound pressure experienced)

Algorithm-based programs are preferred because they cover more situations. However, assumptions are an essential ingredient of algorithms. These basic algorithms, along with sound data from the acoustics laboratories of equipment manufacturers, are incorporated to various degrees in acoustics programs. Whenever possible, equipment sound power data by octave band (including 32 and 63 Hz) should be obtained for the path under study. A good sound prediction program relates all performance data.

Many other more specialized acoustics programs are available. Various manufacturers provide equipment selection programs that not only select the optimum equipment for a specific application but also provide associated sound power data by octave bands. These programs can help in the design of a specific aspect of a job. Data from these programs should be incorporated in the general acoustic analysis. For example, duct design programs may contain sound predictions for discharge airborne sound based on the discharge sound power of the fans, noise generation/attenuation of duct fittings, attenuation and end reflections of variable air volume (VAV) terminals, attenuation of ceiling tile, and room effect. VAV terminal selection programs generally contain subprograms that estimate the space NC level near the VAV unit in the occupied space. However, projected space NC levels alone are not acceptable substitutes for octave-band data. The designer/analyst should be aware of assumptions, such as room effect, made by the manufacturer in the presentation of acoustical data.

Predictive acoustic software allows system designers to look at HVAC-generated sound in a realistic, affordable time frame. HVAC-oriented acoustic consultants generally assist designers by providing cost-effective sound control ideas for sound-critical applications. A well-executed analysis of the various components and sound paths enables the designer to assess the relative importance of each and to direct corrective measures, where necessary, to the most critical areas. Computer-generated results should supplement the designer's skills, not replace them.

5.6 Solutions

In a normal reverberant field room where sound undergoes many reflections, the sound field caused by any single source reaches a more or less uniform level at some distance from the source. This remote sound level does not then change appreciably from one part of the room to another. On the other hand, sounds created out-of-doors or in a large open space decrease in level by 6 dB each time the distance from the source is doubled. A well-designed acoustic plan approaches this condition. This is achieved by treating sound-reflecting surfaces, especially the ceiling and walls, with absorbing materials to reduce sound reflections to negligible values in the speech frequency range.

Ceiling The largest surface available for specular reflection of sound in the open spaces is the ceiling. For that reason, much attention is paid to its acoustical properties. The ceiling surface must be highly sound absorbent, especially at angles of incidence of 45–60°, for a flat ceiling, lest reflections greatly reduce partition effectiveness. The acoustic system designer should provide an absorbent ceiling that approaches an "open sky" condition as nearly as possible.

A variety of ceiling products and configurations are available to reduce ceiling reflections. They range from flat absorbing panels to coffered systems and baffles. Accessory installations in the ceiling also can cause sound reflections. Flat, hard surfaces, such as light fixture shielding media, are generally unsatisfactory when located uniformly throughout the ceiling area. An exception to this rule can occur with task-oriented lighting or daylighting.

Walls Wall carpeting and drapes, although aesthetically pleasing and durable, may not be sufficiently sound absorbing unless extremely thick. Also, they may not have proper fire resistance. It is suggested that acoustically absorptive wall panels specifically designed for acoustic application be utilized. Windows around the perimeter of the building present a problem since most acoustical absorbers will impair vision. Draperies do not provide enough sound absorption unless they are of heavy textile material, fuller than usual, and closed. One possible solution is to tilt the window glass out of the vertical plane, thus deflecting the unwanted sound reflections to the ceiling or floor. Acceptable results may also be obtained by using vertical louvers that do not seriously impede vision but adequately diffuse the otherwise specular reflections.

Floors The acoustical efficiency of carpeted floors for reduction of impact sounds is well known and highly desirable. The sound absorption of carpeting is of limited value for open-space applications. Since sounds reflected from carpeted floors are easily absorbed by the ceiling system and otherwise broadly diffused by the partitions and furniture, no great problems result.

5.6.1 *Static Noise Control*

The most effective noise control step is to locate noisy equipment and its related services as far as possible from noise-sensitive areas. Nonsensitive areas (such as corridors, storage rooms, toilets, shafts, etc.) can be used effectively as buffer zones to shield noise-sensitive areas from mechanical equipment rooms in a typical multistory office building core area. Optimal mechanical room placement relative to other core areas permits the routing of the return air path and the supply air trunk ducts over areas with low acoustical sensitivity (such as restrooms).

To minimize the chances of noise problems, fan and air handling unit rooms should have floor areas of at least 10–15 ft^2 for each 1000 cfm of equipment air flow. This usually allows adequate space for proper air flow into the fan, low noise supply, and return air duct fittings and duct silencers, if required (Fig. 5.3). To minimize further the acoustical coupling between equipment housings and equipment room walls, all building system equipment room should have a floor area large enough to allow a clearance of at least 2 ft (600 mm) around all equipment. In some cases, electrical codes mandate even larger clearances.

Electrical transformer noise can be minimized by these steps:

1. Mount unit on vibration isolators.
2. If transformer is wall hung, use resilient hanging. If it is floor mounted, place on as massive a slab as possible.
3. Locate the unit so that reflections do not amplify the sound. Sound-absorbent material on the walls behind the units is not useful at 120 Hz. Only cavity resonators will absorb appreciable amounts of sound at that frequency.

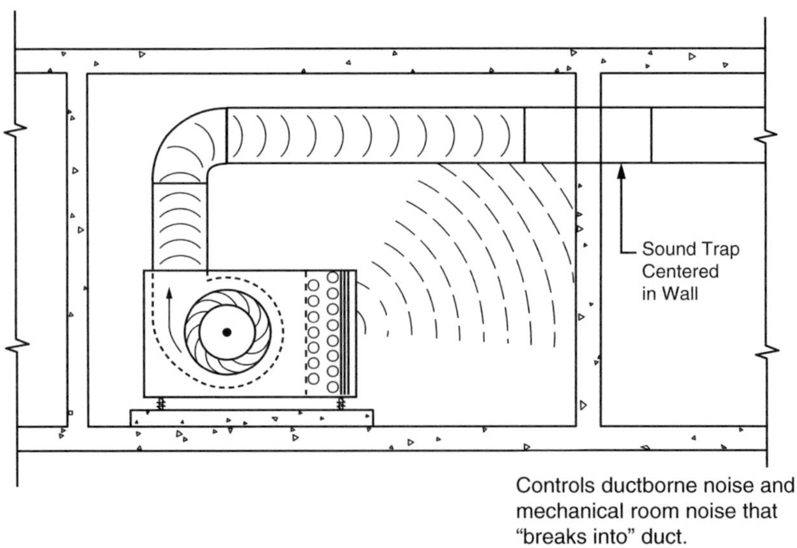

Fig. 5.3 Passive sound attenuation system

4. Use only flexible conduit connections.
5. Avoid locating transformers adjacent to, or immediately outside, quiet areas. A common error in this regard is placing a transformer pad immediately below the window of an NC 15–25 area.

5.6.2 Active Noise Control

Active noise solutions (ANSs) use artificially produced "noise" to counteract unwanted noise. It works by mixing antiphase sounds with the existing in-phase noise to produce "destructive interference" throughout the sound field. The level of attenuation achieved depends on whether the amplitude and phase error of the noise and antinoise signals are matched. When carefully controlled, noise reduction can be as high as 20 dB down.

There are two main approaches to the application of active noise cancellation (ANC). The first is to synthesize a secondary waveform with or without prior information on the original noise signal, which is, however, ineffective against random broadband noise. To reduce both tones and broadband noise, a second approach senses the original noise and produces an antinoise signal by filtering and injecting it back into the noise field.

The ANC system consists of three basic components: a controller to produce the antinoise signal, two sensors to measure the combined noise and antinoise sound field, and an actuator to introduce the antinoise into the sound field. Figure 5.4 shows a typical layout of ANC components in a duct. The first sensor

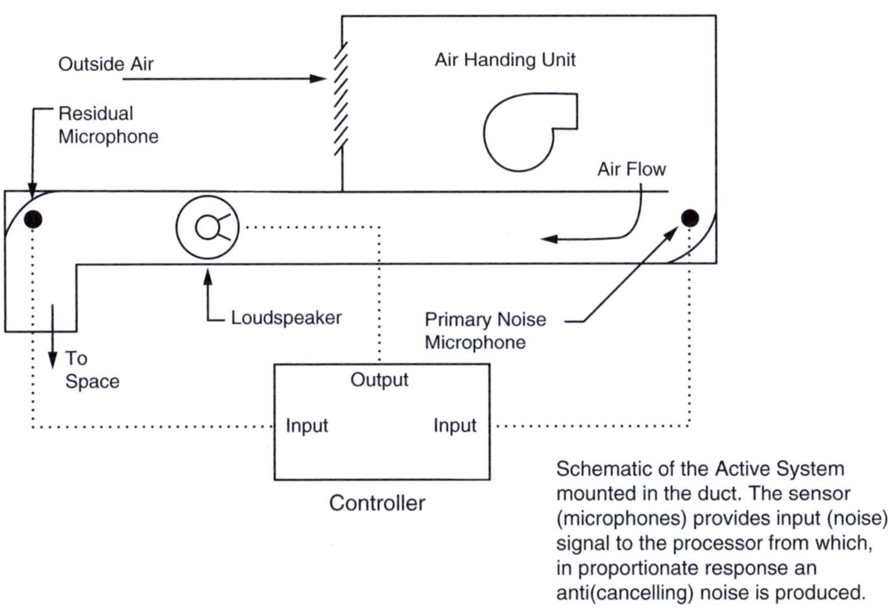

Fig. 5.4 Active sound cancellation system

microphone is located at a position close to the primary source of the noise, the motor/blower. A second microphone, called the residual sensor, is near the discharge of the duct and an actuator, actually a small loudspeaker, is mounted on the duct wall near the residual sensor. The sensors provide input signal to the controller and the actuator receives controller output. The system's controller is based on a digital signal processor (DSP) chip with appropriate analog input/output circuits for signal conditioning. Software that carries the adaptive feed forward (AFF) algorithm allows this unit to attenuate both broadband random noise and "periodic" or tonal noise for a variety of applications. The ANC process can be described by explaining how much each component functions during the cancellation cycle. During each DSP clock period, the controller samples both microphone inputs and provides a speaker output. The primary noise sensor sample is filtered and delayed so that when the noise propagates down the duct to the vicinity of the loudspeaker, the signal outputs to the loudspeaker at 180° out of phase. The residual microphone sample provides the means for the controller to monitor and improve its performance by sensing and minimizing the combined sound field of noise and antinoise. The AFF algorithm adaptively computes the cancellation signal and the antinoise signal continuously changes (active) as the noise changes. Errors (residual noise after the introduction of the antinoise signal) are corrected on the following noise cycle.

Several criteria must be satisfied for the ANC system to perform well. The sensors must be at an adequate distance from the actuator to provide a time delay for filtering of the primary noise signal and for acoustic mixing of the combined noise and antinoise field. Second, as the controller is minimizing the error at the residual sensor (using "knowledge" of the primary noise and the cancellation signal), it is necessary that the former's signal characteristics do not change as the noise travels down the duct. Thus, without the loudspeaker signal, both sensors should hear the same noise. This property is called coherence, which is not affected by the time delay between the positions of the primary and residual sensors. However, coherence is greatly affected by turbulent airflow near the microphones. Turbulent aerodynamic noise degrades coherence because it is localized and differs at each microphone. The coherence between microphones sets a theoretical limit on the maximum amount of attenuation that an active cancellation system can achieve. For the motor/blower test platform, this means that duct design, microphone placement, and microphone cover are aimed at preserving coherence across the frequencies where active attenuation would be most effective. (Perfect coherence has a value of 1.0.)

A third criterion for effective noise attenuation is that the controller must have accurate models of the acoustic propagation paths between the actuator and both sensors. These models are obtained during self-calibration wherein the AFF algorithm computes the transfer functions (the time, phase, and amplitude relationship between the output and input of each sensor and the actuator). In the process, two of the digital filters that the AFF algorithm uses are defined. The filter that models the feed-forward electrical and acoustical paths from controller output, through the actuator, to the residual sensor, and back to the controller input is vital to the monitoring and improvement of the adaptive nature of the AFF algorithm. A second filter models the feedback path from the actuator to the primary noise sensor. Thus, any antinoise "heard" by the primary sensor does not find its way into succeeding antinoise outputs, causing an unstable control loop.

5.7 Vibrations and Buildings

Another acoustical system to consider is that of vibrating equipment, its vibration isolators, and the building structure that supports the equipment. The primary design goal is to mount the equipment on resilient isolators that are attached to very stiff, massive parts of the building structure.

Vibration in a building originates from both outside and inside the building. Sources outside a building include blasting, operations, road traffic, overhead aircraft, underground railways, earth movements, and weather conditions. Sources inside a building include doors closing, foot traffic, moving machinery, elevators, building HVAC and electrical systems, and other building services. Vibration is an omnipresent, integral part of the building environment. The effects of the vibration on building occupants depend on whether it is perceived by those persons and factors related to the building, the location of the building, the activities of the occupants in the building, and the perceived source and magnitude of the vibration. Factors influencing the acceptability of building vibration are presented in Fig. 5.5.

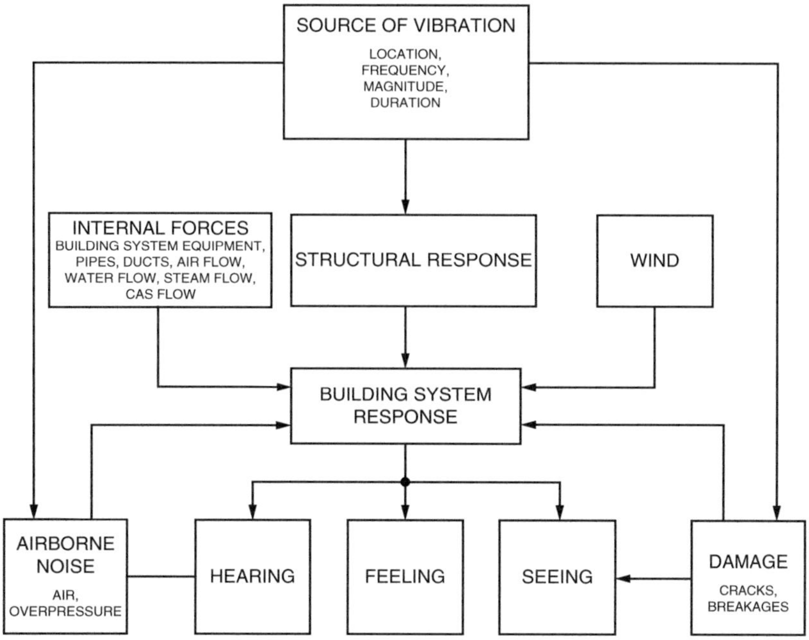

Fig. 5.5 Factors affecting building vibration

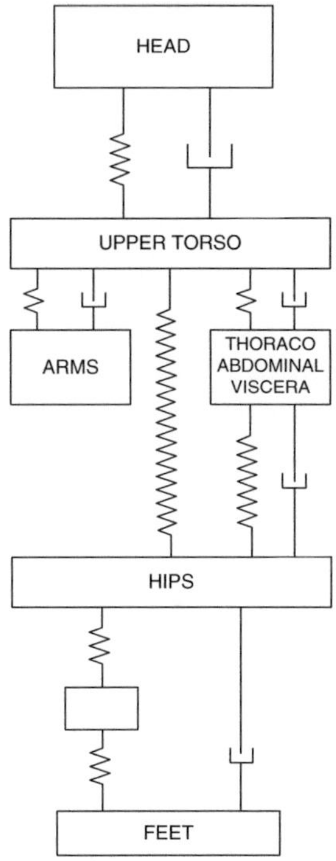

Fig. 5.6 Simple mechanical analog of the human body—a complex resonator. Bent knees and slouching shoulders damp out vibrations received via the feet. The response of body to other vibrations depends on direction, force, intensity, and distribution of vibration

Whatever the source of vibration, a person will perceive its effect by hearing it, seeing it, or feeling it. It is the combination of these perceptions that will determine human response. Because of this and the nature of vibration sources and building responses, it is convenient to consider building vibration in two categories: low-frequency vibrations less than 1 Hz and high-frequency vibrations of 1–80 Hz.

Human response to vibration depends on the vibration of the body. The main vibrational characteristics are vibration level, frequency, axis (and area of the body), and exposure time. See Fig. 5.6 for a simple mechanical analog of the human body, considered as a vibrating system. Tall buildings always oscillate at their natural frequency, but the deflection is small and the motion undetectable. In general, short buildings have a higher natural frequency of vibration than taller ones. However, strong wind forces energize the oscillation and increase the horizontal deflection, speed, and accelerations of the structure.

Higher-frequency vibrations in buildings are caused by building system machinery, elevators, foot traffic, fans, pumps, and other process equipment. Further, the steel structures of modern buildings are good transmitters of high-frequency vibrations. Vibration does not reach us through the air, but audible sound does. A variety of insulations are barriers to sound transmission to a human space and reduce reflections within a human space. This art is part of the field of acoustics. One of the problems of mechanical design is to prevent structural parts from conducting sound, bypassing insulation.

Chapter 6
Lightning, Electrostatic Discharge, and Buildings

Anil Ahuja

6.1 Introduction

What causes lightning? Striking 100 times a second somewhere on the planet and packing temperatures up to 50,000 °F, lightning is both a common and a fearsome occurrence. Atmospheric electrical discharges known as *lightning* or *thunderbolts* (from cloud to cloud or cloud to ground) have captured the imagination and fear of the human race since ancient times. The ancient Greeks believed that lightning was Zeus' tool to punish human misbehavior or to demonstrate his anger.

Meteorologists recognize several types of clouds. When hot, humid air rises, it forms cumulus clouds. Large cumulus clouds rising to the top of the troposphere expand horizontally and form cumulonimbus clouds. These are the familiar "thunderheads" of the summer. Typically, electrical flashes are found in cumulonimbus clouds, but they may also occur in nimbostratus clouds (thick dark clouds at 2–12 km altitude covering the entire sky), in snowstorms and dust storms, and even in the gases of an active volcano.

Thunderstorms occur when a cloud becomes electrically charged. This can happen when ice crystals, water droplets, and other particles collide within the rising and falling air currents in clouds, producing electricity. The atmosphere usually works as an insulator to prevent this electricity from escaping. However, when the electricity stored in the thundercloud reaches a certain level, the insulation effect breaks down and allows the instant formation of an enormous electrical current known as lightning.

Ball lightning is one of the most bizarre of all atmospheric phenomena. While ordinary lightning comes down in bolts from the sky, ball lightning is a fireball with a tail that moves close to the ground. Blazing with the brightness of several hundred light bulbs, ball lightning ranges in size from 6 in. to 2 ft in diameter and in shape from spherical to oval to wispy. Its color is usually white, sometimes with an orange or blue cast. But as varied as fireballs are in their appearance, they differ much more widely in behavior. Following paths that are straight or wildly curved, fireballs can

A. Ahuja, *Integration of Nature and Technology for Smart Cities*,
DOI 10.1007/978-3-319-25715-0_6

float lazily over the ground for minutes or race by at thousands of miles per hour. They typically follow telephone or power lines for a few seconds, then vanish. According to some reports, fireballs also enter homes. Slipping through doors or windows, or diving down chimneys, they dart about furiously, then make a quick exit, leaving scorch marks in their wakes. Fireballs tend to occur in places where the air is stagnant, as over marshes and valleys. In cities, they usually form near high-tension wires, telephone lines, and the corners of metal buildings and towers. Nearly half of all reported fireballs appear inside buildings, with the fireball usually entering through a door or window.

People have reported seeing fireballs for centuries. But because ball lightning is so rare—and so fleeting when it occurs—it was not until the mid-twentieth century that scientists began studying it seriously. Three theories relate fireballs to lightning strikes. In the electricity theory, lightning strips hydrogen atoms from airborne water molecules. The hydrogen bonds with molecules containing carbon form a ball of hydrocarbon molecules that emit light as they release excess energy. In the aerosol plasma theory, electrically charged particles called aerosols form a sphere. When lightning strikes, the sphere becomes energized and glows. In the electromagnetic wave theory, static electricity builds up in clouds and trees, generating electromagnetic waves that bounce off the ground. Where the waves meet, electromagnetic energy exits the air, forming a lightning ball.

6.2 Mechanisms and Characteristics of Lightning

The real cause of lightning is separation and accumulation of electrical charges in clouds via certain microphysical and macrophysical phenomena. To explain these phenomena, initially precipitation and convection theory was developed, but today's most complete theory for lightning phenomena has established the fact that the structure of a thundercloud is *tripolar* (see Fig. 6.1). This is the so-called charge-reversal hypothesis which states that when graupel particles (precipitation consisting of densely packed balls of snow or snow pellets) collide with ice crystals, the charge is transferred to a graupel particle. The polarity of the charge is dependent on the temperature.

At temperatures above a certain value, called the *charge-reversal temperature,* the transferred charge is positive. The exact value of the charge-reversal temperature is being debated, but it is believed to be around −15 °C. Considering that the temperature of the atmosphere is −15 °C (5 °F) at an approximate altitude of 6 km (3.75 miles), this means that owing to collisions of large snow pellets and ice crystals, the thundercloud will be, on aggregate, negatively charged for altitudes about 6 km and positively charged below 6 km. This hypothesis has been verified in the laboratory and explains the levels of negative and positive charges in a thundercloud. Yet the exact microphysics of this phenomenon is practically unknown.

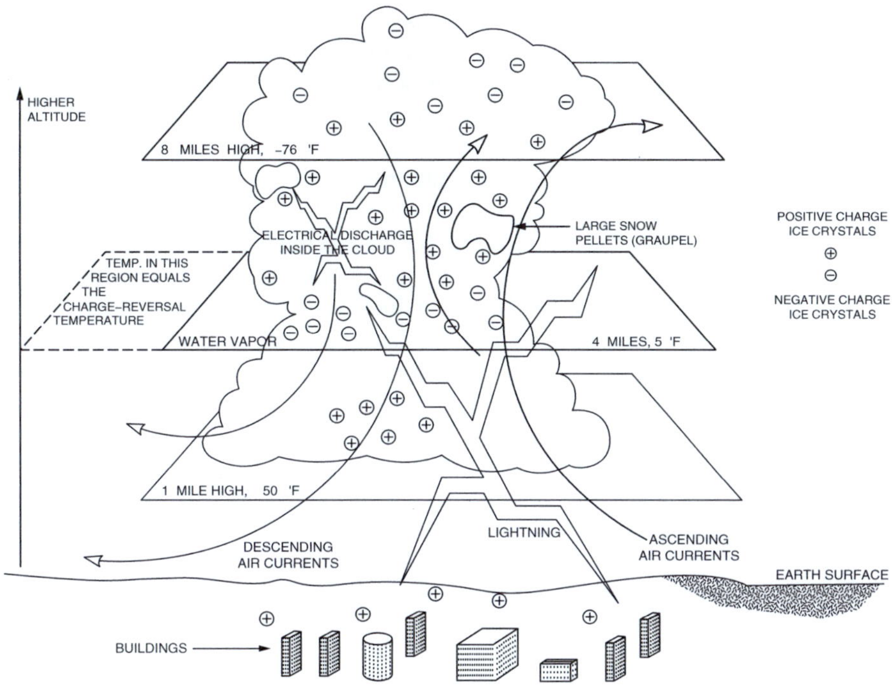

Fig. 6.1 Lightning

Mechanisms of Lightning Lightning begins whenever the charge accumulation in a thundercloud is such that the electric field between charge centers inside the cloud or between cloud and earth is very high. For building engineering purposes, only cloud to earth lightning strokes (ground flashes) are of importance.

Typically, cloud to earth lightning stroke involves three stages. In the first stage, the high electric field intensity may generate local ionization and electric discharges known as *pilot streamers*. A pilot streamer is followed by the so-called *stepped leaden*. The stepped leader is a sequence of electric discharges that are luminous; they propagate with a speed approximately 15–20 % of the speed of light, and they are discrete, progressing approximately 50 m at a time. The time between *steps* is a few microseconds to several tens of microseconds. The stepped leader will eventually reach the surface of the earth and will strike an object on the earth. However, where it will strike is not determined until the stepped leader is within a *striking distance* from the object.

The second stage initiates when the stepped leader reaches an object on the earth or meets an upward moving stepped leader. Specifically, a high-intensity discharge occurs through the channel established by the stepped leader. This discharge is extremely luminous and therefore visible. It propagates with a speed of about 10–50 % of the speed of light. The return stroke carries an electric current of anywhere from few thousands of amperes to 200 thousands of amperes. The current magnitude rises fast, within 1–10 μs to the peak value, and then decreases rapidly. The discharge is known as the *return stroke* or simply the *lightning stroke*. The return stroke transfers a substantial amount of positive charge from the earth to the cloud and specifically to the charge center where the lightning originated. This transfer results in a significant lowering of the potential of the charge center. This phenomenon initiates the third stage of lightning.

In the third stage, discharges may occur from other charge centers within the thundercloud to the depleted charge center because of the increased potential difference between them. This discharge will trigger another stroke between cloud and ground through the already established conductive channel with the first stroke. This process may be repeated several times depending on the electrification status of the thundercloud, resulting in multiple strokes.

Characteristics of Lightning Strokes The parameters of lightning ground strokes are very important in the design of protection schemes against lightning. The most important parameters are:

- Voltage
- Electric current
- Waveform
- Frequency of occurrence

The voltage between a thundercloud and earth prior to a ground stroke has been estimated to be from 10 to 1000 MV. For design work, however, the building protection engineer is interested in the voltage appearing on the stricken power apparatus. This voltage will be equal to the product of the impedance times the stroke current. It is generally accepted that the ground stroke current is independent from the terminating impedance. The reason is that the terminating impedance is much lower than the resistance of the lightning discharge channel, which is on the order of few thousand ohms. Thus a ground stroke is normally considered as an ideal current source at the point of strike. The crest of the stroke electric current can vary over a wide range: 1–200 kA.

6.3 Building System Lightning Exposure and Protection

Building electric power systems are exposed to weather and therefore they are subjected to lightning strikes which result in overvoltages. Lightning overvoltages are generated by direct lightning strikes on a power system apparatus or indirect strikes to nearby objects, from which subsequent overvoltage is transferred to the system via inductive, capacitive, and conductive coupling.

Lightning overvoltages are independent of system voltage but depend on system impedances. For example, a direct lightning hit to a phase conductor of an overhead transmission line will generate an overvoltage proportional to the characteristic impedance of the line and proportional to the current magnitude of the lightning stroke. This overvoltage may be several million volts. It is a practical and economical impossibility to insulate distribution or lower-kilovolt-level transmission lines (i.e., 230 kV and below) to withstand this type of overvoltage. An integrated design procedure is applied to minimize the effects of lightning, which involves, among other things, (1) shielding of lines and equipment, (2) effective grounding, and (3) application of protective devices (surge arresters). The presence of the shielding system ensures that lightning strike, which otherwise will discharge to a line conductor, will terminate on a wire shield, air terminal, etc., that is electrically connected to the grounding system. A well-designed ground system will divert the majority of the lightning overvoltages.

Building distribution circuits are typically not insulated to withstand direct lightning strokes. As a result, direct strikes will cause a flashover. Direct strikes on building distribution lines are not frequent since the poles are not as high and therefore shielded from trees and structures. On the other hand, distribution lines may be vulnerable to overvoltages resulting from lightning strokes to nearby trees, ground, or other objects. Lightning strokes to nearby trees, ground, or other objects can result in voltage surges into the power system through coupling. The coupling can be conductive (through the conductive soil and the power system grounding structures), inductive, or capacitive. In a typical situation, all the coupling mechanisms may be present, resulting in a voltage surge to the power system. These voltages are called *induced voltage surges* and are generally much lower than those occurring after a direct strike. Specifically, they rarely exceed 400 kV.

It is therefore necessary to insulate distribution lines to withstand these surges. This translates into the requirement of a 300 kV basic lightning impulse insulation level (BIL) for distribution lines. In addition, power apparatus, connected to distribution lines and with BIL lower than the induced voltage surges, such as distribution transformers (typically the BIL of distribution transformers is 100 kV), must be protected. It is practical to protect transformers with surge arresters of appropriate ratings. Higher-kilovolt-level lines (i.e., 69 kV and above) have sufficient insulation withstand so that induced lightning voltages do not present the risk of flashover.

By far, lightning overvoltages on building power systems are the most stressful. Since these systems are interconnected to utility power systems, disturbances on the utility power system will be transmitted to them. These systems are also subjected to direct lightning. Therefore, these systems also require shielding against lightning.

Typically, a grounding system will be installed as well as a lightning protection system (shielding) to divert any direct lightning strokes to ground. This grounding system is referred to as *external grounding* to distinguish it from the so-called internal grounding, which refers to the grounding system of various types of equipment in the facility. The external grounding systems of industrial/commercial power systems are interconnected to the power system, as shown in Figs. 6.2 and 6.3. The lightning protection system is basically a shielding system designed to route the lightning surges into the external grounding system for the purpose of minimizing

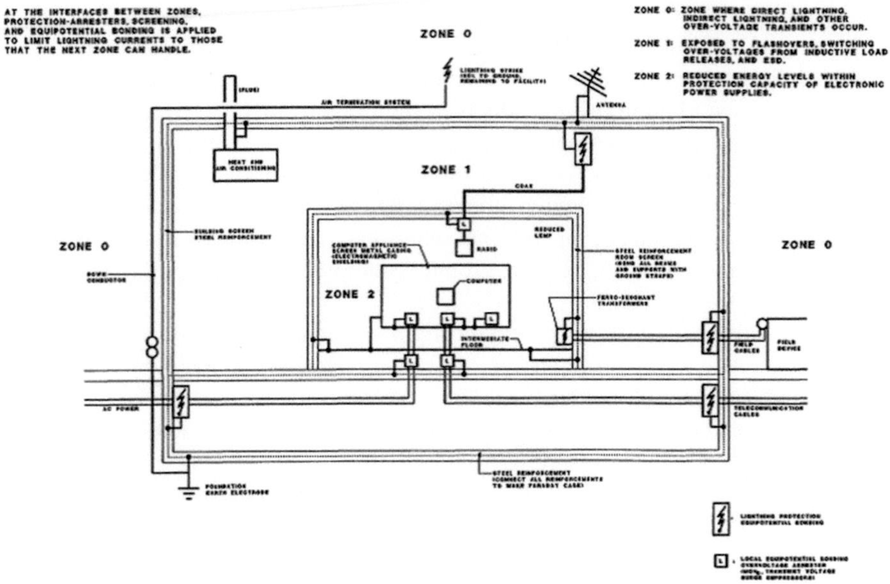

Fig. 6.2 Zoned electromagnetic shielding and lightning protection diagram

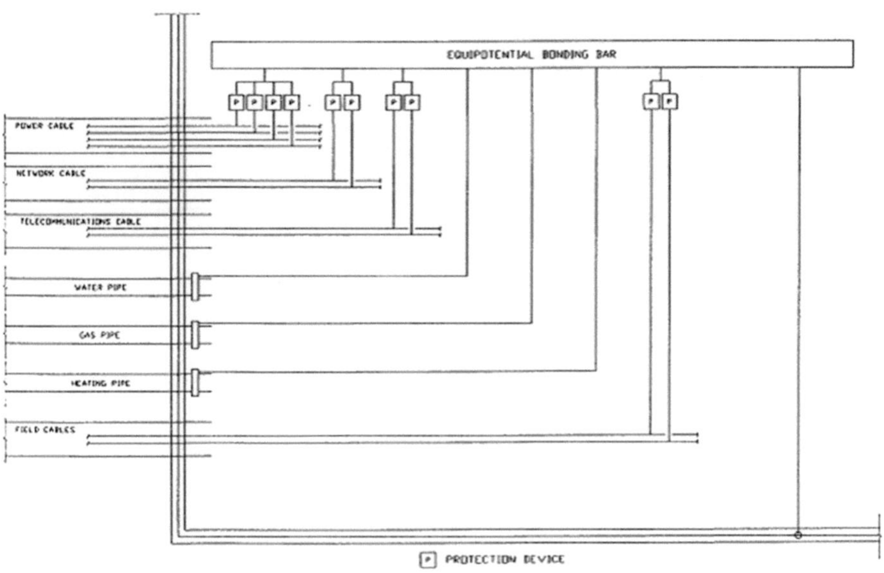

Fig. 6.3 Lightning and transient protection

potential differences within the facility. A system like this is subjected to lightning overvoltages which may enter from a number of points such as:

- Air terminals
- Communication towers (if present)
- Power system grounding
- Fences

An integrated design of the building external grounding system, lightning protection system, and internal grounding system can provide a system that is hardened against lightning and other sources of overvoltages.

Figure 6.3 shows lightning and transient protection facility grounding diagram. Grounding of a facility goes a lot further than just the electrical systems. A full treatment of lightning and transient protection includes a grounding or equipotential system to shunt the high voltage and currents to ground potential in all zones. The total treatment would include current-carrying conductors from lightning rods, located on the structure to rebar and metal inlay within the concrete, to zone protection conductors far within the structure.

6.4 Electrostatic Discharge (ESD): An Introduction

To the average person, the words "static electricity" can mean either a noise in the radio receiver that interferes with good reception or the electric shock experienced when touching a metal object after walking across a carpeted floor or sliding across the plastic seat cover of furnishings inside the building. Some people also have experienced mysterious crackling noises and a tendency for some of their clothing to cling or stick tightly together when wool, silk, or synthetic fiber garments are worn. Nearly everyone recognizes that this phenomenon occurs mainly when the atmosphere is very dry.

The word "electricity" is derived from the ancient Greek work *elektron*, meaning amber, for it was with this substance that the phenomenon of electrification was first observed. When the properties of flowing electricity were discovered, the word "static" came into use as a means of distinguishing the old from the new. The implication that such electricity is always at rest is erroneous; it is when it ceases to rest that it causes the most concern.

For the sake of simplicity, one might imagine electricity to be a weightless and indestructible fluid that can move freely through some substances, such as metals, which are called "conductors," but can flow with difficulty or not at all through or over the surface of a class of substances called "nonconductors" or "insulators." This latter group includes gases, glass, rubber, amber, resin, sulfur, paraffin, and most dry petroleum oils and many plastic materials.

When electricity is present on the surface of a nonconductive body, where it is trapped or prevented from escaping, it is called *static electricity.* Electricity on a conducting body that is in contact only with nonconductors is also prevented from escaping and is therefore nonmobile or "static." In either case, the body on which this electricity is evident is said to be "charged." The charge may be either positive (+) or negative (−). At one time it was thought that the two charges were two kinds of electricity and that in a neutral (uncharged) body they were present in exactly equal amounts. Now it is known that there is actually only one kind of electricity. It is true, however, that in a neutral or uncharged body, the two entities are present in exactly equal amounts.

Work is required to separate positive and negative charges. Electricity, therefore, is sometimes referred to as a form of energy produced by expenditure of energy in some other form, such as mechanical, chemical, or thermal. Likewise, when electrical energy is expended, its equivalent appears in one of these other forms. These entities are components of all atoms, the outer electrons (−) and the inner (nuclear) protons (+). Electrons are free to move from one molecule to another in conductors, but the proton, in the nucleus of the atom, cannot move appreciably unless the atom moves. Therefore, in solids, only the electrons are mobile; in gases and liquids, both are free to move. Curiously, a surface that has an excess or deficiency of one electron in every 100,000 atoms is very strongly charged.

The stable structure of the atom shows that unlike charges attract and, conversely, like charges repel. It follows that a separated charge will be self-repellent and will reside only on the surface of a charged body. If the body were a perfect insulator, the charge would remain indefinitely. However, there are no perfect insulators, and isolated charges soon leak away to join their counterparts and thus bring about neutralization, the normal state.

Static electricity then is the set of phenomena associated with the appearance of an electric charge on the surface of an imperfect insulator. It is "liberated" or usually made alive by the expenditure of mechanical work since electricity cannot be created. The charge on the surface of an insulator can thus attract or "bound" an equal and opposite charge on the nearest surface of any conducting body close to it. If the conducting body is now moved away from the originally charged body, the bound charge is now freed and will redistribute itself over the whole surface of the conducting body. In turn, it can be released in the form of a spark.

6.5 Electrostatic Charge: Charging Mechanism

Whenever two substances of different composition are brought into contact, one of the substances will surrender some of the electrons from its atoms to the other along the contact surface. Although the total (net) charge upon the two substances remains unchanged (and may be zero), the redistribution of charge resulting from this transfer of electrons results in the formation of an "electrical double layer" along the contact surface. One substance will have an increased abundance of electrons (and be negatively charged), while the other will be somewhat depleted of electrons (and be positively charged). Since these equal and opposite charges are strongly attracted to each other, they remain intimately related to the opposing surfaces and are not externally sensible so long as the surfaces remain in contact.

If the substances are nonconductive and are pulled apart, however, much of the charge disparity will remain with the individual substances, resulting in one being charged positively and one negatively. This charging mechanism is referred as "contact/separation or frictional charging." It can be enhanced by increased speed of separation, by lowered conductivity of substances, and by increased disparity in work function of the substances.

The surface of a substance that is subjected to bombardment by an ion shower (such as one originating at a corona point) will become charged by attachment of the ions or by surrender of charge to the surface by the ions. Charging by this mechanism is referred to as "bombardment charging."

When an uncharged object is brought into contact with another object that is charged, some charge will be transferred to the previously uncharged object. This charging mechanism is called "contact charging."

ESD is a charge-driven phenomenon. At any time, the net charge on a body is the difference between charge generated and charge dissipated.

$$Q_{\text{net}} = Q_{\text{generated}} - Q_{\text{dissipated}}$$

In many situations, induced charges are far more dangerous than the initially separated ones upon which they are dependent. Since a spark from the surface of an insulator can release a charge from only a small area, all the charge on the conducting body can be released in a single spark. In effect, a metal plate in close proximity to a charged surface can be considered one plate of a capacitor, and its ability to store energy is described as its capacitance. When a potential difference is applied between the two plates of a capacitor, electricity can be stored. In some instances one of the plates is the earth, the insulating medium is the air, and the other plate is some body or object insulated from the earth to which the charge has been transferred by induction or otherwise. Then a conducting path is made available, and the stored energy is released (the capacitor is discharged), possibly producing a spark. The energy so stored and released by the spark is related to the capacitance (C) and the voltage (V) in accordance with the following:

$$\text{Energy} = \frac{C(V)^2}{2}$$

If the object close to the highly charged nonconductor is itself a nonconductor, it will be polarized; that is, its constituent molecules will be oriented to some degree in the direction of the lines of force since their electrons have no true migratory freedom. Because of their polarizable nature, insulators and nonconductors are often called dielectrics. Their presence as separating media enhances the accumulation of charge. This is the reason dry area enhances accumulation and electrostatic discharge (ESD).

A person walking on an isolated floor covering can charge up to 10 kV. The typical capacitance to earth is 100 pF. When a person is close enough to a grounded structure (a computer, metal door, building steel, etc.), an ESD or arc is formed. Using the above figures, the charge the person carries is

$$Q = CV = \left(100 \times 10^{-12}\right)\left(10 \times 10^{3}\right) = 1\text{microcoulomb}$$

The current flowing is limited by the surge resistance of the person and varies from person to person, but it is significant enough to cause damage to electrical or electronic building system equipment.

6.6 ESD Solutions

One solution is to maintain a humid environment to reduce the dielectric resistance or use ionizers to dissipate or neutralize charge. Second, place a metal barrier between the person and building system electronic components. This does not make the problem go away but does protect the equipment. To solve the problem as well as prevent charge injection, it is necessary for the metal barrier either to completely surround the entire system (including cables) or to connect it to earth ground. An earth connection will bleed off the metal barrier charge and thus eliminate the electrostatic field, as well as prevent charge accumulation. In some cases, ESD can be reduced by very inexpensive means: use of static free carpeting, static sensitive building materials, and humidity control. ESD electromagnetic radiation usually can be prevented from causing malfunctions by use of sound maintenance practices. Replace missing items such as screws, shields, ground straps, and so on. Some degree of ESD potential control can also be obtained by momentarily grounding the human body upon entry into protection zone. This requires good coupling between charged sources associated with the human body (clothing and shoes) at the time of initial grounding and maintaining this coupling as the human body enters protected zone.

Chapter 7
Electromagnetic Waves, Noise, and Systems Susceptibility

Anil Ahuja

7.1 Introduction

To understand electromagnetic interference (EMI), electromagnetic compatibility (EMC), electromagnetic shielding (EMS), and their relation to building system-sensitive equipment, it is important to understand how electromagnetic waves are created, their characteristics, and how they travel.

Any time an electric current varies its speed or course, it generates electromagnetic waves—which are nothing more than fluctuations of electrical and magnetic forces. Electric and magnetic fields are physical phenomena: invisible lines of force that occur whenever electrical power is used. Electric and magnetic fields are produced in energized conductors, electrical office equipment, internal wiring, and other items.

Electric and Magnetic Fields Electric and magnetic fields are decoupled below radiofrequencies and are considered separate fields at low frequencies.

Electromagnetic Fields Electric and magnetic fields at radiofrequencies and above are coupled together and referred to as electromagnetic fields.

The energy of an electromagnetic wave is related to its wavelength—the distance from one wave crest to the next. The shorter the wavelength, the more energetic the wave. In the order of decreasing wavelength, electromagnetic waves include radio waves, infrared rays, visible light, ultraviolet light, X-rays, and gamma rays. Gamma rays are a mere one hundred-billionth of a meter or 3.3 hundred billionths of a foot long, while radio waves can be a few miles long. As electromagnetic waves spread outward at the speed of light, their electric and magnetic fields radiate at right angles to each other and to the direction of wave flow (see Fig. 7.1).

All electromagnetic radiation is transported exactly the same way through empty space. The atmosphere behaves almost like empty space except that particles and vapors selectively absorb radiation differently at different wavelengths. The electric field is a function of *only* the electric system voltage level. The higher the voltage,

A. Ahuja, *Integration of Nature and Technology for Smart Cities*,
DOI 10.1007/978-3-319-25715-0_7

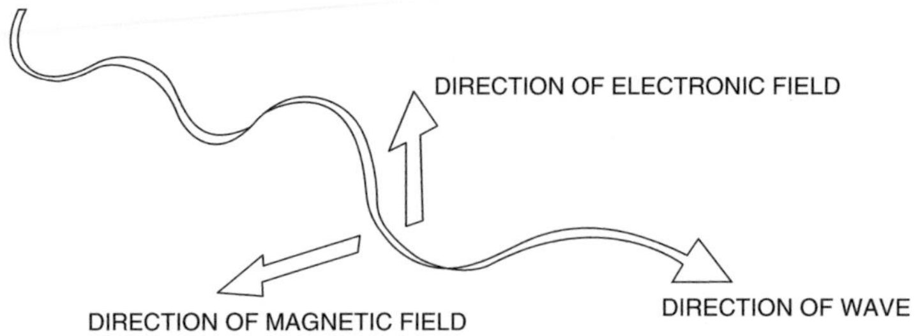

Fig. 7.1 Electromagnetic radiation

the stronger the electric field will be. The unit of measurement of the electric field is volts per meter. At higher levels of the spectrum, it is electron volt/meter (eV/m).

The strength of the electric field decreases as an approximate function of the inverse square of the distance from the equipment or conductor containing the voltage. Thus, the strength of the electric field decreases rapidly as the distance from the source increases. Since a magnetic field is a function of moving charges (current), the magnet field can be looked on as closed loops of force lines such as ripples from a pebble dropping in the middle of a pond. With alternating current, the ripples would reverse with each change in current flow. The magnetic field is a direct function of the amount of electric current flowing in a piece of equipment or conductor. The larger the amount of current, the stronger the magnetic field will be. The strength of the magnetic field around a source decreases as an approximate function of the inverse cube of the distance from the electrical equipment carrying current. Thus, the strength of the magnetic field also decreases rapidly as the distance from the source increases.

7.2 Geomagnetic Fields and Buildings

A naturally occurring magnetic field exists and surrounds the earth. Humans and animals conceived, nurtured, and grew under the influence of magnetic fields: Building systems operate in the company of natural fields. The SI unit of magnetic field intensity is the tesla (T), equal to 1 newton/ampere meter. Other units in common use in geophysics are the gauss (G), equal to 10^{-4} T, and the ganna (y), equal to 10^{-5} G, or 1 nanotesla (nT). The magnetic north and south poles are defined on the basis of the compass needle—the end of the needle that points toward geographic north is called the *north magnetic pole*, and the opposite pole of the needle is called the *south magnetic pole.* The north pole of a compass needle points north because it is attracted by the earth's north magnetic pole, located in eastern Canada (77′18″N.101 48′W). However, a north magnetic pole attracts a south magnetic pole and vice versa.

The intensity of the magnetic field of the earth at sea level ranges from 0.25 G at the magnetic equator to 0.60 G at the magnetic poles. By comparison, that of a small pocket magnet may range around 100 G.

The origin and dynamics of the earth's magnetic field still are not clear. The electron flow needed to support the magnetic field at its present intensity is about 44.10^9 amperes if it is in the form of a toroidal current in the outer core. In order to maintain the flow, an energy source is needed. Heat is the most obvious choice, but mechanical torques related to the precessional motions of the earth's axis have also been suggested. The trick is to transform heat and/or mechanical energy into an electron flow. It is obvious that the earth's rotation has something to do with the orientation of the magnetic field. There are a lot of free electrons in both the inner core and the outer core; there is a temperature gradient; and there is convection, at least in the outer core. The system is highly complicated in terms of electrical properties. It is likely that there are electrons flowing in all directions, except that there may be a minute excess flowing in a direction related to the earth's rotation, that is either overtaking the earth (and producing a normal magnetic epoch) or lagging behind (and producing a reversed magnetic epoch).

Electromagnetic interaction is mediated by photons. Four basic points need to be remembered in order to understand magnetism. The first point is that a charged particle moving with *uniform* motions creates around itself a magnetic field owing to continuous emission and reabsorption of virtual photons; the second point is that a magnetic field forces a change of direction in a charged particle moving at an angle through it; the third point is that a charged particle moving with *accelerated* motion (except electrons within orbitals) produces an electromagnetic wave; the fourth point is that an electromagnetic wave accelerates a charged particle.

If electrons flow with uniform motion through a straight wire, a circular magnetic field is created around the wire that has a counterclockwise direction when viewed along the direction of motion of the electrons. Conversely, if the electrons move with uniform motion around a loop on a counterclockwise direction, a magnetic field is created along the axis of the loop that is directed along the line of sight when viewing the electron flow in the loop as counterclockwise.

7.3 Electromagnetic Compatibility (EMC)

There are three elements to any EMC system: the source of the EMI called emitters; the media through which it is transmitted, conducted, or radiated; and the receptors or susceptors. Susceptibility is the tendency of a piece of equipment or a system to respond to undesired electromagnetic energy which otherwise affects the system adversely owing to the received EMI. Therefore, EMC can be achieved by reducing the EMI emissions levels in the source, by blocking the propagation path of the EMI signals, or by making the receiver less susceptible to the received EMI signals.

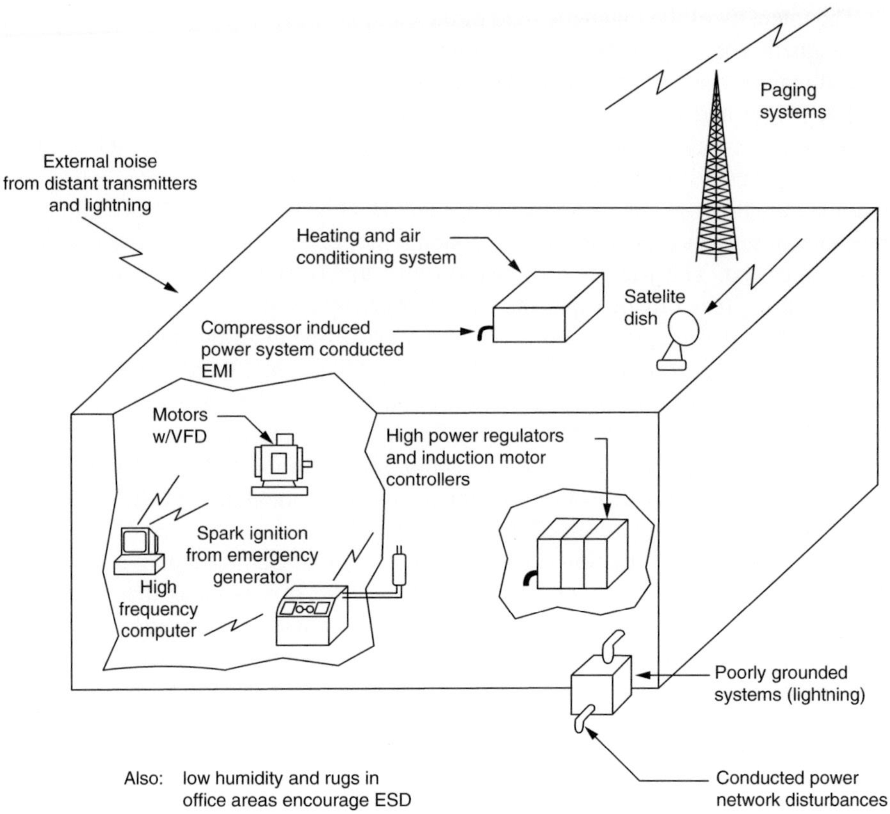

Fig. 7.2 Interference can come from sources both inside and outside the building and range from nearby radio and radar transmission to power line disturbances and lightning

Figure 7.2 shows some of the emitters outside and inside of buildings. The source of the EMI is primarily any system where the current or voltage changes rapidly. Generally, external electrical noise is from the following sources:

1. Ignition noise from gasoline emergency generator engines
2. Static electric-type discharge from high-voltage power lines
3. The endless collection of business automation equipment

EMI sources can also be broadly divided into two categories, natural and man-made. Naturally caused EMI below 10 MHz is mainly due to atmospheric noise resulting from electrical storms. Above 10 MHz, they are primarily a result of cosmic noise and solar radiation. A man-made noise common to all systems in buildings is power line hum. Building system operators used to determine if their telephones were working by noting the hum before attempting to use it. In addition there were the occasional pops and snaps from the power company doing some line switching. From the system point of view, it is necessary to have a way to determine how much noise each component will add to the system.

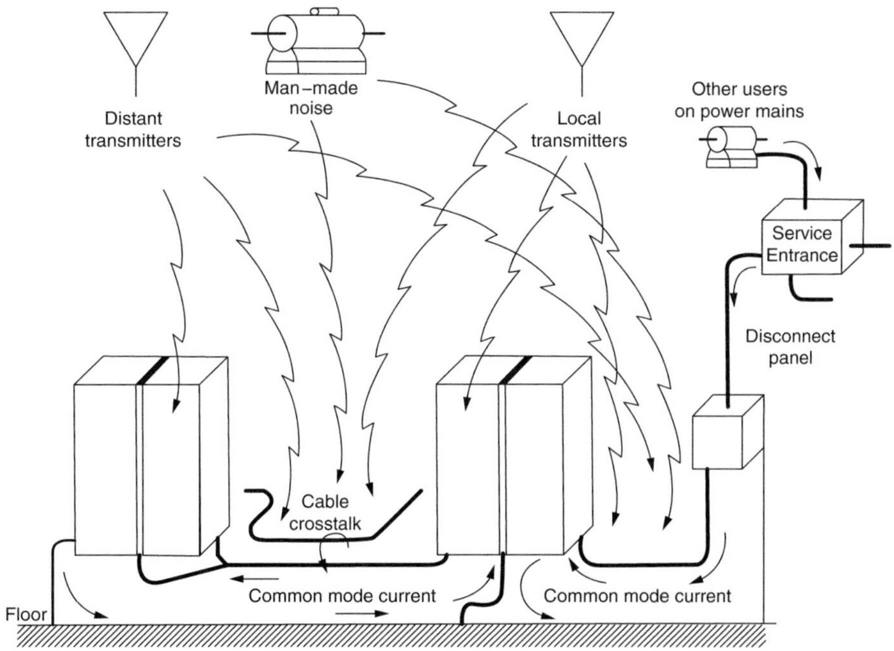

Fig. 7.3 Electromagnetic interference can reach the victim through radiation or conduction

Power electronic circuits used in large building systems, by switching large amounts of current at high voltages, can generate electrical signals that affect other electronic systems. The electric field strength from these radiators is measured in units of volts per meter. When the field exceeds 100 V/m, EMI is almost certain to exist. When it is below 1 V/m, EMI is unlikely. These unwanted signals give rise to electromagnetic interference (EMI), also known as radiofrequency interference (RFI), since they occur at higher frequencies. The signals can be transmitted by radiation through space or by conduction along a cable (see Fig. 7.3).

Apart from emitting EMI, the control circuit of power systems can also be affected by EMI generated by its own power circuitry, by other circuits, or by natural phenomena. When this occurs, the system is said to be susceptible to EMI. Any system that does not emit EMI above a given level, and is not affected by EMI, is stated to have achieved electromagnetic compatibility (EMC).

Naturally caused EMI through lightning can result in interference to ground or airborne equipment and damage if a direct hit occurs. Peak currents can exceed 50 kA with rate of rise in the region of 100 kA/μs, *giving field strengths in nearby conductors of greater than 200 kV/m. The voltage induced in antenna systems, having a large physical length, is especially high and these are also prone to direct strikes.*

Man-made EMI can be intentional or unintentional. In both cases, it is the variation of the voltage and current that produces EMI, whose magnitude depends on the value of the current, the length of the conductors, the rate of change of voltage and current, and the physical position of the conductors relative to each other and any earth plains. Examples of intentional EMI are buildings built near airfields which suffer most from EMI resulting from radar, where field strengths approach 200 V/m, with buildings giving only a low level of shielding.

Unintentional man-made interference is caused by sources such as switches, relays, motors, and fluorescent lights—equipment commonly found in buildings. The inrush current of transformers during turn-on is another source of interference, as is the rapid collapse of current in inductive elements, resulting in transient voltages. Integrated circuits also generate EMI owing to their high operating speeds and the close proximity of circuit elements on a silicon die, giving stray capacitive coupling elements.

Electromagnetic interference is generated in power circuits owing to rapid transitions and ringing. Oscillations can be damped by introducing resistance if the source of resonance is isolated. Harmonics generated by saturating transformers can be minimized by using high-permeability material for the core, although this would cause the device to operate at high flux densities and result in large inrush current. Electrostatic shielding is often used in transformers to minimize coupling between primary and secondary windings.

Interfering signals can often be bypassed by high-frequency capacitors or metal screens used around circuitry to protect them from these signals. Twisted signal leads, or leads that are shielded, can be used to reduce coupling of interference signals. The collapse of flux in inductive circuits often results in high-voltage transients, causing interference in connecting circuitry. This is prevented by providing a path for the inductive current to flow, such as through a diode, Zener diode, or voltage-dependent resistor.

EMI can be radiated through space, as electromagnetic waves, or it can be conducted as a current along a cable. Conduction can take the form of common-mode or differential-mode currents. For differential mode, the currents are equal and opposite on the two wires and are caused primarily by other users on the same lines. These currents are mainly caused by coupling of radiated EMI to the power lines and by stray capacitive coupling to the body of the equipment (see Fig. 7.3).

7.4 Electromagnetic Shielding (EMS)

The usual barrier to long waves is a surface of conductive material, either solid or mesh. The conductive surface is a short circuit to the electrical component of radiation. A sheet of high-permeability magnetic alloy is a barrier to both the magnetic component of radiation and to dc and ac magnetic fields. Refer to Fig. 6.2.

Shielding Enclosures The choice of which type of shielding system to use in a given building is a function of several factors. The first factor is performance. Only performance levels that are actually *needed* in terms of shielding effectiveness for each type of electromagnetic field and frequency range should be *specified.* The required shielding)effectiveness is primarily based on the operational purpose of the facility. Hospital and secure communication facilities generally require shielding for different reasons. The operational purpose also determines the other physical factors required for the shielded enclosure such as appearance, HVAC, physical layout, and environmental requirements. All of these factors must be taken into account when selecting the type of shielding system for a given application.

A shielded enclosure is a complete structure with doors, vents, and other items. All penetrations to the shielded enclosure are treated to maintain shielding integrity. It may be a freestanding structure, such as the)modular clamp-up structure, or a welded steel room built in place on the site. A more recent advent in shielding is the architectural shielding system, where metal foil or sheet metal is built into the walls, floor, and ceiling of existing or new construction.

The most common form of modular shielded enclosure is the clamp-up system using galvanized sheet metal bonded to both sides of a wood core and tied together by a framework made of plated steel. The galvanized panel system is the most common because it is readily available from a large number of suppliers and provides good performance over a broad frequency range. The magnetic field shielding effectiveness generally exceeds 100 dB for electric and plane-wave fields. The actual measured performance of these enclosures is a function of the installation workmanship.

As the size of the space to be shielded increases, the modular system becomes less cost effective. Finishing the shielded room with conventional wall coverings must be done very carefully to ensure that the integrity of the shield is not compromised. Painting of the shielding panels is not recommended since the paint forms capillaries under the strapping and negates the shielding properties of the system. The proof of shielding is in the performance test. This should be conducted by an independent testing service and should be performed after all work that has an impact on the shielding is completed.

Water is an absolute enemy of shielding. Normal city water is loaded with minerals, and if spilled (Coffee, Mineral water, Soda, Beverages etc.), the runoff can get under the strapping and form a deposit that would insulate the strap from the shielding panel and degrade the shielding effectiveness of the system. Typically a sealing material called "kobakoat" is used to seal the floor of shielding systems, preventing damage in the case of a sprinkler discharge. For critical mission facilities, it is recommended that a dry pipe system be installed. This system is activated by smoke detectors, with built-in delays, and alarms so that if any personnel are available, it can be verified that water is needed. A non-CFC gas extinguishing agent system is best for computer centers.

7.5 Electromagnetic Interference and Signal Protection

Signal and power conductors provide the simplest means of interconnecting the different elements in a system. It is not uncommon for these lines to be hundreds or even thousands of feet long. As these lines wind their way from source to destination, they often pass through areas with high electric and magnetic fields, which can severely distort the intended signals. Another threat to signal integrity is interference caused by ground loops and differences in ground potentials. Just keep in mind that, in addition to being a conductive path for noise, signal and power wiring can be a source and receptor of noise.

Crosstalk is a common noise problem caused by one system either radiating or coupling unwanted energy into another system. Refer to Fig. 7.3. At lower frequencies, coupling is the major concern whereas stray radiation becomes dominant at higher frequencies. Just how undesirable crosstalk is depends on whether it is intelligible or not. If an interfering signal can be detected and has many of the same characteristics as the desired signal, it is much more disturbing than an interfering signal of the same power but with different characteristics from the desired signal. An example is the crosstalk sometimes heard on the telephone. If the interference consists of a syllabic pattern, the hearer is considerably more distracted than if the interference is, say, a hum.

Coupled crosstalk is most often found in adjacent pairs of a multiple-pair cable. There will be a capacitance coupling between the wires owing to their proximity to each other and the dielectric between them. There is also the magnetic coupling due to a varying current flowing in one wire, causing a varying magnetic field that intersects the adjacent wire. The magnetic field intersection causes a voltage in the second wire. This magnetic coupling effect is the basic principle of a transformer.

The goal in signal protection from EMI is to minimize, divert, or eliminate one of the three elements necessary for an interference/noise problem. Usually, the element over which the system designer has the most control is the coupling path. Coupling can be capacitive or inductive or conductive (through a common element).

Capacitive Coupling Any piece of building equipment or wiring can develop an electric charge, or potential, which can be expressed as a voltage. If this charge changes, an electric field is generated that can couple capacitively to other equipment or wiring. This type of noise can be significant when a circuit or termination has a high impedance, because the noise voltage that's generated at the receiver is the product of the noise current and the receiver impedance.

An easy and effective way to minimize capacitively coupled interference is to use cable shielding. The shield is a Gaussian or equipotential surface on which electric fields can terminate and return to the ground without affecting the internal conductors. Solid shields provide the best theoretical noise reduction solutions, but they're more difficult to manufacture and apply. Therefore, most cables are shielded with a braid for improved flexibility, strength, and ease of termination. Braided shields are less effective than solid shields because they provide only 60–98 % coverage of the cable. Decreased effectiveness is more prevalent at high frequencies where the holes in the braid are large compared to a wavelength. For maximum shielding, reliability, and ease of use, cables with combined shields are available that use both a solid layer and a braided layer.

Shielding is effective against electric fields only if it provides a low-impedance path to the ground. A floating shield provides no protection against interference.

Grounding of shields can be a controversial subject because there are several ways to do it. *Pigtail* connections from a shield to the ground have inductance, resulting in an impedance that increases with frequency. This type of connection will work at frequencies below 10 kHz, but will cause problems at higher frequencies. The use of short connections with large cross-sectional area minimizes the inductance of a pigtail, but the best connection is a 360° contact between the shield and connector. To complicate matters, some manufacturers of transmitters and receivers supply devices with cables whose shields have been internally connected and cannot be altered.

The correct place to connect an electrostatic shield is at the reference potential of the circuitry inside the shield. This point will vary depending upon whether the source and receiver are both grounded or whether one or the other is floating. It is important to ground the shield at only one point to ensure that ground currents do not flow through the shield. In most applications, the shield ground should not be at a different voltage with respect to the reference potential of the circuitry. If it is, this voltage can be coupled to the shielded conductor through the capacitance.

The capacitance between conductors is inversely proportional to the distance between them. Therefore, another simple way to reduce capacitive coupling is to increase the distance between the victim cable and source cable. It is always a good idea to route *noisy* cables, such as power input wiring, motor control wiring, and relay control wiring, separate from *quiet* cables such as analog I/O lines, digital I/O lines, or LAN connections.

Magnetic Coupling When a cable carries current, a magnetic field is generated. Magnetic coupling is much more difficult to reduce than capacitive coupling because magnetic fields can penetrate conductive shields. The amount of penetration is dependent on the frequency of the incident magnetic field and can be related to the skin depth of the shield material. At one skin depth distance into a material, an incident wave will be attenuated by *1/e* or 37 % of its original value. In general, copper, aluminum, and steel are more effective shielding materials at higher frequencies, with steel providing about an order of magnitude increase in effectiveness over copper and aluminum.

The magnetic field can also be reduced by separating the source of the field from the receiving loop, or by reducing the length (loop area) of the conductors, or by twisting the source wires if it is determined that the current producing the field flows in a wire pair. Twisting causes the magnetic fluxes from each wire to cancel each other out if equal currents flow in the wires, so the net field is ideally reduced to zero. If some of the return current were to flow through another path, such as a ground loop, a magnetic field would be generated by the imbalance in currents.

Twisting applies to both shielded and unshielded cables and to interference caused by shield currents or from other sources. Twisting the wires forces them close together, reducing the loop area and, therefore, reducing magnetic field generation and induced voltage. The effectiveness of twisted pair wire increases with the number of twists per unit length. In addition to reducing magnetic coupling, twisted pairs also act to reduce capacitive coupling. Each exhibits an equal capacitance to a noise source, causing equal and opposite charges to appear along the leads. This results in a net induced charge of zero and, ideally, no capacitively coupled noise. Twisted pairs are beneficial for use at frequencies below about 1 MHz. Above this limit, losses in the cable become a concern.

Some other common couplings encountered are the following: common-impedance coupling, in which two or more units or systems are connected to the same safety wire, ground grid, or plane at more than one place (multipoint grounding); common-mode and ground-loop coupling, in which radiated fields couple into ground loops that convert interference to undesited common-mode currents and to differential-mode currents; and differential-mode coupling, in which radiated fields penetrate signal and control cables to develop interfering voltages at the victim.

Unless all coupling paths are made sufficiently immune to EMI, electromagnetic compatibility may not result.

7.6 Grounding Systems of Buildings

In building systems environments, ground systems carry signal and power return currents, form references for electronic control analog and digital circuits, bleed off charge buildup, and protect people and equipment from faults and lightning. Because of these many requirements, the ground is an elusive and often misunderstood concept.

The term *ground* implies that the soil we walk on is the place to which all currents and voltages levels of system are referred. This equipotential view of the ground is not representative of practical grounding systems because two physically separated ground points are seldom at the same potential. Any current flow in a ground system can cause differences in potential. Lightning strikes or other transient events can generate hundreds to thousands of volts of potential difference.

It is not enough to design a system and leave grounding as a secondary issue. A ground system must be considered from the beginning in order for the system to work in the intended environment and pass interference, emissions, and safety requirements. There is no magic ground system that will work for all applications, so an understanding of the underlying principles is necessary for successful designs.

Proper grounding is dependent on many factors, such as the frequencies and impedances involved, the length of cabling required, and safety issues. When designing a ground or troubleshooting a ground problem, it is first necessary to determine where the current is flowing. If several kinds of grounds coexist, the current may not return by the assumed path.

To maintain high power quality, a proper grounding and bonding is a must. The signal ground should have a low impedance to handle large signal currents, and this is usually done by making the ground plane large. The inductance is reduced by placing the signal current-carrying conductor close to its ground return. Single-point grounds are difficult to maintain at a low impedance and are not suitable for frequencies above about 10 MHz. Generally, above these frequencies, multipoint

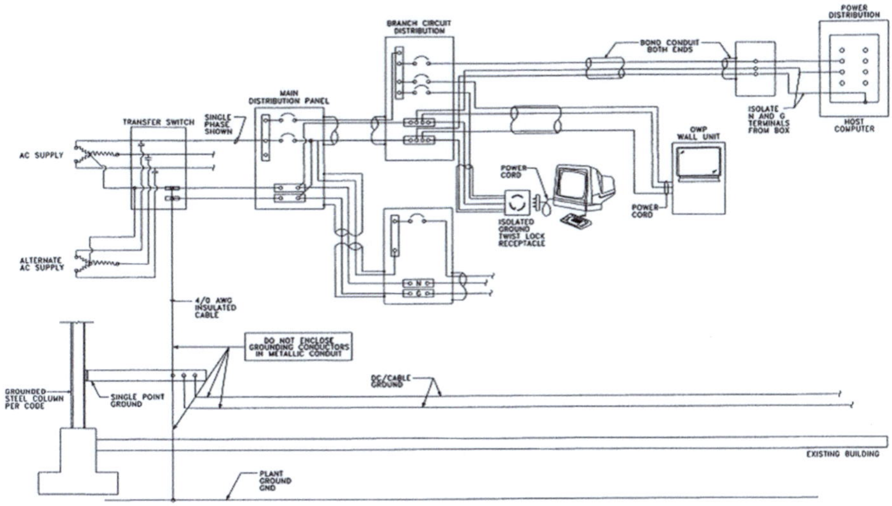

Fig. 7.4 Single-point ground (SPG) dedicated to control system wiring diagram

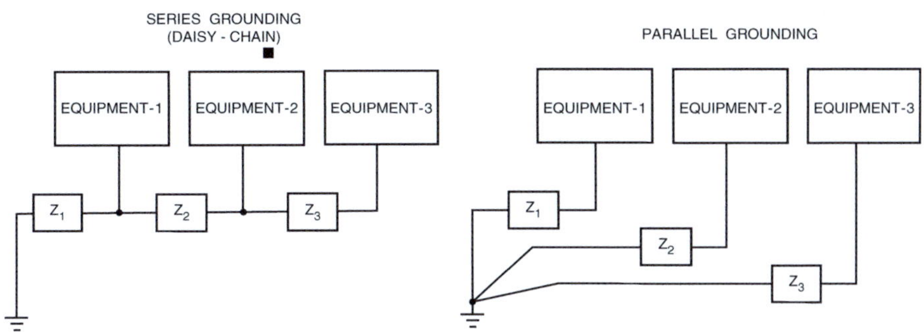

Fig. 7.5 Single-point equipment grounding methods

grounds are used. However, care is now needed to prevent the occurrence of ground loops, which can generate fields that interfere with the signal.

The most desirable type of ground for low-frequency applications is the single-point ground (see Fig. 7.4). Two examples are shown in Fig. 7.5. Avoid the series connection, or daisy chain, when sensitive equipment is involved, because return currents from all the equipment circuits flow through the common ground impedances linking the circuits. The ground potential of the equipment-1 circuit is determined not only by its return current through impedance Z_1 but also by the return currents from equipment-2 and equipment-3 circuits through the same impedance. This effect, which is called common-impedance coupling, is a primary means of noise coupling. Equipment circuits were used in the example, but any grounding scheme follows the same guidelines. Beware of common-impedance coupling when two signal conductors share one ground.

The preferred ground is the parallel connection. It is usually more difficult and more costly to implement because of the amount of wiring involved. When choosing between these ground configurations, first determine the circuit common-impedance coupled noise immunity. In practice, most systems will use a combination of both topologies.

7.6.1 *High-Frequency Interference and Shield Grounding*

The proliferation of electronics in building systems is generating an environment with increasing electromagnetic pollution. As the speeds of electronic circuits continue to increase, electromagnetic interference will present a continuing challenge to building systems engineers. Shielding and grounding concepts have remained constant throughout all the new developments in electronics, but now more than ever, it is important to understand these concepts and to implement them in circuits and systems from the initial design stages.

Single-point grounds for cable shields work well up to about 1 MHz depending on system size. At higher frequencies, parasitic capacitances provide sneak paths for shield currents that form ground loops. Also, standing waves present on a shield cause the impedance to vary along the shield. So even if there is a low-impedance connection to the ground at one end of the shield, the impedance will go to infinity at points that are odd multiples of $^1/_4$ wavelength apart. For digital equipment cabling, or when the shield length exceeds $^1/_{20}$ wavelengths of the highest frequency or harmonic present, a cable shield is often grounded at both ends and possibly at several points in between, depending on cable length and frequencies present.

Multiple shield grounds may seem contradictory to previous statements, but the noise induced by ground loops is generally at lower frequencies and can be filtered out, preserving the high-frequency signals. Also, at high frequencies, the skin effect causes signal current to flow on the inside surface of the shield and noise current to flow on the outside surface of the shield; thus, the shielding benefits of a triaxial cable are realized. Most systems have a hybrid ground system where the shield will have a single-point ground at low frequencies and multiple grounds at high frequencies.

7.6.2 *Ground Loops*

Ground loops exist in a system when there are multiple current return paths or multiple connections to *earth ground.* Current flowing in a ground loop generates a noise voltage in the circuit. No matter when you might think, buildings do have many ground loops, some of which may include cable shields (Figs. 7.6 and 7.7). Here are a few causes associated with building systems:

- Commercial and UPS power ground conductors entering cabinet and one connected to cabinet
- Cabinet grounded to building ground grid and/or steel and indirectly grounded to other ground points

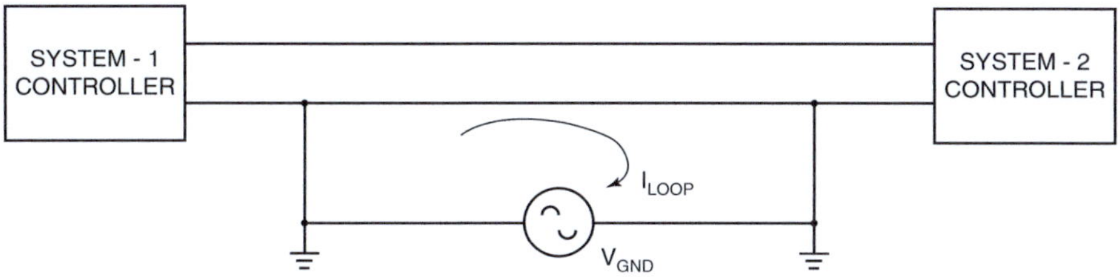

Fig. 7.6 Ground loop between two systems

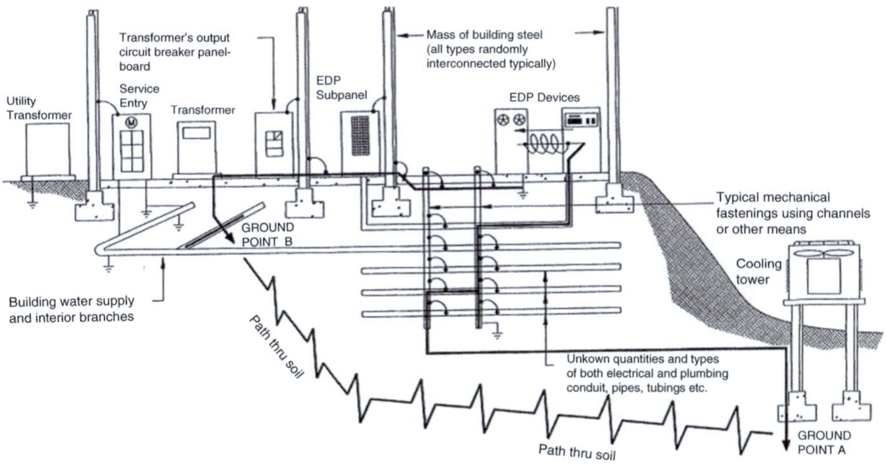

Fig. 7.7 Shown is an example of a ground loop. Here ground point A may have a different potential with respect to ground point B. With some impedance between the two points, we have a potential driving a current around the loop

- Internal dedicated analog reference (returns) grounded via long external runs to building ground
- Analog and logic returns connected together (often indirectly) and grounded to the cabinet
- Uncovered cable trays multipoint grounded to building steel

The most obvious way to eliminate the loop is to break the loop connection between the equipment and ground. When this isn't possible, isolation of the two circuits is a universal way to break the loop. Isolation prevents ground-loop currents from flowing and rejects ground voltage differences. A popular method of isolation involves the use of signal conditioners based on transformers or optical couplers. Signal conditioners often are rated to withstand transient events, thus providing a level of protection for the host system against harsh building systems environments and costly damage in fault situations.

Chapter 8
From Microelectronic Technology to Information Technology

Anil Ahuja

8.1 Intelligent Buildings

Although the term "intelligent building" seems to attribute a human characteristic to an inanimate object, the term's actual meaning, at least as defined by the Intelligent Buildings Institute, is "a building which provides a productive and cost-effective environment through optimization of its four basic elements—envelope, systems, services and management—and the interrelationships between them ... Optimal building intelligence is the synchronizing of solutions to occupant and business need." In the framework of such a definition, the intelligence of a building resides essentially in its system design. An intelligent building is, therefore, not necessarily packed with electronic systems, nor even one with an extensive building automation system, unless it is necessitated by the building program and can demonstrate its cost-effectiveness. Rather, it is a building designed with thought toward integration and perhaps some inspiration, both of which are necessary to satisfy immediate and predictable need and yet also to anticipate the occupants' future requirements. Since modem building usage is computer oriented and rarely static, it can now be said that "It is possible to turn off the computer but, it is not possible to turn off the computer environment." Figure 8.1 shows how intelligence shifts in a mechanical workhorse just as in humans and how a motor junction box carries not only wire terminals but also "brains" to control motor speed.

Experience has shown clearly, however, that in the very fast-moving building systems market, today's high tech can become tomorrow's albatross, and therefore planning for future requirements means intelligent preparation as much as immediate provision. The integrated building is essentially the building that is designed with foresight. Hindsight will determine the degree of intelligence.

A. Ahuja, *Integration of Nature and Technology for Smart Cities*,
DOI 10.1007/978-3-319-25715-0_8

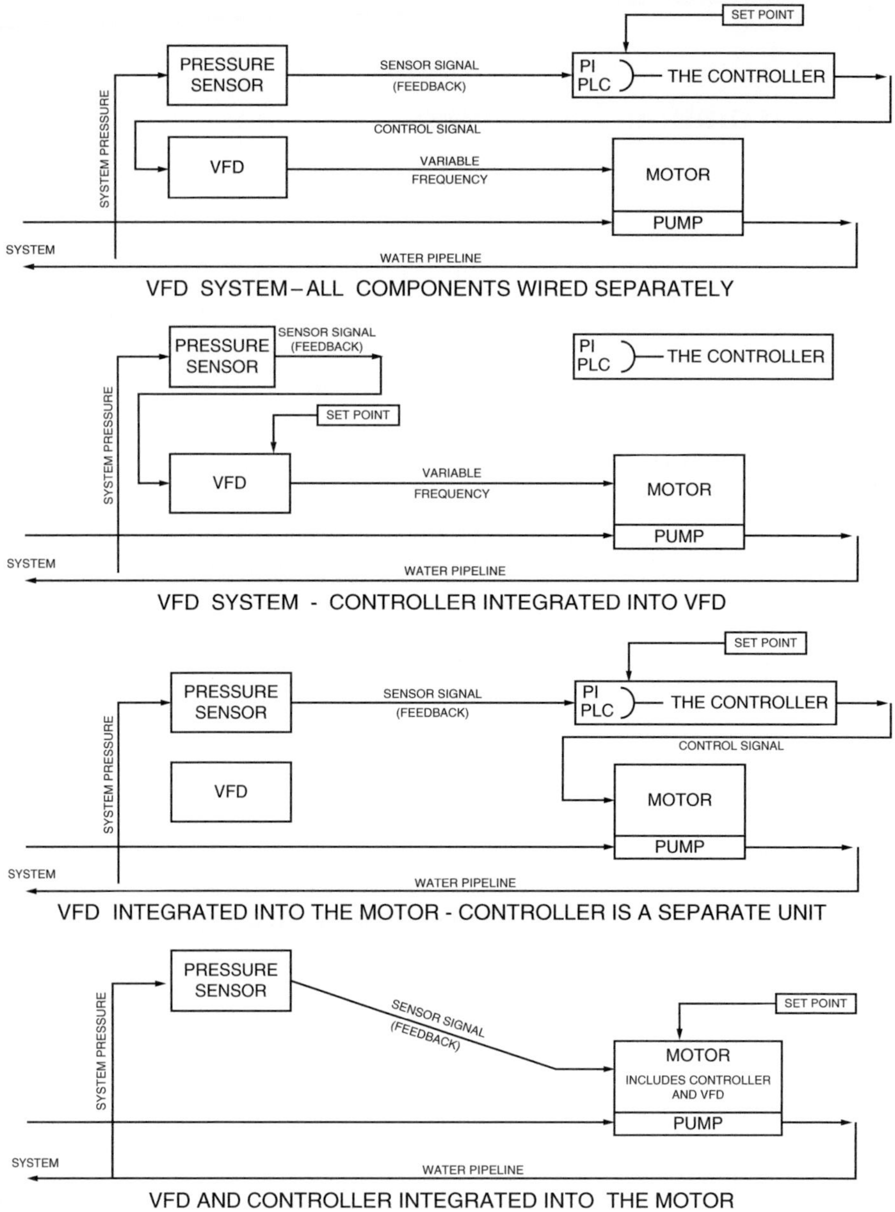

Fig. 8.1 Integrated variable frequency drive motors

8.2 Intelligence Communication and Dissonance

From smoke signals and Samuel F. B. Morse to Alexander Graham Bell and Guglielmo Marconi to talking around the world and to the moon, the history of communications has been one of overcoming distance. The decibel unit is of special interest because it is so commonly used in communication work and because it has several applications. The first application is from the original definition of the Bel, named after Alexander Graham Bell. The Bel is the log (base 10) of a power ratio. This number is generally too small, so the decibel came into being, namely, 10 times the log (base 10) of a power ratio.

$$\mathrm{dB} = 10\log_{10}\left(P_{\mathrm{o}} / P_{\mathrm{i}}\right)$$

It is generally used to define the gain or loss of a device. If P_{o} is the power *out* and P_{i} is the power *in*, then the ratio $P_{\mathrm{o}}/P_{\mathrm{i}}$ is the response of the system. The log of the response allows very large or very small responses to be easily written.

A common error is to read a large system gain number in a specification without the input specified and then to assume that the output power will also be large. However, system inputs are usually limited to a fixed maximum, so the output may not be as large as suspected or desired.

Example: If $P_{\mathrm{o}} = 10$ watts and $P_{\mathrm{i}} = 1$ watts, the system gain in dB is

$$\mathrm{db} = 10\log\left(10/1\right) = 10\mathrm{dB\ gain}$$

However, if $P_{\mathrm{o}} = 100$ watts and $P_{\mathrm{i}} = 10$ watts, the system gain is

$$\mathrm{db} = 10\log\left(100/10\right) = 10\mathrm{dB\ gain}$$

Note: the system gains are the same with a 10 times difference in output power.

Today technology has minimized the practical distance limitation, although for specific communication requirements distance is still the major key to finding the proper system. The large number of choices alone makes it very difficult for the non-communication-type building systems engineer to choose confidently. All intelligence communication systems can be divided into five basic pieces. The information source and the information user mark the beginning and the end. Between these are the two signal conditioners and the medium that separates the source from the user.

Information sources of building systems are typically pressure, temperature, current, flow, voltage, voice, and data. Typical information users are charts, computers, printers, video display, alarms, and loudspeakers. The signal-conditioning requirements and the transmission medium determine the communication system itself. Common conditioning devices are modulators, analog-to-digital or digital-to-analog converters, modems, and isolators. The medium may be a simple wire (twisted/untwisted), a coaxial cable, a free space, a radio system, a satellite, or a fiber optic cable.

An electrical signal is often the reasonable choice to transfer information when the distance or obstacles preclude directly coupling the information source and the information user. An example of a directly coupled system is two people talking, the human voice (the source) acoustically coupled to the human ear (the user).

Dissonance in intelligence is caused by electromagnetic interference and covers the frequency spectrum from de to the optical frequencies (30,000 Ghz). In recent years, many buildings have been upgraded to intelligent buildings, resulting in controls upgrade from analog type to digital type. Yet, many systems have both analog and digital components. The low-level, narrowband analog has given way to the high-level, high-noise immunity logic families. But the bandwidth of the logic is orders of magnitude wider than that of the analog circuits and therefore more sensitive to transients and electrostatic discharge (ESD). The electromagnetic interference (EMI) problem continues because of the system grounding, large number of ground loops, and long cable runs.

The most frequent EMI violations in grounding of control systems today are the following: both commercial power and UPS safety grounds entering the cabinet, cabinet grounded to building ground grid and/or steel and indirectly grounded to other ground points, internal dedicated analog reference (returns) grounded via long external runs to building ground, analog and logic returns connected together (often indirectly) and grounded to the cabinet, uncovered cable trays multipoint grounded to building steel, cable shields that are left unterminated at the point where they enter the cabinet, unsuppressed and unfiltered power mains at the cabinet entry point, and analog sensor and control device cables located in the same cable tray (see Fig. 8.2).

If analog sensor and control device cables must share the same cable tray (a poor practice), at least position the sensor cables on one side of the tray and the controlled-device cables on the other side. Use tray separators. Since most cable trays, unfortunately, are open at the top, ferrites can be added every 10 or 20 feet.

8.3 Intelligence Source Objects, Quality, and Transmission

The device required to transform the primary source—pressure, flow, temperature, rpm, vibration, whatever—into an electrical signal is called a transducer. The definition of a transducer given by the Instrument Society of America (ISA) is "a device which provides a usable output in response to a specified measurand." For our purpose, the output will be the electrical signal that carries the desired information.

The transducer will almost always require an external power source, even if the sensing portion is "self-generating," for example, a thermocouple. This power requirement is often overlooked if the systems are not integrated and later requires a system "add-on." In addition, this power source must be considered in any reliability calculations made for the system. Three important characteristics to consider about the transducer are the output signal's type, power, and range. Equally important is how they interface with the next stage, that is, signal-conditioning equipment.

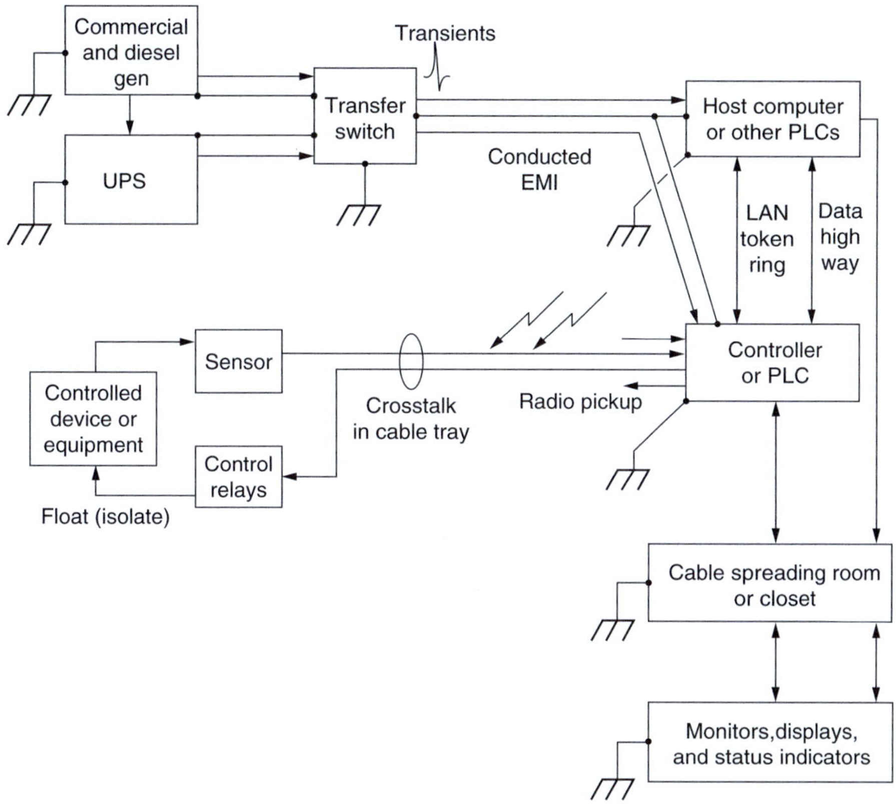

Fig. 8.2 A typical building system automation control installation contains many ground loops and other ways for EMI to couple into equipment

Intelligence Quality The object of the signal-conditioning equipment is to prepare the signal to overcome the medium loss, overpower any undesired interference, and arrive at the user input terminals with less than or equal to acceptable distortion. Many times the conditioning done on the intelligent source end must be undone at the user end. Thus, the pairing is often seen in manufacturers' literature. Examples of these pairs are (digital to analog–analog to digital) D/A–AID convertors, modulators–demodulators, and rectifiers–inverters. A good example of this type of system would be a computer that is used to monitor the pressure of a pipeline. The pressure transducer is the source and the computer is the user.

Intelligence Transmission By definition the transmitting medium will include everything between the two signal conditioners. This can be any system from a single wire to a satellite circuit. Therefore, the term "intelligence loss" can be misleading in that there could be signal gain devices included in the medium, so the user receives signal level that could be higher than the source signal output level. Whether the medium presents a gain or a loss to the overall system, it is a piece of the problem that can be accounted for easily. The hidden problem, and one that is much harder to account for, is the medium's susceptibility to interference. The interference can be noise in any or all of its forms or a variety of unwanted signals. For simplicity, noise will be considered as any signal interference, including the desired signal interfering with itself as in the case of an echo. A large intelligence loss that requires a large signal output power is no different than a low medium loss with interference that requires a large signal output to override the interference.

Some building systems are more susceptible to a particular noise or distortion, and of course, the longer the distance between system components and controllers, the greater the exposure to these conditions. Because of the importance of noise considerations in system design, many volumes have been written and will continue to be written describing noise. Acoustically, common descriptions are hum, pop, rumble, click, etc.—typically, those things that are heard that interfere with the communication process. Visually noise is not so common, yet it is present. Glare and spots are considered visual noise to the eye, and on a video display ghosts, snow, and tearing are caused by noise.

All mediums will distort the signal to some degree. Two things can be predicted about noise. The first is that it will always be present and the second is that its waveform is unpredictable. If the noise waveform could be predicted, then a 180° out-of-phase duplicate waveform could be generated and added, thus achieving a noise-free system. This is the ultimate ideal, in the same category as perpetual motion.

Noise, although difficult to describe, can be measured. The measurements vary as do the instruments used, but generally root mean square (rms), average, rectified average, or peak voltage determines the quantity of noise present. These measurements depend on the filtering and weighting applied and, of course, represent only the amount of noise present at the time of measurement. The most important design criterion for most building intelligence communication systems is the signal-to-noise (S/N) ratio received at the user's terminal. Often it is used as a measure of quality of an overall system. In data systems, bit error rate (ber) is the normal design criteria. In reality ber is also dependent on S/N.

Noise in intelligent buildings can be divided into two major groupings. The first is noise generated internally to the building system by design problems, integration difficulties, or a system operating at parameter thresholds. In any system design or device specification, there are trade-offs between the technical ideal and costs. The second category is noise encountered from independent external sources.

8.4 Intelligence Availability

In intelligence communication systems, failure should also include external source interference as well as actual equipment failure if the interference is sufficient to make the information transmitted unacceptable. Thus, intelligent building communication can be lost as a result of (1) equipment failure, (2) loss of usable signal, or (3) loss of power. Although related, each of the three can to some extent be considered separately. Individual unit reliability or power reliability is covered under a separate chapter; therefore, further discussion is not necessary here. Outside influences such as corrosion, propagation failure, or noise problems, as a special case, are the focus here.

Corrosion protection, like lightning, is a study in itself. One of the causes is the galvanic corrosion occurring between two dissimilar metals touching each other in the presence of an electrolyte. The source of the electrolyte could be the humid atmosphere, soil, or the metal surfaces themselves. A lead-shielded communication cable buried in an electrolyte of soil and water carrying a current of 1 amp from the lead to ground can lose as much as 75 pounds of lead a year as corrosion to the ground.

A common problem in intelligence communication circuits is the aluminum junction box terminal strips connected to copper wires. Sometimes aluminum is used as grounding wire connected to a dissimilar copper-jacketed steel grounding rod. The typical method of protection is to maintain the junction as free from any electrolyte as possible. Coat the junction with corrosion-resistant material: paint, organic coating, or electrodeposits. This coating must be free of pores and discontinuities.

Corrosion can also be caused by stray currents passing through metal to ground. These currents can be coming from local power systems, other cathodic protection, etc. A common method for protection is to apply an external voltage of the opposite polarity to cancel the stray current. The buried metal is forced to be a cathode (negative relative to ground) instead of an anode. And the other metallic ground connection would be the sacrificial anode.

When currents are caused by fluids flowing through a pipe or other ground potential differences that causes a current flow from the metal to ground, corrosion takes place. The protection against this action is called impressed current. A voltage is applied to the system so the impressed current opposes the natural current. The applied voltage will cause a net current to flow from a sacrificial anode to the metal system, thus adding metal to the system instead of metal corroding away.

In intelligent buildings, the basic principles of external interference protection are:

Distance: can be controlled by construction and installation techniques.

Barrier: can be provided by insulation/insulators, sectioning, shielding, and metallic raceway.

Training: to ensure maintenance personnel are familiar with the system so they are aware of potential failure to the operation.

Interference: Exposure protection often is the least expensive and may in some cases be mandatory. The National Electrical Code deals extensively with spacing requirements between communication circuits and power lines. Usually there are governmental regulations that determine minimum clearance for cables and wire that cross.

Maintenance: The building systems engineer putting together an intelligence communication system to meet some need that is a small part of an overall building system will almost invariably forget maintenance—until the system falls in some area. Maintenance plays a large role. It is discussed in detail in the reliability section. Maintenance will often be the controlling factor in making an economic choice among different types of systems.

Three major approaches are taken regarding systems maintenance in today's industrial world. Leasing versus ownership is one approach. If the system can be leased, then the routine and emergency maintenance will come as part of the system lease costs. A time factor may have to be considered by both parties; that is, the mean time to repair (MTTR) should be agreed to in the leasing contract. A lease versus ownership study would have to be tailored to each system requirement and is outside the scope of this book.

Another approach is to purchase the intelligence system units and, after the system is tested and commissioned, utilize a maintenance agreement with either the vendor or a service organization. The terms of such a contract are normally discussed in general with potential suppliers before the final selection is made. Along with the list of equipment is a description of the system, usually with a block diagram showing the overall intelligence signal flow.

8.5 Optical Fibers and Intelligent Buildings

Optical fibers have been used for years for guiding light in noncommunication applications such as medical illumination and image transmission. Introduced to the building systems communication industry in the mid-1980s, fiber optics received a tentative reception. That now has given way to the point that fiber cables are the medium of choice for many new and retrofit intelligent building information exchange or communication systems.

The term "fiber optics" refers to the technology of transporting information using guided light waves in an optical fiber (see Fig. 8.3). The fiber's capacity for total internal reflection enables a beam of light from an optical source to travel very long distances and around bends. The transmission mechanism of fiber optics is simple. Light flows only in the center portion, or core, of multilayer glass or plastic fibers, which are contained by an outer layer or cladding. The flow rate (or power throughput) depends on the source power, input-coupling efficiency, size and physical condition of the fiber, leakage at connectors and splices, constrictions at bends, and output-coupling efficiency.

Optical transmission system components include transmitters, receivers, and cables. A laser diode (LD), known as a coherent transmitter, provides a high degree of spectral purity and high power densities. LD signals exhibit nearly perfect beam collimation, meaning there is virtually no spread to the light beam. The alternative is a light emitter, which can be used for short transmission lines.

LDs and LEDs are currently the only practical light sources for fiber optic communication systems. Generally, LDs perform better due to their more tightly focused spectrum, efficient coupling into fibers, and high modulation speeds. On the other hand, LEDs are less expensive and simpler to operate, are less temperature sensitive, and last longer. Whichever light source is specified, transmitters should produce more than a few milliwatts of monochromatic power efficiently and operate at or near room temperature with no special cooling. Long life, compactness, and handling convenience are also important considerations.

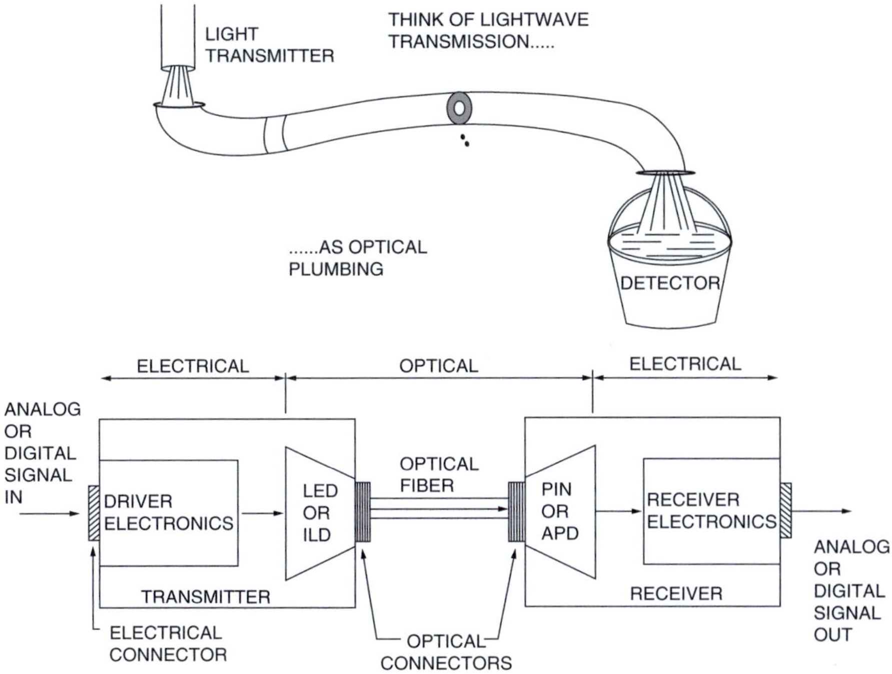

Fig. 8.3 (*Top*) Example of light-wave transmission leakage. (*Bottom*) Basic light-wave system

Receivers incorporate photodetectors, which convert optical radiation into electrical signals, and square-law detectors that respond to light intensity averaged over time. Avalanche photodiodes, one type of photodetector, have internal gain resulting in high sensitivity but also require a relatively large power supply. The alternative, positive-intrinsic-negative diodes are less sensitive but may be a better choice when high sensitivity is not required.

Cables consist of optical fiber cores covered with cladding and sheathed in protective jackets. Fibers are available in various combinations of glass and plastic. Both core and cladding may be plastic, yielding a fiber with high losses but one that is easy to handle and economical for short links. Although plastic fibers do not have the dimensional stability and environmental durability of glass, they are acceptable for some applications. A glass core also may be clad in plastic for special applications, such as communication in high-radiation environments. All-glass fiber features low signal attenuation and probably is the best choice for long transmission paths.

Optical fibers may be either single- or multimode. Only one light mode can propagate in single-mode cables and there is no modal dispersion. These fibers, which have core diameters of 5–10 μm, also require single-mode light sources, that is, lasers, and are relatively expensive and difficult to couple. Multimode fibers allow more than one light mode to propagate and, with core diameters typically 50 μm or more, are easier to work with.

Splicing fiber optic cables provides reliable permanent connections with low signal loss but requires reasonable skill and special tools. Couplers are available to join or split optical signals. Plastic fiber ends can be cut with a razor blade, but glass must be polished to be effective. However, because glass accepts a better polish than plastic and shows little or no deterioration with time or wear, it usually is specified for applications requiring high reliability. As in wire-based systems, minimizing the number of connectors and splices reduces signal loss.

Multilane Fiber optic technology can be applied to bidirectional and multidirectional communication networks. Two fibers, carrying signals in opposite directions, are used in the most straightforward scheme—referred to as half-duplex—for transmitting and receiving at both ends of a point-to-point link. A full-duplex system, permitting simultaneous transmission in both directions on the same fiber, conserves fiber, offering a significant advantage on long links. By applying special equipment, large amounts of data also can be multiplexed, or combined for transmission and separated at receipt, on fiber optic systems.

The advantages of using fiber optic cables over conventional copper cable, coaxial cable, or twisted-paid systems for intelligence communications are numerous:

- Optical systems do not generate electromagnetic interference and are not susceptible to outside interference.
- Complete electrical ground isolation is achieved between transmitter and receiver on an optical system, eliminating ground loops and common ground shifts in data circuits and permitting safe operation in hazardous environments while providing high-voltage isolation.
- Data security is increased because fiber optic transmissions do not radiate electronically detectable signals and are thus difficult to intercept. Building systems integrated with building security systems require this attribute.
- Optical transmission cable and equipment are generally smaller and lighter than what is required for comparable electronic systems. Furthermore, a single line can provide multiple control functions and carry audio/video signals.
- Crosstalk is eliminated on optical multichannel systems.

Disadvantages of fiber optic systems, while few, nevertheless should be considered:

- Not all building system equipment enclosures are able to accept fiber optic transmitters. A practical solution is to wire input/output points within a given perimeter zone with copper to a junction box near the zone equipped with a fiber optic multiplexer. The multiplexer accepts individual inputs and transmits them over a single fiber optic line.
- Each fiber optic transmission link requires a transmitter and receiver with corresponding power supplies, which means power must be available locally at all devices using fiber optic communications.
- Fiber optic splices require trained technicians and more sophisticated equipment than do metallic conductor systems. Fiber preparation for outdoor terminations

and splices can be difficult and may require a protective enclosure around the work area.

- Fiber optic equipment, including cables and connectors, is not as readily available as the components for metallic conductor systems. This, however, is changing rapidly as equipment suppliers are beginning to stock high-quality components with more variety to ensure off-the-shelf availability.

Chapter 9
Energy Storage

Anil Ahuja

9.1 Introduction

A general principle applicable to all physical systems in which mass is neither created nor destroyed is the *law of conservation of energy*, which states that energy is neither created nor destroyed; it is merely changed in form. This principle, together with the laws of electric and magnetic fields, thermodynamics, electric and hydraulic circuits, and Newtonian mechanics, is a convenient means for finding the characteristic relationships of electromechanical energy coupling. Energy required to operate building systems can be stored in thermal storage; electrochemical, passive electric elements; and coupling fields.

Irreversible conversion of electric energy to heat arises from three causes: Part of the electrical energy is converted directly to heat in the resistances of the current paths, part of the mechanical energy developed within the device is absorbed in friction and windage and converted to heat, and part of the energy absorbed by the coupling field is converted to heat in magnetic core loss (for magnetic coupling) or dielectric loss (for electric coupling). If the energy losses in the electrical system, the mechanical system, and the coupling field are grouped, the energy balance may be written in the following form:

Electrical energy input minus resistance losses = mechanical energy output plus friction and windage losses + increase in energy stored in coupling field plus associated losses

Basic property and characteristics desired in a building system energy storage mechanism are that it should be convertible and storage system charging and discharging should be part of base system configuration.

A. Ahuja, *Integration of Nature and Technology for Smart Cities*,
DOI 10.1007/978-3-319-25715-0_9

9.2 Thermal Storage

Thermal storage systems remove heat from or add heat to a storage medium for use at another time. Thermal storage for building system applications can involve storage at various temperatures associated with heating or cooling processes. High-temperature storage is typically associated with solar energy or high-temperature heating processes and cool storage with air-conditioning and refrigeration. Energy may be charged, stored, and discharged daily, weekly, annually, or in seasonal or rapid batch process cycles.

Thermal storage may be an economically attractive approach to meeting heating or cooling loads in the building if one or more of the following conditions apply:

- Loads are of short duration.
- Loads occur infrequently.
- Loads are cyclical in nature.
- Loads are not well matched to the availability of the energy source.
- Energy costs are time dependent (e.g., time-of-use energy rates or demand charges for peak energy consumption).
- Utility rebates, tax credits, or other economic incentives are provided for the use of load-shifting equipment.
- Energy supply from the utility is limited, thus preventing the use of full-size nonstorage systems.

The following is a list of some benefits of thermal storage:

Reduced equipment size
Capital cost savings
Energy savings
Improved system operation

9.2.1 Applications and Requirements

Thermal storage can take many forms to suit a variety of applications. Some groups of thermal storage applications are off-peak air-conditioning, industrial/process cooling, off-peak heating, and other applications. Whether for heat storage or cool storage, and whether for storing sensible or latent heat, storage designs follow one of two control strategies: full storage or partial storage.

Full storage (load-shifting) designs are those that use storage to fully decouple the operation of the heating or cooling generating equipment from the peak heating or cooling load. The peak heating or cooling load is met through the use (i.e., discharging) of storage, while the heating or cooling generating equipment is idle. Full storage systems are likely to be economically advantageous only under one or more of the following conditions:

- Spikes in the peak load curve are of short duration.
- Time-of-use energy rates are based on short-duration on-peak periods.

- There are short overlaps between peak loads and peak energy periods.
- High cash incentives are offered for using thermal storage.
- High peak demand charges apply.

For example, a school or business in which electrical demand drops dramatically after 5 P.M. in an electric utility territory where on-peak energy and demand charges apply between 1 and 9 P.M. can economically apply a full cool storage system. The brief 4-h overlap between 5 and 9 P.M. allows the full load to be shifted with a relatively small and cost-effective storage system and without oversizing the chiller equipment.

Partial storage more often provides the best economics and, therefore, represents the majority of thermal storage installations. Although they do not shift as much load (on a design day) as full storage systems, partial storage systems can have lower initial costs, particularly if the design incorporates smaller equipment by using low-temperature water and cold air distribution systems. For many applications, a form of partial storage known as load leveling can be used with minimum capital cost.

A load-leveling system is designed with the heating or cooling equipment sized to operate continuously at or near its full capacity to meet design-day loads; thus, equipment of the minimum capacity (and minimum cost) can be used. During operation at less than peak design loads, partial storage designs may function as full storage systems. For example, a system designed as a load-leveling partial storage system for space heating at winter design temperatures may function as a full-shift system on mild spring or autumn days.

Storage Media A wide range of materials can be used as storage media.

Desirable characteristics include the following:

- Commonly available
- Low cost
- Environmentally benign
- Nonflammable
- Nonexplosive
- Nontoxic
- Compatible with common building system materials
- Noncorrosive
- Inert
- Well-documented physical properties
- High density
- High specific heat (for sensible heat storage)
- High heat of fusion (for latent heat storage)
- High heat transfer characteristics
- Storage at ambient pressure
- Characteristics unchanged over long use

9.2.2 Sensible Heat Storage

Common storage media for sensible heat storage in buildings include water, soil, rock, brick, ceramics, concrete, and various portions of the building structure being heated or cooled. In HVAC applications such as air-conditioning, space heating, and water heating, water is often the chosen thermal storage medium, providing virtually all of the desirable characteristics when kept between its freezing and boiling points. In lower temperature applications, aqueous secondary coolants (typically glycol solutions) are often used as the heat transfer medium, enabling certain storage media to be used below their freezing or phase-change points.

Sensible thermal energy can also be stored in electrically charged, thermally discharged storage devices. For devices that use a solid mass as the storage medium, equipment size is typically specified by the nominal power rating (to the nearest kilowatt) of the internal heating elements, and the nominal storage capacity is taken as the amount of energy supplied (to the nearest kilowatt-hour) during an 8-h charge period. For example, a 10-kW heater would have a nominal storage capacity of 80 kWh. Some commonly used electrical storage equipment includes room storage heaters, central storage air heaters, pressurized water storage heaters, and underfloor heat storage.

For high-temperature heat storage, the storage medium is often rock, brick, or ceramic materials for residential or small commercial applications. The use of the building structure itself as passive thermal storage offers advantages under some circumstances. Building elements such as floor slabs and internal walls form the storage and the materials (concrete, brick, etc.) are the storage media.

The objective for building mass thermal storage is not to maintain a constant temperature in the conditioned space but rather to limit the rise in temperature over a normal working day. In air-conditioned buildings, running the systems overnight to cool the building mass limits the peak and total cooling loads the following day. The method usually takes advantage of "free cooling" with outside air, which is available during the night in most climates, but the free cooling can be supplemented or replaced by other means of cooling.

9.2.3 Water Thermal Storage

Water is commonly used in both heat storage and cool storage applications. Water has the highest specific heat (Btu/lb.°F) of all common materials and is well suited for thermal storage. The amount of energy stored in a chilled water storage is directly related to the volume of water in the storage times the temperature differential between the entering and leaving water. For example, chillers that cool water in storage to 40 °F (4.4 °C) and coils that return water to storage at 60 °F (15.5 °C) provide a range of 20 °F. The cost of extra coil surface required to provide this range can be offset by savings in pipe size, insulation, and pumping energy. Further, fan energy costs are reduced if the extra coil surface is added as face area rather than as extra rows of coils.

Water Storage Enclosure A perfect storage tank would deliver water at the same temperature at which it was stored. It would also require that the water returning to storage neither mix nor exchange heat with the stored water or the tank. In practice, however, both types of heat exchange occur. Water storage is based on maintaining a state of thermal separation between cool water and warm water.

Tanks are available in many shapes; however, vertical cylinders are the most common. Tanks can be located above ground, partially buried, or completely buried. Tanks are usually at atmospheric pressure and unpressurized and may have clear-span spherical dome roofs or column-supported flat roofs. Exposed tank surfaces should be insulated to help maintain the temperature differential in the tank. Insulation is especially important for smaller storage tanks because the ratio of surface area to stored volume is relatively high.

Heat transfer between the stored water and the tank contact surfaces (including divider walls) is a primary source of capacity loss. Not only does the stored fluid lose heat to (or gain heat from) the ambient by conduction through the floor and wall, but heat flows vertically along the tank walls from the warmer to the cooler region. Exterior insulation of the tank walls does not inhibit this heat transfer.

Space Requirements The location and space required by a thermal storage system are functions of the type of storage and the architecture of the building and site. Building or site constraints often shift the selection from one option to another.

Chilled water systems are associated with large volume. As a result, many stratified chilled water storage systems are located outdoors (such as in industrial plants or suburban campus locations). A tall tank is desirable for stratification, but a buried tank may be required for architectural or zoning reasons. Tanks are traditionally constructed of steel or prestressed concrete. A systems supplier who assumes full responsibility for the complete system performance often constructs the tank at the site and installs the entire distribution system.

9.2.4 Latent Heat Storage

Common storage media for latent heat storage include water-ice, aqueous brine-ice solutions, and other phase-change materials such as hydrated salts and polymers. Water has the highest latent heat of fusion of all common materials—144 Btu/lb (80 cal/g) at the melting or freezing point of 32 °F (0 °C). Other than water the most commonly used material for latent heat thermal storage is a hydrated salt with a latent heat of fusion of 41 Btu/lb (22.7 cal/g) at the melting or freezing point of 47 °F (8.3 °C) and a density of 93 lb/ft^3 (1.5 g/cm^3). Hydrated salts have been in use for many decades, often encapsulated in plastic containers. This material is a mixture of inorganic salts, water, and nucleating and stabilizing agents.

For building systems air-conditioning applications, water-ice is the most common storage medium. The challenge is to find an efficient and economical means of achieving the heat transfer necessary to alternately freeze and thaw the storage medium. Various methods have been developed to limit or deal with the heat transfer approach temperatures associated with freezing and melting; however, leaving fluid temperatures (from storage during melting) must be higher than the freezing point, while entering fluid temperatures (to storage during freezing) must be lower than the freezing point. Ice storage can provide leaving temperatures well below those normally used for comfort and nonstorage air-conditioning applications. However, entering temperatures are also much lower than normal.

9.2.5 Ice Storage

The latent heat of fusion in ice storage systems requires a capacity of about 5.5 ft^3/ton-hour (0.055 m^3/kWh) for the entire tank assembly, including piping headers and water in the tank. Buildings commonly use following ice storage systems:

Ice-on-Coil Systems (External and Internal Melt) Ice-on-coil systems are available in many configurations with differing space and installation requirements. Because of the wide variety available, these systems often best meet the unique requirements of many types of buildings. Some key performance data requirements of an external melt system are:

- Evaporator and suction temperatures at the start of ice build
- Evaporator and suction temperatures at the end of ice build
- Ice thickness at the end of ice build
- Time to build ice

From these data, determine actual efficiency. This way, refrigeration capacity deviation from published ratings can indicate refrigerant loss or surface fouling.

In the case of a internal melt-ice-on-coil storage system, the following performance data are important:

- Secondary coolant temperature and suction temperature at the start
- Secondary coolant temperature and suction temperature at the end
- Secondary coolant flow
- Tank water level at the start
- Tank water level at the end
- Time to build ice

Based on the above measured flow data and heat balance, determine efficiency at start versus theoretical and again efficiency at end versus theoretical. Compare this with published ratings.

Ice storage can be integrated as a part of the building structure, by installing bare coils in concrete wall cells. The bare steel coil concept can be used with direct cooling, in which the refrigerant is circulated through the coils, and the water is circulated over the coils to be chilled or frozen. This external melt system has very stringent installation requirements. Coil manufacturers do not normally design or furnish the tank, but they do provide design assistance, which covers distribution and air agitation design as well as side and end clearance requirements. These recommendations must be followed exactly to ensure success.

Encapsulated Ice This system uses plastic containers filled with deionized water and an ice-nucleating agent. The plastic containers must be flexible to allow for change of shape during ice formation. During discharge as ice melts, the plastic containers return to their original shape.

Ice Harvesting Systems This system separates ice formation from ice storage. Ice harvesting system can melt the stored ice very quickly. During the ice generation mode, the system is energized if the ice is below the high ice level. Challenge is to accurately quantify remaining ice in the tank. When ice is floating in the tank, the water level will always be constant, so it is impossible to measure ice inventory by measuring water level.

The conductivity of the water-ice solution increases as water freezes, and dissolved solids are forced out of the ice into the liquid water, thus increasing their concentration in the water. Accurate ice inventory information can be maintained by measuring conductivity. The following data are important in commissioning an ice-harvesting system:

- Suction temperature at the start
- Suction temperature at harvest
- Harvest time/condensing temperature
- Time from start to full signal
- Tank water level at start
- Tank water level at bin full signal

From the above data, determine efficiency and capacity and compare with published ratings.

9.2.6 System Configurations

Thermal storage can be configured for various operating modes. Optimum configuration is related to storage technology, discharge temperature needed, and operating mode flexibility required. Common operating modes used in facilities are shown in Fig. 9.1.

Open Systems Chilled water, external melt-ice-on-coil, and ice-harvesting systems are all open chilled water piping systems. Due to the potential for drain-down, the open nature of the system, and the fact that the water being pumped may be saturated with air, the design must provide the piping details carefully to prevent

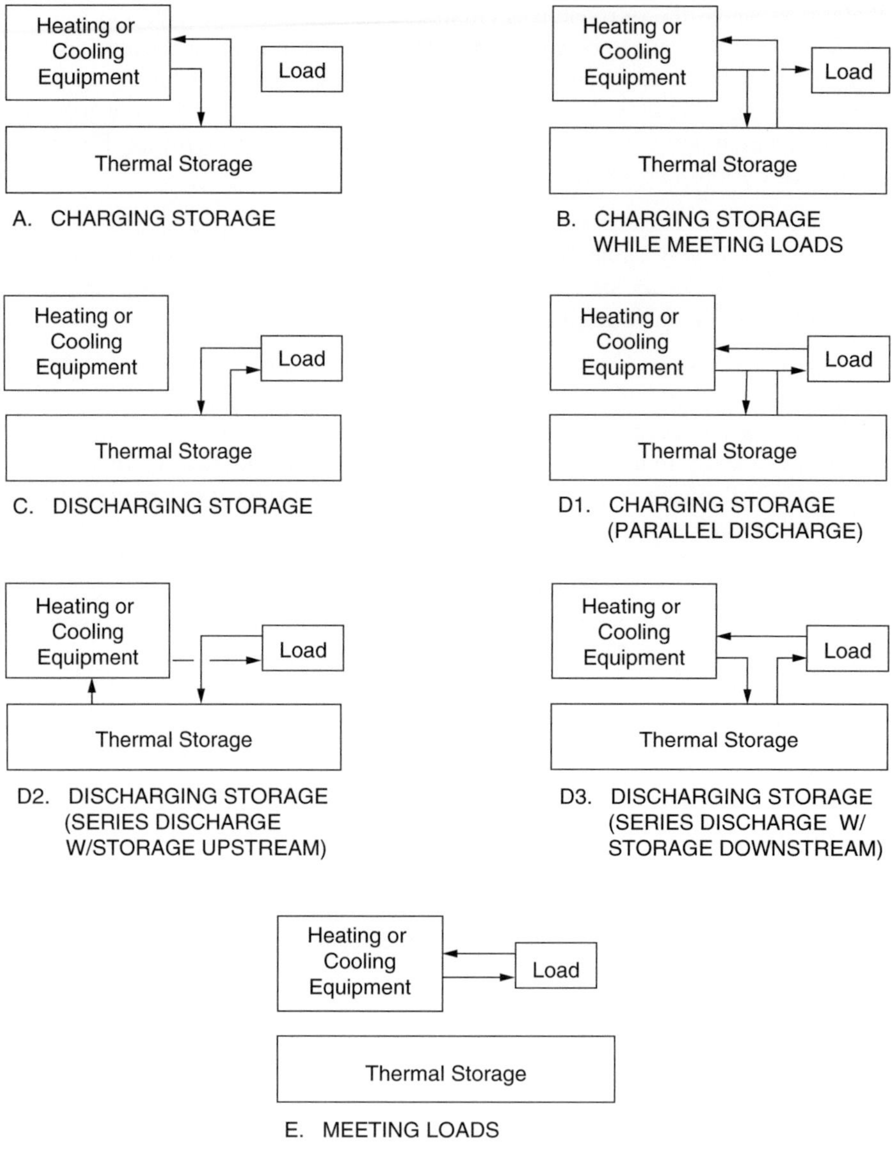

Fig. 9.1 Thermal storage operating modes (Reprinted by permission of the American Society of Heating, Refrigerating and Airconditioning Engineers, Atlanta, Georgia, from the *1995 ASHRAE Handbook—Applications*)

pumping, drain-down, or piping problems. Drain-down must be prevented by isolation valves, pressure-sustaining valves, or heat exchangers.

Closed Systems Closed systems normally circulate a glycol solution coolant either directly to the cooling coils or to a heat exchanger interface to the chilled water system. A domestic water makeup system should not be the automatic makeup to the glycol solution coolant system. An automatic makeup system that pumps a premixed solution into the system is recommended, along with an alarm signal to the building automation system to indicate makeup operation. An accurate estimate of volume is required. The coolant must contain inhibitors to protect the steel and copper found in the piping system. The water should be deionized.

Piping for ice storage systems can be configured in a variety of ways. The chiller may be located upstream or downstream of the building load. See Fig. 9.2 for an upstream chiller arrangement.

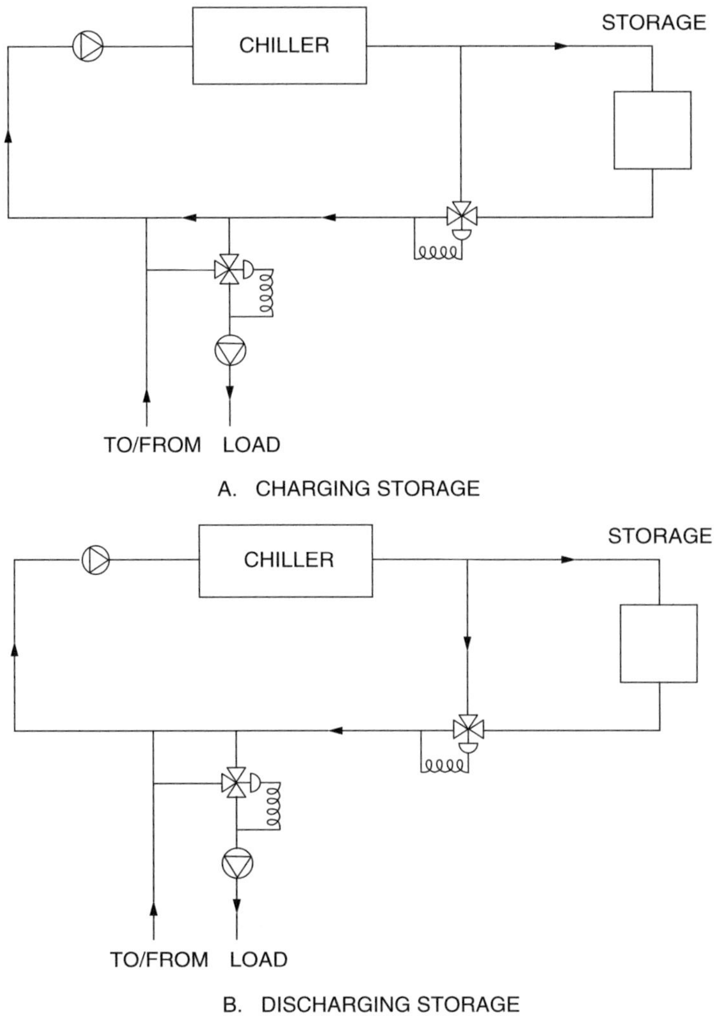

Fig. 9.2 Thermal storage system with chiller upstream (Reprinted with permission of the American Society of Heating, Refrigerating and Airconditioning Engineers, Atlanta, Georgia, from the *1995 ASHRAE Handbook—Applications*)

9.3 Electrochemical Storage

First we consider some basics like *What is chemical energy?* When gasoline mixes with air and ignites, a rapid and violent transformation takes place. This chemical reaction produces carbon dioxide, water, and enough energy to turn a generator shaft. Chemical reactions inside the human body change the compound glucose into carbon dioxide and water. Although these reactions are less violent than the one involving gasoline, they produce enough energy to power the human body. Countless different types of chemical reactions occur throughout the universe, and they all involve changing the composition of matter. This change consists of a regrouping of the atoms in one set of molecules and breaking the complex bonds that hold them together to produce a different arrangement. Some reactions happen slowly and release only a small amount of energy. Other reactions require a separate source of energy before they can occur. For example, in chemical reactions known as photosynthesis, plants use the energy in sunlight to join together molecules of carbon dioxide and molecules of water to produce glucose and oxygen.

Electrochemical energy is converted into electric energy in batteries and fuel cells. The former are energy storage devices and store the chemical energy within the battery. Fuel cells are energy conversion devices and produce energy all the time. The required chemicals (fuel and oxidant) are supplied in the cell.

9.3.1 Batteries

A battery is an assembly of electrochemical cells that convert the energy produced by chemical reactions into low-voltage, direct current electricity. Similar to other electrochemical cells, each cell contains three major components: a positive electrode or cathode, a negative electrode or anode, and the electrolyte.

Primary Battery Primary batteries convert chemical energy into electric energy once only, and the original chemical state cannot be regained by passing electric energy through the battery. They are discharged once and are discarded after use.

Secondary Battery After discharge of a secondary battery, their original chemical state may be regained by passing through the battery a quantity of electric energy equivalent to that drawn during discharge. Secondary batteries are, therefore, distinguished from primary batteries by the feature of rechargeability. Secondary batteries are commonly used in building system storage applications.

Battery capacity to deliver a certain amount of energy is generally provided in terms of energy density. *Energy density* is the energy delivered per unit weight (or unit volume).

$$\text{Energy density} = \text{ampere hour capacity} \times \text{cell voltage}$$

Energy density is expressed in Wh/lb (Wh./kg) or Wh./in^3 addition to the energy density of a cell, another important parameter which the energy, that is, the power, can be delivered. *Power density* delivered by the battery per unit weight or volume:

$$\text{Power density} = \text{current} \times \text{voltage/battery weight}$$

Units are W/lb or W/in.^{3}

Batteries rarely use 100 % of the theoretical ampere-hour capacity. For primary cells, the reason is that mass transport and resistance effects limit the *utilization* of reactants, whereas for secondary cells, the energy used in one cycle is maintained below 100 % in order to maximize battery life. For secondary battery applications, high-energy density is useful only if the battery can be repetitively discharged, that is, long cycle life at high reactant utilizations. Depth of discharge, or utilization, is defined as the ratio of the number of ampere hours delivered to the theoretical number of ampere hours calculated. Utilization depends on the battery discharge rate, and higher utilizations are achieved at slower discharge rates (i.e., lower currents).

The ampere-hour capacity available from the battery to a preselected voltage depends on the value of the current. At high load (i.e., high currents), the battery delivers far less capacity than at low load. A *capacity/time quotient* is, therefore, defined to normalize the discharge rate for cells of differing capacity. For example, C/5 is the current that completely discharges (or charges) the battery in 5 h; 4C is the current that discharges the cell in ¼ h. These values are independent of the battery size and weight, but are characteristic only for the type of battery. For example, a radio battery has a discharge rate of about C/100, while a building system requires a C/3 discharge rate. *Shelf life* is the loss of capacity of the cell during storage. The period beyond which it becomes uneconomical to store the battery is known as its shelf life.

Cycle life is the number of times a secondary battery can be discharged (or charged) before the battery voltage falls below a prescribed value.

Discharge Voltage Characteristics When a cell is discharged, that is, a current is drawn from the cell, the voltage immediately decreases owing to electrode and ohmic polarization. There is then usually a period during which the cell voltage remains essentially constant. Subsequently, there is a sudden change in voltage with time. This point of inflection indicates that little useful energy can be obtained by continuing discharge beyond this point.

The shape of this curve depends on the battery and the discharge rate. There is a similar, though opposite, curve during charge, at which further application of the charge current results only in gas evolution at the electrodes rather than reforming the chemical reactants.

9.3.2 Battery Design and Types

A large number of primary and secondary batteries are available or are being developed to satisfy the building systems. There is no one universal battery that can satisfy all these requirements, and each has its particular market niche. See Table 9.1 for types of commercial batteries applied in building systems.

Battery Design Features As with other electrochemical cells, care is taken in the design of the battery to minimize cell polarization. This is achieved in several ways. First, the electrodes are fabricated to maximize their surface area to volume ratio; this is achieved by using porous electrodes. Thus, a high specific area is formed at the interface between the active materials of the anodes and cathodes and of the electrolyte phase in order to obtain high currents during charge and discharge. At the same time, good electronic conduction must be achieved between the active battery material and the current collector which, in turn, must be stable and noncorrodible. Several types of electrode structures have been developed that satisfy this condition depending on the nature of the battery.

Ohmic polarization is reduced by using electrolyte concentrations corresponding to maximum specific conductivity and by minimizing the anode-to-cathode distance. In order to have small anode-to-cathode separations, care must be taken that the electrodes are not in contact. This is particularly necessary for secondary batteries where electrode shape changes occur during cycling with a resultant change in

Table 9.1 Commercial battery types

		Commercial primary batteries	
Type	Cathode	Anode	Electrolyte
LeClanche	MnO_2	Zn	NH, Cl/$ZnCl_2$
Alkaline manganese	MnO_2	Zn	KOH
Rubens (or Duracell)	HgO	Zn	KOH
Panasonic high energy	$(Cf_x)_n$	Li	Propylene carbonate with $LiC1O_4$
Lithium (GTE)	$SOCl_2$	Li	$SOCl_2$ with LiCi-$AlCl_3$
Lithium (Mallory)	$SOCl_2$	Li	Propylene carbonate with $LiC1O_4$
Lithium (Saft)	CuO	Li	Propylene carbonate with $LiC1O_4$

		Secondary batteries	
Type	Cathode	Anode	Electrolyte
Lead acid	PbO_2	Pb	H_2SO_4
NiCad	NiOOH	Cd	KOH
Nickel–iron	NiOOH	Fe	KOH
Silver–iron	Ag_2O_2	Zn	KOH
Zinc–halogen	Br_2 (or CI_2)	Zn	$ZnBr_2$ (or $ZnCi_2$)
Zinc–air	O_2	Zn	KOH
Nickel–hydrogen	NiOOH	H_2	KOH

this distance. As a result, a separator is placed in the electrolyte between the two electrodes. This material is selected for maximum stability in the electrolyte between the two electrodes. This material is selected for maximum stability in the electrolyte, low cost, and optimum porosity to minimize concentration gradients (and, hence, concentration polarization) in the electrolyte.

The remainder of the design considerations involves electrode configurations, low cost, stable case materials, methods of sealing, and low-cost manufacturing techniques, all of which are specific for a particular battery system. However, as a result of added materials to achieve a practical battery, practical energy densities are only about one quarter to one fifth of the theoretical value based on the weight of electroactive materials and electrolyte.

Battery design criteria are related to cost, performance, durability, safety, and environmental considerations. Hence, there is the requirement for high cell voltages, reversible electrode reactions, high electrolyte conductivities, and minimal volumetric differences between reactants and products. Operating costs are influenced by cycle life, maintenance requirements, and energy efficiency of the charge–discharge cycle.

9.3.3 *Fuel Cells*

A fuel cell is an electrochemical device that continuously converts the chemical energy of the fuel and the oxidant directly to electric energy without an intermediate combustion process. Unlike a battery, a fuel cell does not run down or require recharging. It will continue to operate as long as the fuel (hydrogen) is fed to the anode and the oxidant (air) is fed to the cathode. The electrodes, which are solid material of high internal area, act as reaction sites where the electrochemical transformation of the fuel and oxidant occurs. These reactions produce electrons that flow from one electrode to the other when the electrodes are connected together through an external circuit. Several types of electrolyte may be used; the particular type is determined by the type of fuel and oxidant and by the cell operating temperature. The electrode reactions are

$$2H_2 \rightarrow 4H^+ + 4e \quad (\text{anode})$$

$$O_2 + 4H^+ + 4e \rightarrow 2H_2O \quad (\text{cathode})$$

giving as the overall cell reaction

$$2H_2 + O_2 \rightarrow 2H_2O$$

Fuel Cell Systems A practical fuel cell power system is composed of three major subsystems: fuel processor, fuel cell, and dc-to-ac inverter. Fuel, which may be basically any fossil fuel (oil, natural gas), is fed into a fuel processor. The fuel processor

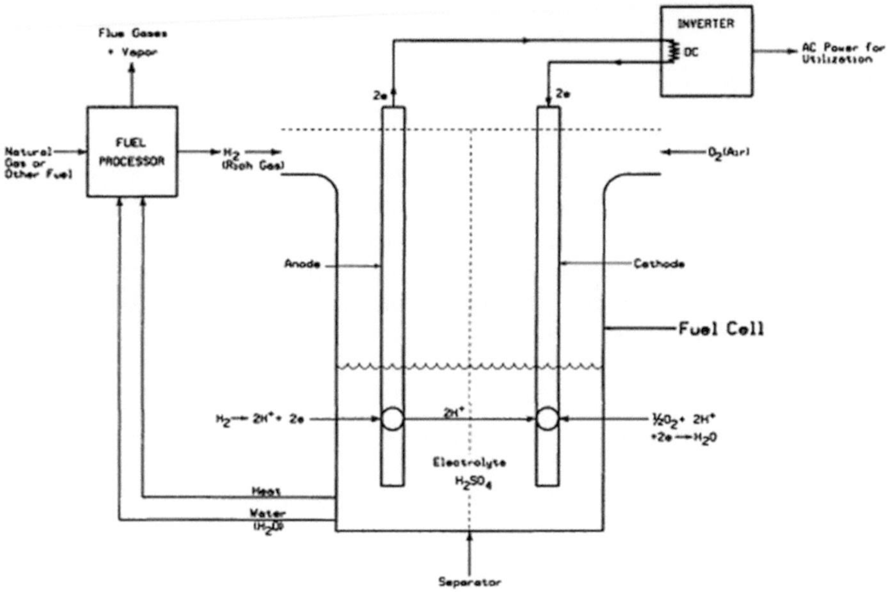

Fig. 9.3 Schematic of fuel cell power generation

converts the fossil fuel into a gas suitable for the fuel cell (i.e., mainly hydrogen); the gas entering the fuel cell also contains other constituents such as carbon monoxide and carbon dioxide depending on the fuel processor system, coal gasifier, steam reformer, or direct gas fired. The third subsystem of the fuel cell power plant is the inverter to convert the direct current of the fuel cell into alternating current. See Fig. 9.3 for a schematic of such an arrangement.

Several types of fuel cells are being developed. The cells, characterized by the nature of the electrolyte, are alkaline, acid, and molten carbonate fuel cells.

Fuel Cell Characteristics The fuel cell is an environmentally acceptable device since it is quiet and has no moving parts. Because it is not a combustion device, emissions such as NO_x, CO, and unburned hydrocarbons are not a problem. Such low-level emissions, coupled with the fuel cell's quiet, water-conserving operation, result in environmental acceptability and siting flexibility.

A single fuel cell normally generates power at approximately 0.5 to 1 V and can be connected in series stacked with other cells to obtain almost any desired voltage. The current produced is a function of the area of the single cells. The range of sizes, the modularity, and the load-following capabilities make the fuel cell system an attractive candidate for power generation in a variety of applications, including on-site and central plants for commercial, industrial, and residential use.

The major advantage of fuel cells is that they are highly efficient energy conversion devices. In addition, because the system is located at the point of use, the waste heat generated by the irreversibilities of the electrode reactions is recoverable, thus further improving the system efficiency.

9.3.4 *Fuel Cell and System Efficiency*

The efficiency of converting gaseous fuels to dc power in a fuel cell power plant is a function of thermodynamic voltage and current efficiencies and of the heating value of the composition of the fuel which may contain nonelectrochemically active, combustible species.

Thus, the total fuel cell efficiency (EFC) for the conversion of chemical energy in the feed gas to dc power is then defined as follows:

$$E_{\mathrm{fc}} = E_{\mathrm{T}} \times E_{\mathrm{v}} \times E_{\mathrm{I}} \times E_{\mathrm{H}}$$

where

ET = thermodynamic efficiency
EI = current efficiency
Ev = voltage efficiency
EH = heating value efficiency

Fuel Cell System Efficiency The previous equation describes only the fuel cell power section of the fuel cell system; the fuel processor, power conditioner, and waste heat utilization are not included. The overall efficiency for a fuel cell system (E_s) without waste heat utilization is given by

$$E_{\mathrm{s}} = \frac{\text{ac power}}{\text{HHV of raw fuel into fuel proccessor}} = E_{\mathrm{fp}} \times E_{\mathrm{fc}} \times E_{\mathrm{pc}}$$

where fuel processor efficiency

$$E_{\mathrm{fp}} = \frac{\text{LHV gaseous fuel from fuel processor to fuel cell}}{\text{HHV raw fuel into fuel processor}}$$

and power conditioner efficiency

$$E_{\mathrm{pc}} = \frac{\text{ac power}}{\text{dc power}}$$

Additional energy can be recovered in the system for heating or by integrating to produce additional electricity. For a system employing waste heat utilization, the fuel cell system efficiency (ESE) is given as

$$E_{\mathrm{SE}} = \frac{\text{fuel cell} + \text{electricity produced} + \text{cogeneration BTUs}}{\text{HHV raw fuel into fuel processor}}$$

9.4 Passive Electrical Power Storage

Passive electrical storage devices store energy in the coupling of the device to the power system. The coupling fields are electric and magnetic fields. These fields are physical phenomena: invisible lines of force that occur whenever electrical power is used.

1. *Electric and Magnetic Fields*: Electric and magnetic fields are decoupled below radiofrequencies and are considered separate fields at low frequencies.
2. *Electromagnetic Fields*: Electric and magnetic fields at radiofrequencies and above are coupled together and referred to as electromagnetic fields.

This passive storage device operates low frequencies, and thus the separate terms electric or magnetic fields will be used.

An electric field represents a force that electric charges exert on other charges. An electric field begins on a charge and ends on a charge, although we picture lines of force emanating from the source like quills on an excited/alarmed porcupine. The electric field is a function of *only* the electric system voltage level. The higher the voltage, the stronger the electric field will be. The unit of measurement of the electric field is volts per meter. The conductor does not have to have current flowing to generate an electric field. The conductor has to be energized to have an electric field present.

A magnetic field is a physical phenomenon of invisible lines of force. Since a magnetic field is a function of moving charges (current), the magnetic field can be looked on as closed loops of force lines such as ripples from a pebble dropping in the middle of a pond. With alternating current, the ripples would reverse with each change in current flow. The magnetic field is a direct function of the amount of electric current flowing in a conductor. The larger the amount of current, the stronger the magnetic field will be. A popular model is that of the self-exciting dynamo: A conductor rotating in a magnetic field creates an electron flow. If the flow passes through a coil, the coil itself creates a magnetic field that keeps the flow moving as long as the conductor keeps rotating. The components described in the following sections are basic passive power storage devices.

9.4.1 *Inductors*

An inductor is an electrical circuit element whose behavior is described by the fact that it stores electromagnetic energy in its magnetic field. This feature gives it many interesting and valuable characteristics. In electromagnetic radiation, electric field strength is considered the driving force and magnetic field strength the response. In mechanical systems mechanical force is always considered as a driving force and velocity as a response. Four basic points need to be remembered in order to understand electromagnetic interaction. The first point is that a charged particle moving with *uniform* motions creates around itself a magnetic field; the second point is that a magnetic field forces a change of direction in a charged particle moving at an angle through it; the third point is that a charged particle moving with *accelerated* motion produces an electromagnetic wave; the fourth point is that an electromagnetic wave accelerates a charged particle. In its most elementary form, an inductor is formed by winding a coil of wire—often copper—around a form that may or may not contain ferromagnetic materials. Energy storage in inductors is

$$W = \frac{1}{2} LI^2.$$

where

L = inductance
I = current through the coil of wire
W = energy stored in the magnetic field

The unit of inductance is called the henry (H), in honor of the American physicist Joseph Henry.

9.4.2 Capacitors

A capacitor is an electrical circuit element that is described through its principal function, which is to store electric energy. This property is called *capacitance*. Capacitance is that property of a system of conductors and dielectrics that permits the storage of electrically separated charges when a potential difference exists between the conductors. In its simplest form, a capacitor is built with two conducting plates separated by a dielectric.

Energy storage in capacitor is

$$W = \frac{1}{2} CV_{\mathrm{p}}^2$$

where

W = energy stored in the capacitor
C = capacitance in farads
V_{p} = peak voltage across the capacitor terminals

Because the amount of the stored energy varies with the square of the peak voltage, the capacitor's ability to sustain load power during an outage drops off twice as fast as reductions in line voltage. This is why sometimes few cycle outage does not shut down equipment.

9.5 Active Electrical Power Storage (Solar Cells)

Photovoltaics (PV) cells convert sunlight into electric power. PV cells and modules have found many commercial applications ranging from solar powered calculators and watches to arrays making a few megawatts of peak power for utility

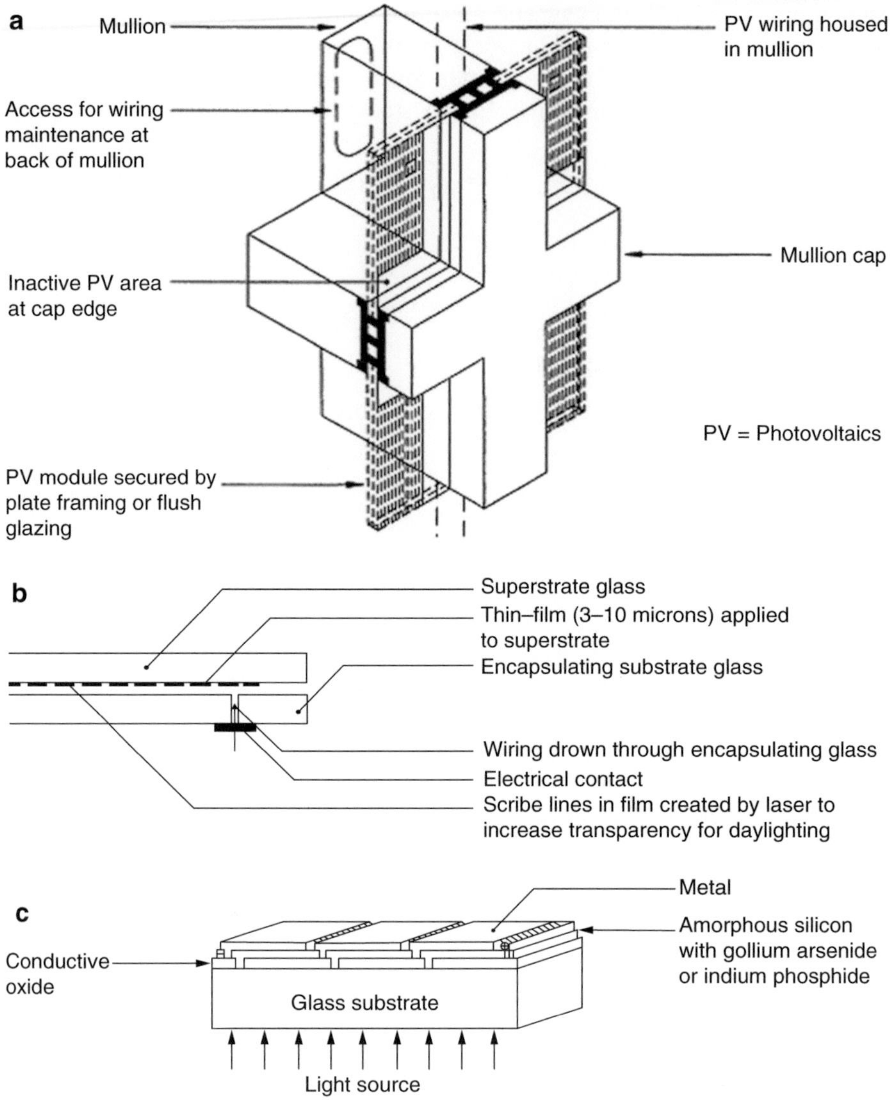

Fig. 9.4 (**a**) Building system integrated, glass-based PV glazing; (**b**) thin-film superstrate-type PV module; (**c**) diagram of an amorphous silicon module with three series-connected cells

generation. The output of solar arrays is direct current; as a result, it must be converted and conditioned if it is intended to feed power into building alternating current system.

One of the easiest way to incorporate photovoltaics into building systems is to use glass-based PVs for certain wall glazing. See Fig. 9.4a, b. Glass-based PVs come in two formats: crystalline silicon or thin-film amorphous silicon. Conversion efficiency ranges from 12.5 % for single-crystal silicon to 7 % for polycrystalline silicon to 5 % for amorphous silicon. The cost of photovoltaic and solar power range from 8 to 15 ¢/kWh with installation cost of \$8–\$15/watt.

A typical solar cell consists of either crystalline or amorphous silicon treated with thin layer of cadmium sulfide, gallium arsenide, indium phosphide, or similar compound. See Fig. 9.4c. This produces semiconductor junction between materials. Sunlight impinging on cell creates electron flow and hence produces electric direct current flow.

A new technology known as "ac PV modules" integrates a dc–ac "micro-inverter" directly into large-area PV modules thus eliminating the separate inverter used to transform the dc electricity generated by the PV array into ac electricity.

The cost calculation for building integrated PV (BIPV) has many components, including design, hardware (PV modules, power electronics, support structure, and wiring), installation, utility interconnection, metering, maintenance, geographical location, and financing. Systems installed in lower latitudes produce more power than higher latitudes due to greater annual solar resources.

9.6 Batteries

Flow batteries are centered around two aqueous electrolytes, which are held in separate tanks when the battery is idle (Fig. 9.5). To get electricity from it, the liquids are pumped into a chamber separated by a membrane, sparking an electron-producing chemical reaction across the membrane. To store energy, an external current is applied across the membrane and the process works in reverse.

Flow batteries are also considered to be very safe, because unlike some lithium-ion designs, they are not prone to thermal runaway, which can cause battery fires (Fig. 9.6).

Technology comparison for Grid-Level applications

Technology	Moving Parts	Operation at Room Temperature	Flammable	Toxic Materials	In production	Rare metals
Flow	Yes	Yes	No	Yes	No	No
Liquid Metal	No	No	Yes	No	No	No
Sodium-Ion	No	No	Yes	No	No	No
Lead-Acid	No	Yes	No	Yes	Yes	No
Sodium-Sulfur Batteries	No	No	No	Yes	Yes	No
Ni-Cd	No	Yes	No	Yes	Yes	Yes
Ai-ion	No	Yes	No	No	No	No
Li-ion	No	Yes	Yes	No	Yes	No

Fig. 9.5 Battery comparison

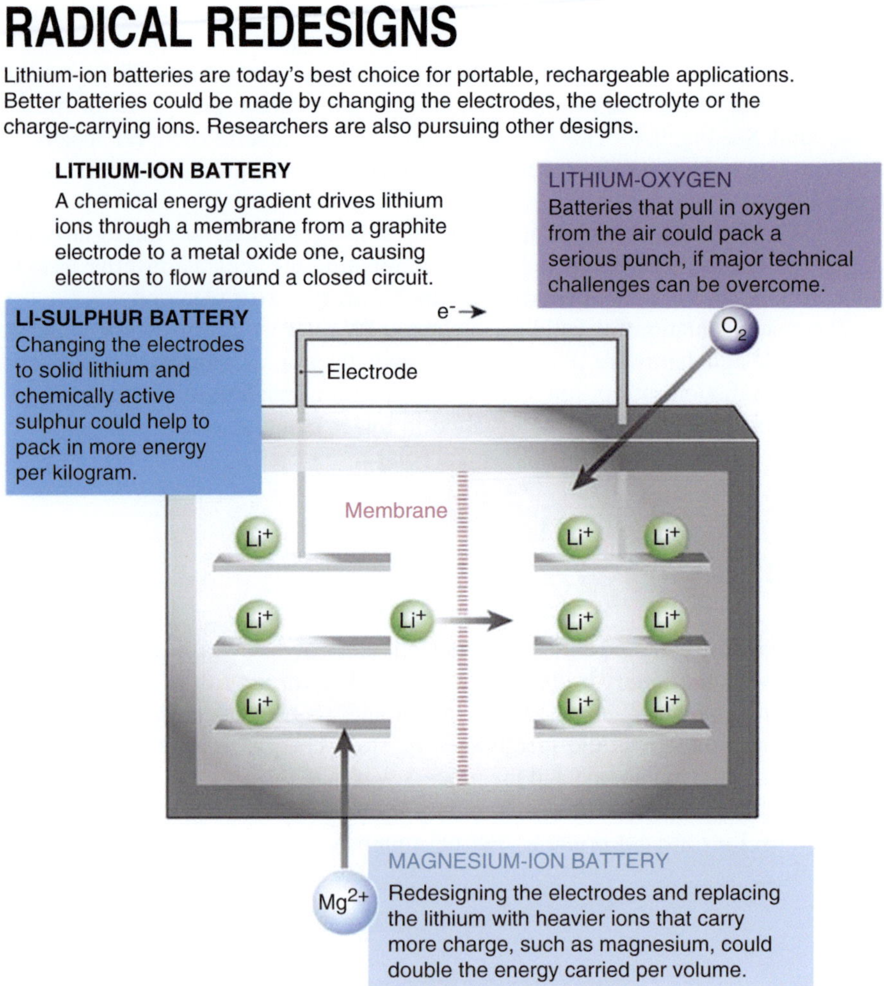

Fig. 9.6 Lithium ion battery technology (Source: http://www.nature.com/news/507026a-i3-jpg-7.15873?article=1.14815)

Chapter 10
Integrated Building Systems Engineering and Automation

Anil Ahuja

10.1 Introduction

With influx of smart grids and smart apps, the building automation systems landscape has developed much faster than MEP products over the last few decades. It is the driving force for optimum operations, improved reliability, and energy savings through the integration of building systems critical to the functionality of a facility. Through the years, occupancy, comfort, safety, reliability, and efficiency have been key factors in the development of the technologies. However, as the global conscious has changed in regard to the impact of humans on our own environment, and global demand and access to energy has risen, a new driving imperative has been placed on facility professionals, engineers, and humanity alike regardless of energy cost: Our goal globally is to reduce our energy consumption in a manner that significantly decreases the negative impacts our consumption has on our environment and thereby on all of nature (including us) alike.

According to the Center for Climate and Energy Solutions, commercial and residential building space represents 39 % of total energy consumption in the United States, more than any other sector, of which 70 % is associated with building HVAC systems depending on locations and environs and 30 % is plug loads and lighting.

It stands to reason integration and operation of these buildings to reduce energy consumption is a top priority. This can only be done through integration of the system and the use of the powerful technologies available to us today and by doing it in a manner ensuring the efficient operation of a facility. Because of this, we now can leverage these technologies to create integrated systems. Combine this with the new frontier that is the Cloud and the Internet of Things, and we have the ability to create more efficient buildings that utilize current technologies to decrease consumption. This can all be done while giving access to analytical tools designed to help pinpoint outlying operation and energy issues before they become a problem (Fig. 10.1).

A. Ahuja, *Integration of Nature and Technology for Smart Cities*,
DOI 10.1007/978-3-319-25715-0_10

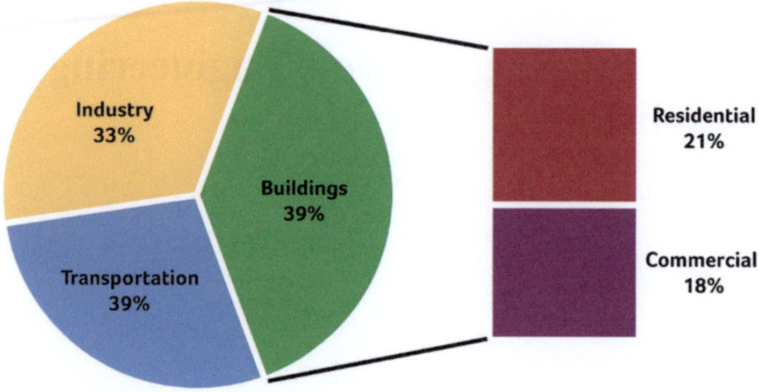

Fig. 10.1 Commercial and residential building space represents 39 % of total energy consumption in the United States, more than any other sector

Building Automation as Been Evolving to Meet Our Energy Reduction Needs In the 1970s and 1980s due to increased energy prices, the demand for a more cost-effective building sent the BAS industry off in a whirl of development to make buildings smarter. Taglines like Smart Building Solutions® started dotting the competitive landscape as building automation companies were able to leverage cheapening computer technologies and bring them into the building space. This was a golden age for the BAS industry. During this peak demand to reduce consumption, BAS manufacturers globally created new and relevant technologies related to controlling a building. Competition increased and the direct digital control solution became the standard expectation in buildings everywhere. However, most systems were disparate and only designed to control specific operations of a building with other key components not being visible or integrated through the controlling mechanisms.

Over the next decade, the building automation landscape began to change with larger leaps as technologies began to evolve at an almost exponential rate. During this time, energy prices inevitably fell, and the larger cost driven model for energy reduction began to wane as the economy strengthened. The pressure was off and therefore the pain soon forgotten. However, the building automation industry was still hard at work developing solutions relevant to the building market and energy reductions. The end user was beginning to ask for disparate systems to link together to allow for competitive selection of products as well as service providers. Open protocols began to become a common topic of discussion as users realized the power of integration. During this time, our world was just beginning to wake to the impact our energy consumption was having on the planet. However, we were still developing the solutions to dramatically reduce our impact: open protocols for fully integrated systems. This was the beginning of the BACnet® Project Standards Committee in 1987 (introduced as an ANS/ASHRAE standard in 1995) and the introduction of LonWorks® in 1990 by the Echelon Corporation (LonTalk® submitted and accepted as an ANSI standard in 1999).

Today, the majority of buildings are designed with direct digital building automation systems. Through the evolution of protocols and technologies, the BAS has become a key component to ensuring the effective function of an entire building. In a perfect world, this would mean all buildings could take advantage of a "plug-and-play" system. However, it is not as easy due to market forces of building low-cost buildings and passing over energy cost to leasing tenants in commercial buildings. Factors such as first cost, disparate manufacturers, legacy control systems, Internet security, and age of the building all come into play when considering the integration. In addition, the ever expanding Cloud and the Internet of Things has opened a whole new world of interconnectivity, allowing smart mobile devices and applications to be applied to these complex systems. Web browsers, SaaS, and mobile applications can be used to create a dynamic system where data storage capacities are virtually limitless and access and control are just a touch away. With all this, the opportunity for efficiency is greater than ever. However, it is only through successful integration this efficiency is achieved.

10.2 Integration and Protocols

A successful integration relies first on understanding the circumstances of the Division 25 specification. Obviously, a retrofit scenario versus a new construction model changes the dynamics of the integration. In new construction, devices can be specified to integrate natively with other solutions through protocols like BACnet and Modbus or by specifying an all-encompassing proprietary solution from a single vendor. However, in the retrofit world, engineers may be dealing with legacy systems that are not being changed out, while newer systems are being expected to be brought in and "bolted" together with the system. In this case, it is important to begin the process of identifying where things can be integrated and where they may not. The good news in these situations is many manufacturers are now starting to make native or gateway devices that allow disparate systems to be brought together through the use of open protocols. In many cases, this can either be accomplished with their existing system or with third-party devices that manage the connection of the legacy protocol; most popularly, a BACnet or LonWorks with some Modbus or a combination of the three (with BACnet and LonWorks rarely being used on the same systems). While integration of these protocols is not always simple, a little understanding by systems integrator can go a long way.

BACnet This protocol has gained significant momentum in the last decade to become arguably the most specified and utilized open protocol in the market today. This is largely due to the fact manufacturers collectively create the outcome of the protocol for integration through working groups, while the protocol is validated through the BACnet Testing Laboratories. The BTL certification mark ensures products developed to utilize the protocol meet the standard as associated to product type. According to www.BACnet.org, BACnet currently has 825 vendor ID's.

In addition, according to BACnetinternational.org, *there are currently 661 active products from 112 distinct manufacturers* (http://www.bacnetinternational.net/btl/) carrying the BTL mark. This includes products for HVAC, lighting, fire and life safety, energy metering, and a wealth of others.

BACnet most popularly comes in two flavors: MS/TP and IP. Both data communication layers have their merits. MS/TP is commonly used for lower level controller devices designed for the control of equipment and uses a twisted shielded pair for wiring with EIA-485 as the physical layer. This comes in handy specifically in retrofit situations where buildings often already have existing wiring in place for their current control systems. However, caution must be taken here as baud rates do vary depending on the revision of the standard the controller was designed to meet. This can be verified in Internet world of information by checking the PIC statement for a device available at www.bacnetinternational.org.

With most BACnet IP, CAT5e or greater (Ethernet) wiring or wireless protocol is used to transfer data over the IP layer between devices and to centralized interface mechanisms. This is a faster means of communication and a more convenient way if existing IP infrastructure is already in place. BACnet IP networks typically reside separate from the standard IT backbone as broadcast messaging can slow general communications. In most cases, a BACnet network will consist of both BACnet MS/TP and IP devices mainly through the use of routers, bridges, and repeaters to bring disparate BACnet networks together to tie into the centralized server for data display, monitoring, and control. This does not mean that a building must have access to the outside world in order to take advantage of IP or MS/TP. These layers can sit internally on an intranet structure.

When specifying BACnet products for integration, it is important to understand there are multiple variations and revisions of the BACnet protocol. Therefore, all BACnet is not created equal and may not be able to communicate if certain qualifications or standards are not met. This is why BACnet tests to various device types and to only specific types of communication layers. Check www.BACnetinternational.org for BTL product listings and more information (Fig. 10.2).

LonWorks This protocol was developed by the Echelon Corporation and is largely considered to be the competitor to BACnet. Its foundation is based on the LonTalk protocol and mostly the use of embedded Neuron® chips in devices created by competitive manufacturers. This single source model of course has at times created the question of whether LonWorks was a truly "open" protocol. However, the use of this methodology does create a stricter standard of interoperability as it is controlled more tightly due to the nature of its development. As a result, testing standards are not required by a consortium to ensure interoperability.

LonWorks gives the ability for both twisted, shielded pair communication bus and IP (Ethernet)-enabled technologies through varying methods and is intended to reside on an intranet or Internet environment. Its protocol can be accessed via chips, routers, network interfaces, software, and controllers on a multitude of devices manufactured by various companies. However, unlike BACnet, it is not expanding into certain markets needed for pure interoperability between systems. Echelon, as well

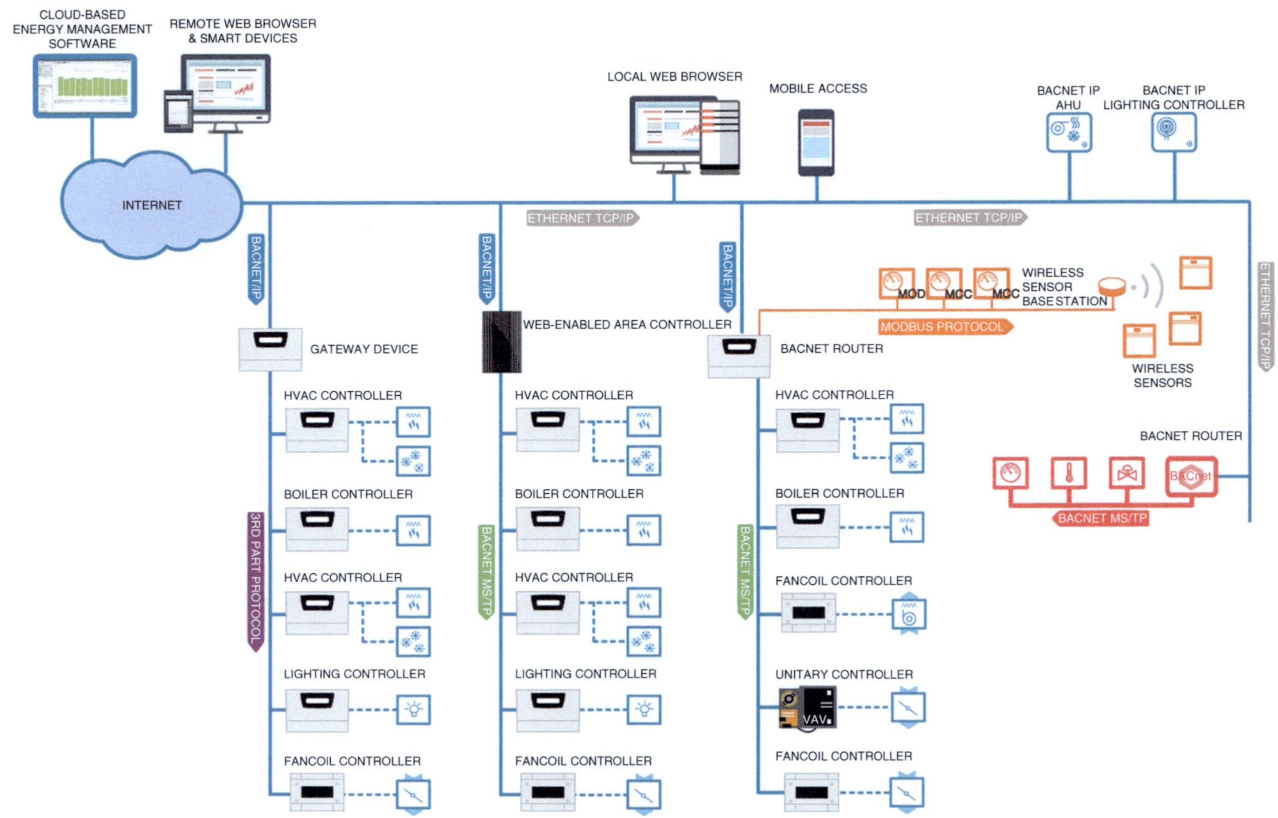

Fig. 10.2 BAS architecture encompassing HVAC and other systems integration through multiple protocols into a BACnet system. Web browser, cloud, and mobile applications are used to create a dynamic system where data storage capacities are virtually limitless and access and control are just a touch away

as a number of control manufacturers, offer devices that can bring LonTalk onto a network that is using other protocols as well as convert it to BACnet. This means products competitively more suited for an application using the LonWorks protocol can be used in a building system to create pure interoperability and efficiency.

Modbus Modbus is an open protocol that is commonly used in some packaged control applications and sensor topologies. It consists of both RTU (EIA-485) and TCP/IP (Ethernet) communication options and is also designed to exist in and intranet or Internet environment. Typical Modbus applications are more popular in industrial market and include energy measurement, motor drives, and automation routines. Integration of this protocol is not necessarily an issue with most devices and can be pulled in through the majority of manufactured systems today or through gateways.

Proprietary Protocols There are a wealth of these protocols existing in buildings today. While the industry is abuzz with open, these protocols should not be left out in consideration for a fully functional building. Costs to integrate to existing systems can be high with third-party devices, and there is always the issue of disparity between the integrator knowing the product they represent and them possibly not knowing the other systems they are integrating to. While the integrator might know their device inside and out, the lower level systems may require additional, competitive service teams be available.

A careful eye should be placed in the market for manufacturers who support their legacy systems through backwards compatibility to their newest and most market relevant solutions. Many of these companies offer ways to bring a legacy system into the modern world through interoperability to their protocols. In addition, many of these manufactures are producing cost-effective ways to bring in open systems one piece at a time to give users a more efficient building while maintaining lower installation and integration costs. With these systems, facilities get the more modern tools they need while still maintaining a familiar system and expanding or replacing localized systems at a pace suited to their needs.

10.3 The Internet of Things

We have all become more connected to products, services, devices, and each other over the years as the Internet has begun to integrate into almost every aspect of our lives. While one could wax philosophic on the merits of this, one thing is for sure, a more connected and accessible building can make for a more efficient building.

Right now a wealth of applications have been developed and are progressing in the building space. Certain companies are leveraging this by taking integration protocols like BACnet, Modbus, and LonWorks and tying them into larger Internet-enabled services that allow for more streamlined access and analytical tools for a building. These services work on simplifying the interaction of the BAS, so the end user can control their building space without the need for understanding the complexities of the building automation system. This is accomplished through integration to IT services already available to these facilities. For example, in this day and age, certain building automation systems can now tie into prominent calendaring tools like Microsoft® Outlook®, Google™ Calendar, and Apple® iCal. This means zones can be scheduled automatically to turn on and off based on a user's schedule set right from their desktop, tablet, or phone just by booking the space. Therefore, reducing the chance of forgotten temperature overrides,or missed chances to change a setting because of a scheduling conflict. (That is, if the meeting is canceled in the executive conference room, the BAS knows not to turn on the lights or cool or heat the zone, all without someone interacting with the BAS.) Other examples include smart device applications that allow for local control of temperature, and native interfaces that allow users to troubleshoot their system, acknowledge alarms, and view trends all from a mobile device.

In addition to BAS systems taking advantage of our Internet-enabled world, many services are starting to emerge that leverage the Cloud in order to promote energy savings and relay important energy, carbon, and financial data to facility professionals, green teams, consumers, and financial managers alike. These services offer tools that can track energy usage and automatically convert it to carbon footprint data. They can create historical energy analysis and use degree days or other conditions to normalize information in order to get accurate representations of performance. These tools can also create targets based on driving factors related to energy consumption and create predictive models that create alarms when something is performing out of range. With these features, and a wealth of others, energy and comfort conditions can be effectively monitored and managed and therefore optimized. In addition, with the almost limitless storage capacity of the Cloud, this data can be stored securely and redundantly and used over a buildings life cycle to effectively track and analyze energy performance, maintenance performance, and savings measures all for a fraction of the cost associated with embedded systems.

We live in an ever advancing real-time world that requires a new perspective on how we consume energy. As we continue to bring technologies that integrate us to each other, we need to consider the impact these same technologies can have on our interaction with our environment. As the history of BAS has advanced to tackle problems of high energy costs, interoperability, and integration, a larger imperative has arisen that can leverage this progression and our now interconnected world.

Our impact on our environment is increasing at a rapid rate. However, we have the gift of being able to do something about it. Through current technologies combined with decades of product evolution, we can create more efficient systems by merging the developments of yesterday with the technologies of today. The future of efficiency is now; all we need to do is integrate and smartize operation with remote management (Fig. 10.3).

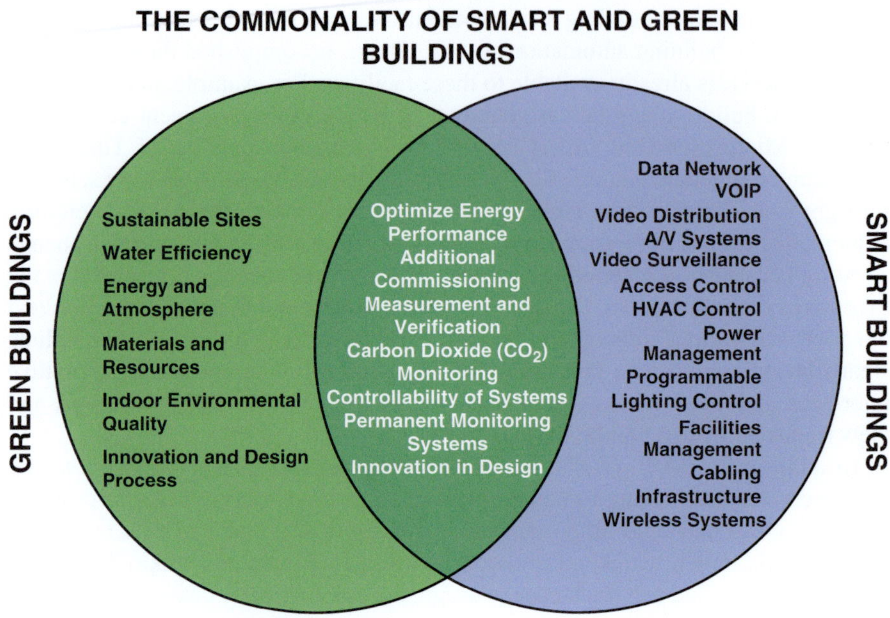

Fig. 10.3 Smart and green building integration

10.4 Active Energy

With smartphones and Internet universal availability, BAS and other software/IT companies are launching smart buildings management platform and marketing SaaS (software as a service).

SaaS is any software application that you run that is not located on the premise being managed. It is a full-blown application, not a component part of something else. It is not a way to build applications. It is not a plug-in to other applications. Instead of having the application running on servers and data storage on building servers, it is running in the vendor's data center.

The way SaaS applications are licensed is different from on-premise applications. Instead of buying the license to use the application, and then paying for software maintenance to support it and keep it current, you "rent" the software over a period of time—usually monthly or yearly. Instead of buying and installing infrastructure and then paying ongoing operating and maintenance costs, the vendor runs the application on their infrastructure. The cost of the SaaS application covers the costs of the software itself and the ongoing operations and infrastructure costs.

When you run a SaaS application, you generally log into your vendor's website and you are on. You can say that SaaS applications are running "in the cloud," and you would be correct. But SaaS applications are not the Cloud.

So what is "the Cloud?" Cloud computing provides computing resources that are not tied to any specific location. Cloud computing basically consists of:

1. Virtual computers/servers.
2. Data storage capacity.
3. Communications and messaging capacity.
4. Network capacity.
5. Development environments

In other words, Cloud computing is for software developers, application vendors, savvy computer users, and corporate IT departments, not for people who use computer applications.

Active energy is cloud-based SaaS and provides Live Access=Active Management.

See Figs. 10.4, 10.5, and 10.6 below.

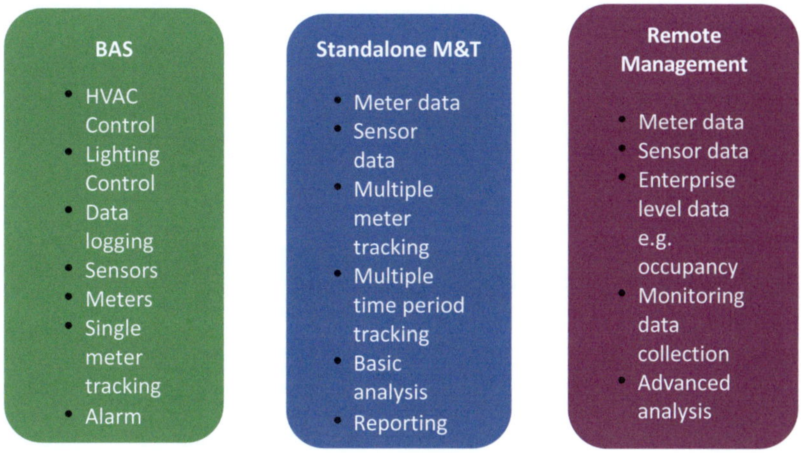

Fig. 10.4 Remote energy monitoring systems bring the power of the Internet of Things to energy reduction. Through using the SaaS cloud-based model, users are able to take advantage of lower upfront costs, continuous upgrades without the need for service calls, and global access to data and reporting features

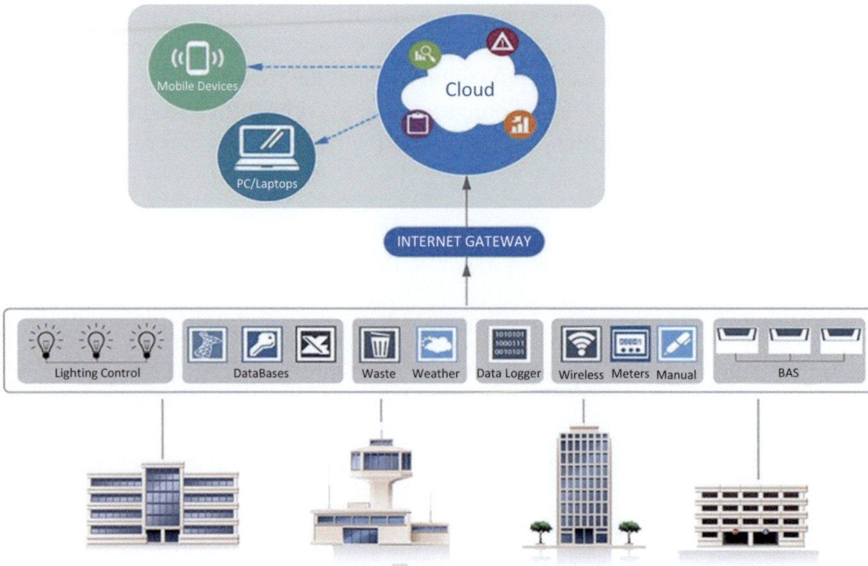

Fig. 10.5 Data collection architecture: through integration protocols that are open, Internet protocols, as well as gateways and routers, information can be brought to a centralized location from disparate systems

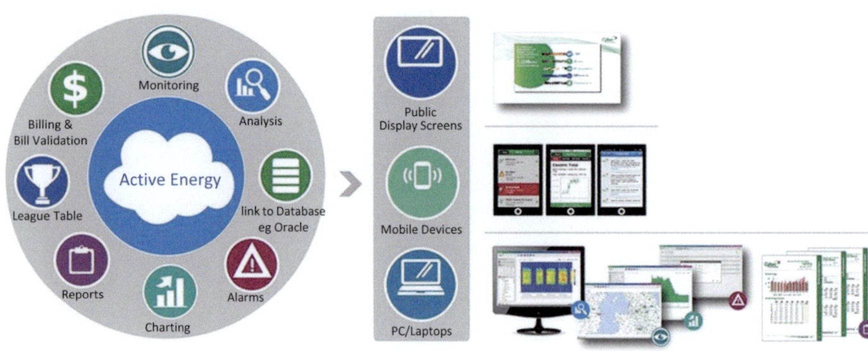

Fig. 10.6 Data output and control: from the centralized SaaS model, analysis, monitoring, alarming, and reporting can all be accomplished, thereby creating an energy saving schema that produces energy savings through prevention and diagnosis

Chapter 11
Atmospheric Radiation Soup and Buildings and Internet of Things

Anil Ahuja

11.1 Introduction

The daylight and the warmth we feel on a fine summer's day reach us as solar radiation. As everyone has known for many thousands of years, it is the sun that makes life possible and that sustains it continually. The radiation we see and feel, however, represents only a small part of the total radiation emitted by the sun.

The sun is a "main sequence" star. This sounds as though our local star is rather ordinary. It is, but that is not what the term means. Stars spend the greater part of their active "lives" burning hydrogen which is why this is called their "main sequence." The sun is a main sequence star because it is still burning hydrogen. So far, it has burned about half of its available stock—not all the hydrogen can be burned—and its present composition is about 90 % hydrogen, 8 % helium, and 2 % consisting of heavier elements. This energy released by the burning of hydrogen makes the center of the sun very hot—the temperature is calculated to be about 21.6 million °F (12 million °C).

11.2 Basics of Radiation

Electromagnetic radiation consists of a stream of photons. Photons are particles. They carry no electrical charge and it appears that they have no mass. You may wonder how it is possible for a particle to exist and yet to have no mass. The answer calls for the suspension of our common sense view of the world, which is necessary when considering the world of subatomic particles. A particle may have energy because it possesses mass and also because it is in motion. If a photon is stopped, however, it is destroyed and disappears, but it moves faster than any other particle. It is this that leads physicists to conclude that its energy consists only of the energy of its motion—it has no mass.

A. Ahuja, *Integration of Nature and Technology for Smart Cities*,
DOI 10.1007/978-3-319-25715-0_11

All photons travel at the same speed, the speed of light—in a vacuum, this is about 186,000 miles per second (300,000 km/s). Not all photons have the same amount of energy, however. Because a photon has no mass and it can move at only constant speed, there is only one way in which the difference in energy between one photon and another can appear—in its wave characteristics. The greater the energy a photon has, the shorter will be the distance between one of its wave peaks and the next. Its energy can be expressed as its wavelength. Conversely, the closer together the waves, the greater the number of wave peaks that will pass a fixed point in a given time.

The number of wave peaks per second is called the "frequency" of the wave, and therefore, the relationship between the wavelength and frequency of any regular wave motion is fixed, and the energy of the photon can also be expressed as its wave frequency. The wavelength is equal to the speed of propagation of the wave (in the case of electromagnetic radiation, the speed of light) divided by the frequency, and the frequency is equal to the speed divided by the wavelength.

The unit of frequency is the hertz (Hz), 1 Hz being equal to one wave cycle per second. This unit is often used to describe radio waves. A brief look into the history of the Hertz: Until approximately 1946, frequency was measured in cycles per second or just cycles for short. After World War II, the United States recruited German engineers and scientists who had developed Hitler's Vl and V2 weapons and brought them to the United States to work on rocket programs. The Germans found that although most electrical units were named after famous scientists (volt after the Italian scientist Volta, ampere after the French scientist, Ampere, and so forth), oscillation frequency was named in simple English. So the Germans declared that cycles per second should be called hertz after the famous German scientist Hertz and they sold the idea.

The range of electromagnetic wavelengths is called its "spectrum," and it is divided into two parts: ionizing and nonionizing radiation. The ionizing radiations have sufficient energy to break down atoms of other materials, causing electrons to be dislodged and leaving atoms positively charged or ionized. Ionizing radiation consists of x-rays, gamma rays, cosmic rays, and other rays with frequencies above 10^{15} Hz. The part of the solar spectrum related to building systems operation—mostly nonionizing with ultraviolet at a frequency of 10^{15} Hz on the upper level proceeding down to visible light, infrared, microwaves, radio frequency, and power frequencies at 50–60 Hz—is quite narrow, but it is the region in which the sun radiates most intensely. Because the spectrum is so large, different parts of it have been given their own common names (Figs. 11.1 and 11.2).

There are several ways to describe the energy of a particular part of the electromagnetic spectrum, each with its own unit. The joule (J) is the scientific unit of energy, and the electronvolt (eV) relates the energy of the photons to that of an electron. Conventionally written as one word, one eV is the energy needed to move one electron through a potential difference of one volt. More commonly, the spectrum is divided into sections according to the wavelength or frequency of the radiation. Sunlight consists of a range of wavelengths each visible as a distinct color, but which we see as white when they are mixed together. Literally, these are the colors

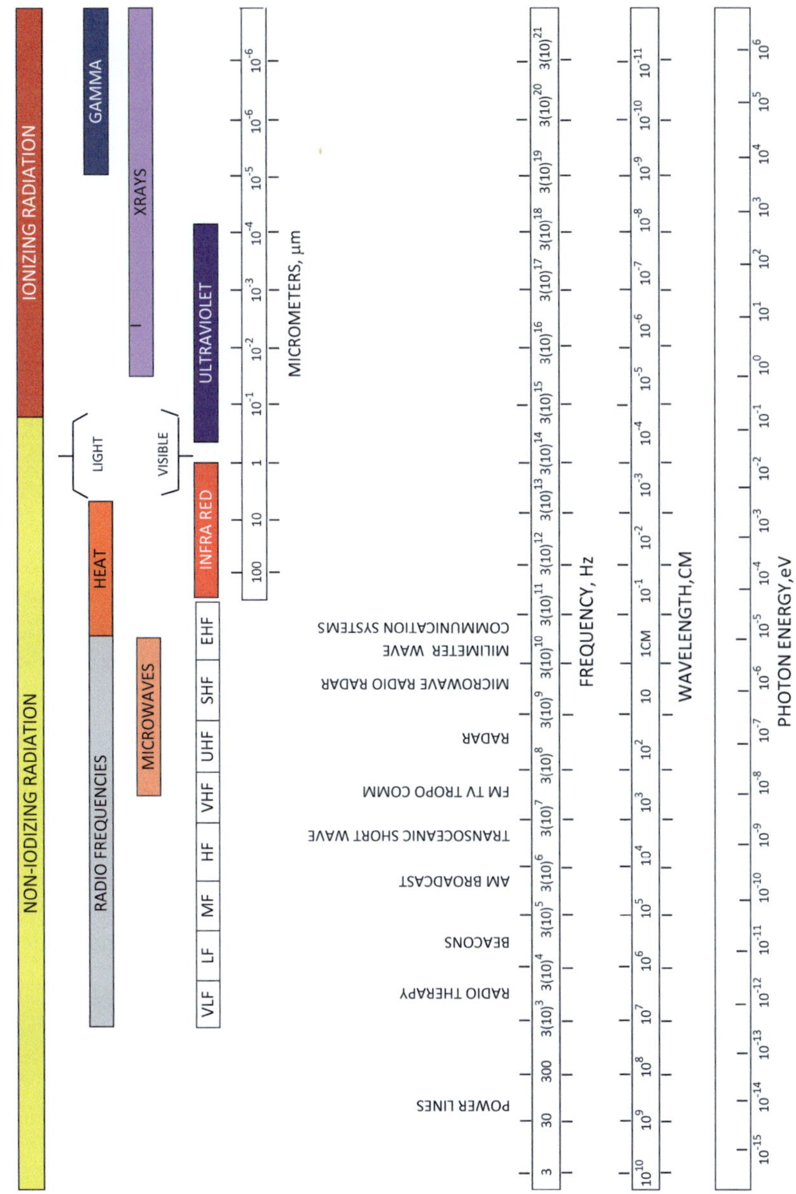

Fig. 11.1 *Electromagnetic wave* frequency spectrums

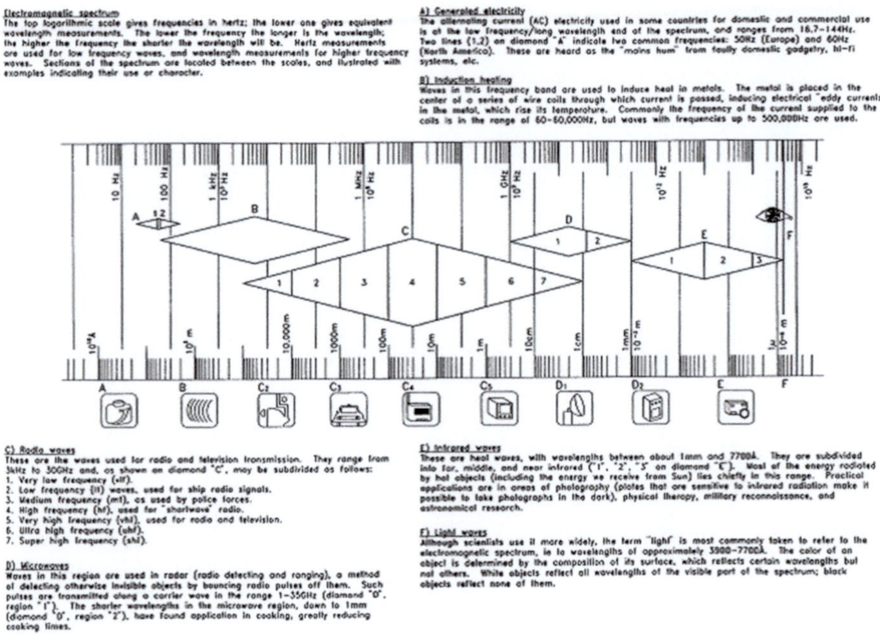

Fig. 11.2 Building system electromagnetic waves

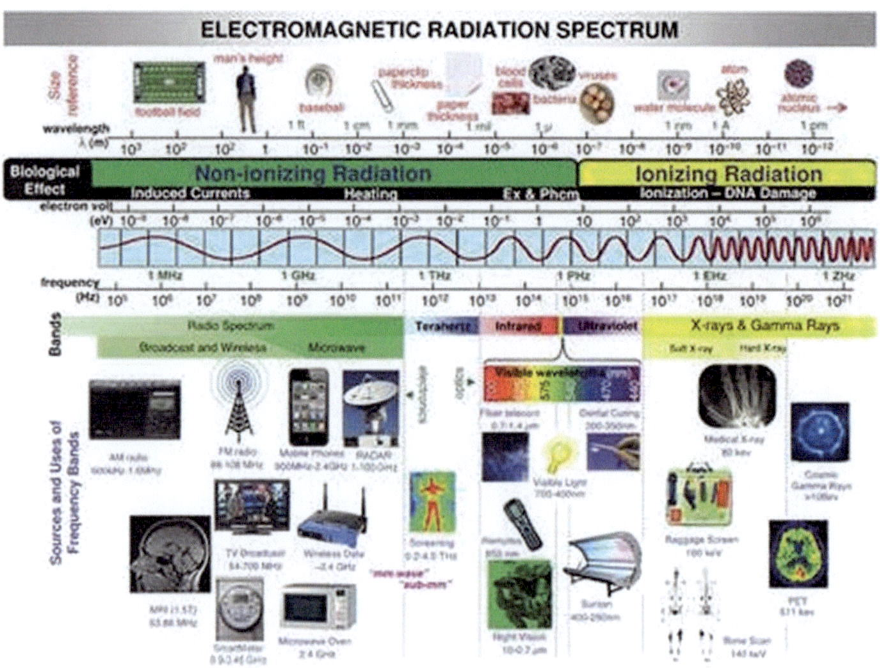

Fig. 11.3 Radiations from cell phones and other electronic devices (Source: Scientific American)

of the rainbow, because the water droplets that break sunlight into the bands of a rainbow are separating it according to its wavelengths (Fig. 11.3).

In modern civilization, everywhere you go there are electronic devices, emitting electromagnetic radiation (EMR). Radiation-emitting devices include electronic products, such as medical and nonmedical equipment, lasers, x-ray systems, ultrasound equipment, microwave ovens, color TVs, laptops, tablets, and PDAs. According to the FDA, a radiation-emitting device is any product that uses electricity to power an electronic circuit.

Significantly, EMR comes in two types: ionizing and nonionizing, which basically refers to the ability of the energy to break chemical bonds into ions. Obviously, EMR of the ionizing radiation type is bad for you as it can rip your molecules apart and cause DNA damage. But, what about EMR that is nonionizing? If it does not break your chemical bonds, does that mean it is safe? Well, the answer is a bit more complicated than that. Although no research to date has definitively linked specific diseases, such as cancer, to nonionizing EMR, there has been a lot of recent studies suggesting exposure to nonionizing radiation emitted from laptops is harmful, too.

11.2.1 US Standards (Fig. 11.4)

Laptops and tablets emit two types of electromagnetic radiation: extremely low frequency (ELF) and radio frequency (RF). Given the prevalence of these electronic gadgets in our daily lives, you may be surprised to learn that in the United

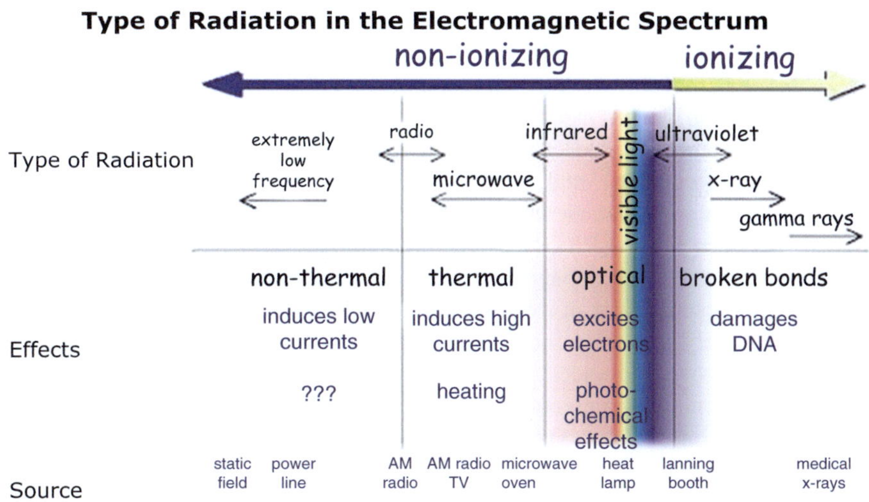

Fig. 11.4 Ionizing versus nonionizing radiation (Source: EPA.gov)

States, there is currently no government-funded organization that actively studies and regulates nonionizing electromagnetic radiation safety of common consumer electronic devices. The FDA largely focuses on medical devices, which exclude laptops and tablets. Trusted authorities, such as the American Cancer Society, even go so far as to suggest that nonionizing radiation is safe due to lack of evidence conclusively linking it to cancer. However, this is a mistake. A lack of information should never be considered a confirmation of safety, especially considering the fact that there has been a resurgence of independent laboratories and academic research institutions conducting studies on such topics as laptop-induced erythema ab igne and evaluation and characterization of fetal exposures to magnetic fields generated by laptop computers. Recent research is very suggestive, indicating long-term nonionizing EMR may, in fact, be harmful. Experts all seem to agree more research needs to be done on the long-term effects of exposure to modern day devices' nonionizing electromagnetic radiation in order to have proof one way or another.

Fortunately, some basic research has been funded by government to understand general human exposure limits to electromagnetic radiation. Experts recommend limiting extremely low frequency (ELF) exposure to below 2 mG and radio frequency (RF) exposure to under .6 V/m. Unfortunately, no laws or definitive guidance documents have been issued to the public.

11.2.2 International Standards

In Europe, the International Commission on Non-Ionizing Radiation Protection (ICNIRP), a publicly funded body of independent scientific experts, aims to disseminate information and advice on the potential health hazards of exposure to nonionizing radiation to anyone who has an interest in the subject. Members include experts in medicine, dermatology, ophthalmology, epidemiology, biology, photobiology, physiology, physics electrical engineering, and dosimetry. In 1998 and 2009, the ICNIRP established and confirmed guidelines for limiting exposure to time-varying electric, magnetic, and electromagnetic fields (EMFs) of up to 300 GHz. (For comparison purposes, the average radio waves emitted by a Wi-Fi network is typically between 2.4 and 5.8 GHz. Note: GHz>kHz.

When it comes to electromagnetic radiation from electronic devices like laptops and tablets, exposure is just an inevitable part of life. To be safe, experts recommend limiting extremely low frequency (ELF) exposure to below 2 mG and radio frequency (RF) exposure to under .6 V/m. As we continue to learn more about health dangers from nonionizing electromagnetic radiation contact, we should do everything we can to limit our personal exposure with EMF protection. While the specific chronic radiation exposure limits need more definition, the best course of action and the one recommended by experts is to avoid extra doses of electromagnetic radiation whenever you can.

11.3 Protective Role of the Atmosphere

The atmosphere is the gaseous envelope of the earth. All gases in the atmosphere exist in a temperature equilibrium that is energized and maintained in balance by the sun. The earth intercepts only about 0.002 % of the total amount of radiation emitted by the sun. This amount is proportional to the distance between the earth and the sun, and at the average distance of about 93,200,000 miles (150,000,000 km), it comes to about 1131 watts per square yard (1353 W/m^2). This is the amount of radiation that would reach the surface if none of it deflected, reflected, or absorbed on its way. It is called the "solar constant." Variations in the output from the sun is not quite so constant as people once believed and, of course, it is an average, it increases as the earth approaches perihelion and decreases as it approaches aphelion.

The figure for the solar constant seems low until you remember that the sun radiates in all directions and the earth is a very small target. Most of the solar output comes nowhere near our planet. Of the solar radiation that does reach us, the solar constant, about 51 %, an average of 595 watts per square yard (690 W/m^2), penetrates all the way to the surface, and about 19 %—215 watts per square yard (257 W/m^2)—is absorbed in the atmosphere and so contributes to the total amount of energy available to us from the sun.

The proportion of sunlight that reaches the surface when the sky is cloudy depends on the extent of the cloud cover but also on the type and thickness of the clouds themselves. Estimates of the extent of cloud cover are included in all reports from meteorological stations. The figure is usually expressed in "oktas"—one okta is equal to one eight of the sky—but sometimes you may see it reported in tenths or as a percent age. The effect on the local radiation budget is calculated by deducting from the incoming radiation an amount determined by the type of cloud and the number of oktas.

Visibility is more severely reduced around dawn and sunset, because the sun is lower in the sky, and therefore, its more oblique radiation must pass through a greater thickness of mist and more of it is lost. It is evident, therefore, that, despite their reflective brilliance, clouds are not opaque. Light can and does penetrate them, and inside a cloud, but near the top, the light can be very bright. This is the reason pilots are taught that, when flying in a cloud, they must rely wholly on their instruments to judge the altitude of the aircraft, because their ordinary senses become very unreliable.

11.4 Radiation Balance

Incoming and outgoing energy can be compared like the credit and debit columns of a balance sheet. A calculation of this kind produces a "radiation budget" that, like a financial budget, must balance. It can also be expressed as a "heat budget." Units of radiant energy can be translated into heat units because all the incoming energy that is absorbed—by the atmosphere as well as at the surface—is converted into heat. "Absorption" means that energy is delivered to the atoms and molecules of the receiving substance, raising their temperatures. Even photochemical processes cause energy to be released as heat or, in the case of photosynthesis, to be stored for later release as heat.

The balances do not have to be achieved immediately or locally. There can be, and are, imbalances according to season and latitude. In high latitudes, for example, the ground surface loses more energy over a year than it receives, and in low latitudes, where the sun is more directly overhead, the ground receives more energy than it loses.

There can also be very local variations. A hillside that faces away from the equator may never be warmed by direct sunshine and may lose more energy than it receives, while the other, equator-facing, side of the same hill receives more than it loses. Elsewhere, there are seasonal imbalances, with gains exceeding losses during the summer and losses exceeding gains during the winter. Even in a particular place and during a particular season, the balance can change from one hour to the next when clouds shade the surface. It is only over a number of years and over the planet as a whole that the totals must balance if the world is not to become hotter or colder.

The sun behaves as an almost perfect "black body." That is to say, it radiates energy very efficiently, at an intensity related to its surface temperature of about 11,000 °F (6000 °C). If the earth were also an efficient black body, promptly radiating back into space all of the energy it receives, the average surface temperature on the planet would be about–9 °F (–23 °C). The earth would be locked into a permanent ice age more severe than any it has ever experienced. Clearly, this is not the average temperature on the surface of the earth; over the whole surface and a full year, it is about 59 °F (15 °C).

The wavelength at which a black body radiates is determined by its temperature, which is exactly what you would expect. The wavelength of radiation represents the amount of energy it conveys, and so the higher the temperature of the radiating body, the more energetic the radiation it emits will be. At 59 °F (15 °C the earth radiates very slightly at less than 0.0004 mm, but mainly between 0.004 mm and about 0.04 mm, and most intensely at around 0.01 mm. The shortwave limit, of 0.004 mm, is taken as the boundary between "short-wave" and "long-wave" radiation.

Where the temperature is higher—over the subtropical deserts or during subtropical summer, for example—the wavelength is toward the shorter end of the wavelength, and where the temperature is lower, over high latitudes, it is toward the longer end. It is the difference in the amount of radiation absorbed in different latitudes that produces our climates, as heat is transferred from warmer to cooler regions through the atmosphere and oceans.

The amount of energy involved is considerable, at least by human standards. A commonplace local shower of rain involves about the same amount of energy as is required to keep a modern airliner airborne for 24 h, and some local thunderstorms release as much energy as the burning of 7000 tons of coal. "Powering" the Asian monsoon requires as much energy as used by all the factories, offices, cars, and homes in the world.

The earth emits long-wave radiation, but this does not leave the planet immediately. Atmospheric carbon dioxide, water vapor, and ozone absorb almost all radiation at wavelengths below 0.008–0.012 mm. At about 0.01 mm, however, which is where the earth radiates most intensely, there are gaps—"windows"—between the absorption bands of water vapor and ozone and between those of ozone and carbon dioxide at about 0.015 mm. Especially when the air is fairly dry, long-wave radiation escapes through these windows.

Of each 100 units of energy arriving at the top of the atmosphere, 34 are reflected. The atmosphere absorbs 19 units and only 47 units are absorbed by earth's surface. See Table 11.1 for debits and credits on the radiation balance sheet. The radiation

Table 11.1 Surface plus atmosphere energy balance

Incoming solar radiation (100 %)	
Absorption by O_3	3
Absorption by H_2O and aerosols	13
Absorption by clouds	3
Backscatter by clouds	25
Backscatter by air and aerosols	7
Backscatter by the surface	2
Total	53
Absorption by the ground	
From the sun	25
From the atmosphere and clouds	22
Total	47
Total incoming radiation	100
Outgoing terrestrial radiation	
Short wave	
Back radiation from H_2O, CO_2, and clouds to outer space	32
Back radiation from ground to outer space	2
Total	34
Long wave	
Back radiation from ground	6
Back radiation from H_2O, CO_2, and clouds	60
Total	66
Total outgoing radiation	100

that is absorbed in the atmosphere, together with latent heat released by the condensation of water vapor, is also reradiated in the same way and in all directions. Some escapes directly into space, some is directed downwards, and some is directed to the sides, where it is absorbed and radiated again or scattered in all directions by atmospheric molecules and particles. See Fig. 11.3. It is because of the absorption of long-wave radiation in the atmosphere—the "greenhouse effect"—that the average temperature on the earth is held above its theoretical black-body value.

11.5 Buildings and Radiation Albedo

The word "albedo" is derived from the Latin *albus,* meaning "white," and, in a sense, it is a numerical measure of whiteness or, more correctly, of the proportion of radiation a surface reflects. In summer, many of us wear white clothes, or at least light-colored ones, and, in winter, we tend to favor darker colors. We believe that white clothes reflect heat and dark clothes absorb it. This is true, although the number of layers of clothes and the materials from which they are made is a more important factor in keeping us warm or cool than their color.

Within regions, albedos change from time to time. Snow cover in winter increases albedo and the spring thaw reduces it again. As you would expect, fresh snow is the surface with the highest albedo. Depending on how smooth and clean it is, its value lies between 0.8 and 0.9—between 80 and 90 % of the radiation falling upon it is reflected.

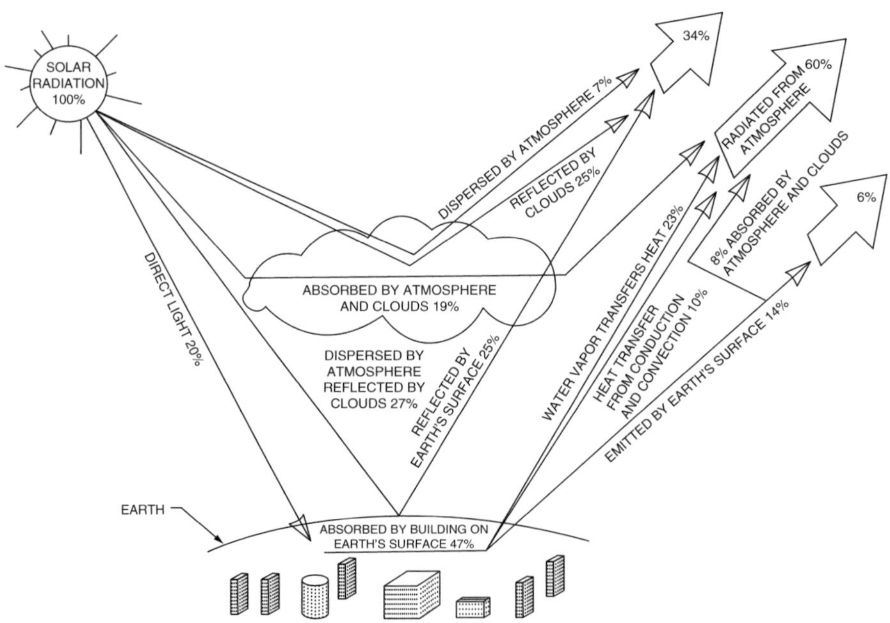

Fig. 11.5 Heat circulation (% division)

As the snow melts, its albedo falls to between 0.4 and 0.6. Sand comes next in the ranking, with an albedo of 0.30–0.35—desert sands tending toward the higher value.

Bare rocks have an albedo of 0.12–0.18. The albedo of urban areas is the same as that of rocks. Buildings and roads are made from rock, after all. The color of a building surface has such a strong influence on the proportion of solar radiation that is reflected and absorbed that it is a major factor in determining the size and capacity of building systems. An increase in its albedo means a surface absorbs less radiation. This will lower its temperature and that of the air above it, which, in turn, will reduce the rate of evaporation of water and cloud formation. Less cloud, however, will partly compensate by allowing more radiation to reach the surface (Fig. 11.5).

Some clouds that is, say, 70 feet (21 m) thick will reflect about 40 % of the incoming radiation (its albedo is 0.40). A cloud 200 feet (61 m) thick has an albedo of about 0.65, one 400 feet (122 m) thick 0.78, and one 600 feet (183 m) thick 0.80. Very deep clouds, such as cumulonimbus storm clouds, may extend for 10,000 feet (3000 m) or more and have an albedo of 0.90.

Such high albedo values suggest that cloud cover cools the ground below, but matters are not quite so simple. Apart from warming the surrounding air by absorbing a small amount of incoming radiation and by the latent heat of condensation, water vapor absorbs long-wave radiation. This radiation, from the surface, is prevented from escaping into space and so clouds have a warming as well as a cooling effect. During the day, when the intensity of incoming radiation exceeds that of outgoing radiation, clouds block incoming radiation and have a cooling effect. At night, when incoming radiation ceases, they block outgoing radiation and so have a warming effect. Their overall influence is to make hot days cooler and cold nights warmer.

11.6 Ozone Layer

Near the *stratopause* of stratosphere zone at a height of about 30 miles (50 km), solar radiation at the short-wave end of the ultraviolet (UV) waveband splits oxygen molecule into their constituent atoms, and the UV energy is absorbed by the atoms. The atoms rejoin to form molecules, some of them in threes to form ozone, and ozone also absorbs UV radiation, the atoms of its three molecules separating. The repeated breaking and reforming of oxygen and ozone molecules blocks virtually all UV radiation at wavelengths below about 0.0004 mm. Mixing of air in the stratosphere carries some upper air, enriched in ozone, to lower levels where the ozone is below the main area of UV absorption and accumulates, especially between about 12 and 16 miles (about 20–25 km). This is the ozone layer.

Ozone is extremely reactive, and therefore, it is destroyed rapidly at lower levels, where it is more likely to encounter compounds it can oxidize. Above the ozone layer, ozone is destroyed by radiation, and below the ozone layer, it is destroyed chemically—although it is also formed in the troposphere by photochemical reactions. There is little exchange of gases or particles across the tropopause but, since the 1960s, there have been fears of a gradual accumulation in the stratosphere of substances that could react with ozone. Were this to happen, ozone might be broken down faster than UV radiation causes it to form, and there would be a proportional increase in the amount of short-wave UV radiation reaching the surface. In 1987, evidence of ozone depletion was reported from Antarctica—the so-called ozone hole.

During the long Antarctic winter, stratospheric air moves in an approximately circular path, with a region of very still air—a vortex—at the center. The temperature in the vortex, about −130 °F (−90 °C), is low enough to cause water vapor—probably produced by the oxidation of methane—to form polar stratospheric clouds made from ice crystals. Various nitrogen compounds adhere to surfaces of the ice crystals and, as the winter draws to an end and the returning sun supplies a little energy, these compounds react with free chlorine. This, in turn, forms a stable oxide with ozone, depleting the ozone layer. As spring advances, the air circulation changes, breaking down the vortex. Air containing ozone is drawn in, and, with the increasing intensity of the sunlight, the formation of ozone is resumed.

Chlorine is the key ingredient in the polar depletion of ozone. More than half of the chlorine that enters the stratosphere arrives in the form of methyl chloride (CH_3Cl). This is released when surface vegetation is burned—in bush and forest fires, for example—and by many wood rotting fungi. It is chemically stable, but, once in the stratosphere, UV radiation breaks the bond linking the methyl (CH_3) and chlorine (Cl) ions, liberating free chlorine. The chlorine from methyl chloride provides a natural background that changes little over the years. The amount of stratospheric chlorine has increased mainly because of the use of chlorofluorocarbon compounds—CFCs in air-conditioning and fire suppression systems. These compounds, of which there are several, are chemically very stable and do not react with the ordinary constituents of the atmosphere. Some CFC molecules are washed to the ground by rain, some enter the oceans, and some adhere to solid particles, such as sand grains, but a proportion remain airborne and seep across the tropopause into the stratosphere.

The ozone layer is thickest in high latitudes in summer. Current predictions, of a possible 10 % depletion within 50–75 years, would increase UV exposure by the equivalent of moving about 100 miles (160 km) toward the equator from high latitudes. There could be a small reduction in yields of sensitive crops and sunbathing could lead to more rapid burning, but the biological consequences are unlikely to be severe. Monitoring suggests, however, that in recent years the intensity of UV radiation over the continents has been decreasing rather than increasing.

11.7 The "Greenhouse" Effect

Long-wave radiation—"black-body" radiation—is absorbed in the atmosphere by a range of gases but escapes through "radiation windows." This temporary retention of heat by the atmosphere is called the "greenhouse effect," and it serves to keep the earth much warmer than it would be otherwise. If the atmospheric concentration of the "greenhouse gases" increases; however, the existing radiation windows may be partly closed. Provided there was no compensating effect, this would lead to a warming of the atmosphere.

Almost any gas molecules that consist of three or more atoms is likely to be of a size comparable to the wavelength of part of the long-wave electromagnetic spectrum, and such a molecule will absorb some outgoing radiation. Not all molecules are of equal importance. It may be, for example, that a particular gas is already so abundant that it absorbs all radiation in its waveband. Adding more of that gas will have no effect, but adding another gas, even in small amounts, may cause radiation to be absorbed in a part of the spectrum that previously was open.

Carbon dioxide is the best known of the greenhouse gases, but this is only because it is the most abundant. It contributes about 50 % of our present greenhouse effect. Methane contributes about 18 %, CFCs 14 %, tropospheric ozone 12 %, and nitrous oxide 6 %. Concentrations of all of them are increasing. Over the past century, the proportion of carbon dioxide in the air has increased from about 290 to 350 parts per million. It is still rising by about 1.5 % a year, mainly because of the burning of carbon-based fuels in facilities and the clearing and burning of forests. Methane, released mainly from cattle and the growing of paddy rice, is increasing at about 1 % a year. Nitrous oxide is released when organic substances are burned, and ozone is produced by photochemical reactions involving nitrogen oxides and unburned hydrocarbons from boiler flues.

The average global temperature today is about 1 °F (0.5 °C) higher than it was a century ago. This may or may not be due to a greenhouse warming, and, beyond predicting a general rise in temperature, calculating the climatic implications of the warming in each region of the world is very difficult. Most scientists agree it would be wise to reduce emissions of greenhouse gases.

Chapter 12
Reliability Requirements, Risk Management, and Associated Building Systems Engineering

Anil Ahuja

12.1 Introduction

A commonly used definition for reliability is the ability of a system to perform a required function under known conditions for a desired period of time or a determined amount of usage. Closely related to reliability, and actually what the systems engineer is interested in, is availability. Again, the definition is as follows: the probability that a system subject to repair will perform a required function under known conditions on demand. Notice that availability has introduced two additional concepts: (1) the probability of something happening and (2) repair.

In overall system reliability, don't overlook the operating time. In building systems, typically the standby generator operates for less time than the chillers. Two terms are useful: MTBF (mean time between failure) and MTTR (mean time to repair).

Quantitative reliability or availability figures are normally used for comparison purposes. Quantitative reliability consists of two major elements: (1) the probability of failure and (2) equally important the magnitude of the consequence of failure. If the system can fail with little consequence, then the quantitative reliability is different than if some failure resulted in major undesirable consequences. An individual building system equipment manufacturer is concerned about balancing the different component failure rates, so that the unit doesn't result in an obvious weak link. The same idea is true for the system designer; the system is no more reliable than the least reliable unit. The good design results in the least reliable link having the least consequence.

Common system protection is normally employed for three reasons: (1) personnel safety, (2) equipment safety, and (3) equipment reliability. Of course, what the designer is protecting against often determines what the protection is for. As an example, protection against accidental contact with high voltages is for personnel and equipment. The major category to be protected will determine the type of protection equipment. Personnel protection will be different from equipment protection

A. Ahuja, *Integration of Nature and Technology for Smart Cities*,
DOI 10.1007/978-3-319-25715-0_12

even though they both may be protecting against high-voltage exposure. It becomes apparent that the designer must first decide what must be protected and what it must be protected from, in order to decide the type of scheme to implement. There are, of course, types of equipment to protect multiple categories as well as multiple causes. For many years when engineers talked about the reliability of building systems, they often used nonspecific quantifiers such as "good, better, best," or "what we propose is better than what you have," or "if you don't do something soon, it will probably fail and cause tremendous problems." If they did try and get more scientific, they usually would drag out the IEEE gold book or military handbook and dig through it, and then try to make heads or tails about how the reliability calculations worked, and then try to figure out which piece of information from the mounds of data in the book should be used to represent the problem at hand.

One of the key points emphasized here is the need to speak the language of the business community. In today's world, this language is centered around such factors as business risk, cost impact, first cost, life-cycle cost, and the cost to operate and maintain. Notice that the language almost always deals with cost and is based on actual data. It may well be a waste of time if reliability is discussed in a language other than cost and in a context based on something other than fact-based data. To have meaningful results from a system reliability evaluation, the answer must be in dollars. Develop a modeling tool that can be applied to both new and existing real-life systems that will help determine the reliability at any point in the system and will determine the cost impact from loss of function. Figure 12.1 shows a reliability evaluation process that speaks engineering and business management language.

For example, if a building systems model is analyzed to achieve 97 % availability and it did not include planned turnarounds, which were estimated to be 2 %, this will result in a 99 % actual availability. Of the remaining 1 %, 0.5 % can be allocated to problems associated with human operating issues. This leaves 0.5 %, half of which (0.25 %) can be allocated to management-related problems and 0.25 % can be allocated to building system problems. Using simple mathematics, 0.25 % of one year is about 22 h (say one day). To meet business requirements, the impact of all building system failures to the operation of building functions had to be less than one day per year. "What-if" analysis could be performed to see the impact of component on system structure reliability.

12.2 Basic Reliability Terminology

Annual risk: The calculated financial losses of building functions due to a building system failure divided by the frequency (MTBF) of the failure.

Availability: A ratio that describes the percentage of time a system component or system can perform its required function.

Component: A piece of electrical or mechanical equipment, an interconnecting pipe/valve or circuit, a section of building system, or a group of items that is viewed as an entity for the purpose of reliability evaluation.

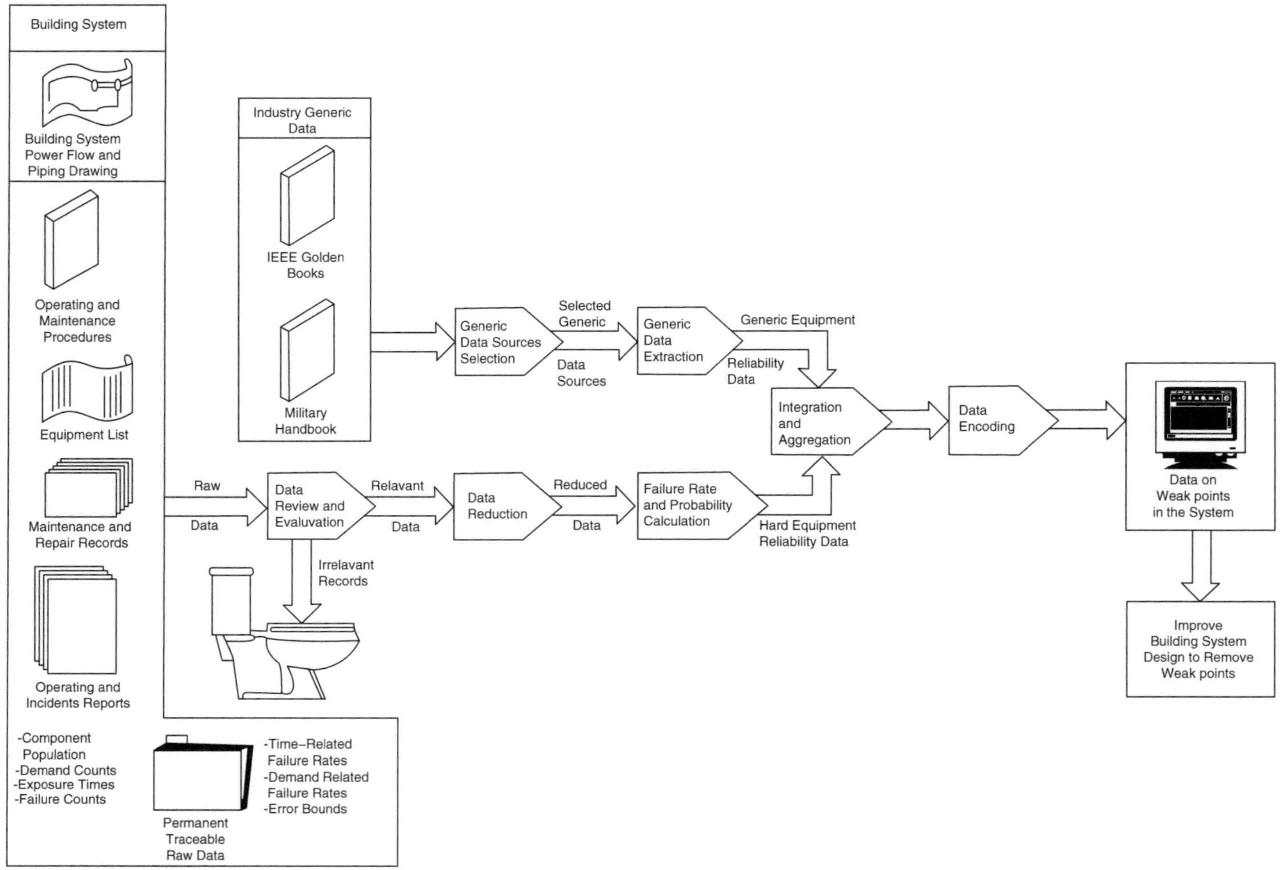

Fig. 12.1 The RAM (reliability, availability, and maintainability) database development process for industrial and commercial facilities

Failure: The termination of the ability of a component to perform a required function.

Failure rate: The mean number of failures of a component per unit exposure time.

Forced downtime: The average time per year a system or equipment is unavailable in between failures and expressed in hours per year.

Lambda (λ): The inverse of the mean exposure time between consecutive failures. λ is typically expressed in failures per year.

MTBF: The mean exposure time between consecutive failures of a component or system. The mean time between failures is usually expressed in years per failure. For some applications, measurement of mean time between repairs (MTBR) rather than mean time between failures may provide more statistically correct information.

MTTR: The mean time to repair a failed component. For a system, it is the total amount of time it is unavailable in between failures and is expressed in hours in both cases.

Point: Any place or location within the building critical system. The name or designation for a point is always the same as the name of the zone in which the point is located.

RAM table: A lookup historical and field data table that displays the MTBF and MTTR for system components.

Reliability: An indication of the ability of a component or system to perform its intended function during a specified time.

Restore time: The time to restore is the sum of the mean time to repair (MTTR) for the failure plus the computed time to restream or restart the connected load.

Building system: A group of components connected or associated in a fixed configuration to perform a specified function. Function could be distributing power or air-conditioning, circulating water, controlling system interoperation, etc.

Zone: A zone is defined as a segment of a power (or a segment of water or steam distribution system) in which a fault (or leak) at any location within the segment or zone would have the common impact of causing the first upstream protective device to isolate the system (or in the case of water/steam create mess and empty the system).

12.3 Building System Reliability Analysis Procedure

Failure rates for series and parallel systems combine in a manner very similar to that of impedances. The only problem with this approach is that while an array of electrical/mechanical components may be physically connected in parallel, its reliability equivalent model may have the components connected in series. For example, if there is a district chilled water line running along several buildings with tap-offs into each building, the flow diagram would show all the building tap-offs in parallel to the central chiller plant. Yet when one building system experiences a major leak, all of the other buildings also lose flow until the problem can be isolated and repaired. Thus, when representing the reliability of all the tap-offs, they end up being in series. To overcome this misrepresentation, the distribution system should be divided into zones.

The concept of developing reliability zones is quite similar to the technique used in protective relaying for describing zones of protection. One of the primary purposes of protective devices from the standpoint of a system is to improve reliability by isolating faulty components while minimizing the impact to the remainder of the system. When configuring a system, a zone can be configured either in series with a single source point immediately above it, or a zone can be configured in parallel with the two sources immediately above it.

The analysis should determine the failure rate and repair rate of all components within a zone without taking into consideration that the zone is connected to any other portion of system. The analysis should then determine the failure rate and repair rate for all the points in the system by taking into consideration the configuration of all the zones. Create a matrix called the zone table in which the configuration of the zones can be entered. Create another matrix table, called the unit impact table, which allows the entry of various interconnected units or loads and the impact that each point has on the unit or load. It should also include data entry to reflect the cost impact for not having the unit available for 24 h. The last data entry field should be the time it will take to get the unit or load reenergized from a cold start.

The analysis should determine if the unit has been down long enough to become "cold." For most critical application units, the time required to restream the unit following a failure is relatively short if the problem can be corrected quickly. However, if the repair persists for an extended period of time, the unit will get "cold" and restreaming of the unit may take much longer. The system analysis should have a result section labeled the "Critical Building System Component Summary." Here the total quantity of each system component is displayed along with the expected failure rate for that population of components. The summary should include the total number of components within the critical system analyzed and the failure statistics for the entire population of components.

The following is a brief description of the computations normally used when combining zones in series and parallel.

12.3.1 Component Analysis

12.3.1.1 Single-Component Analysis

Reliability and availability are necessary to describe the characteristics of the single component shown (see Fig. 12.2a).

$$d_1 = \text{MTBF}(\text{in hours})$$
$$r_1 = \text{MTTR}(\text{in hours}).$$

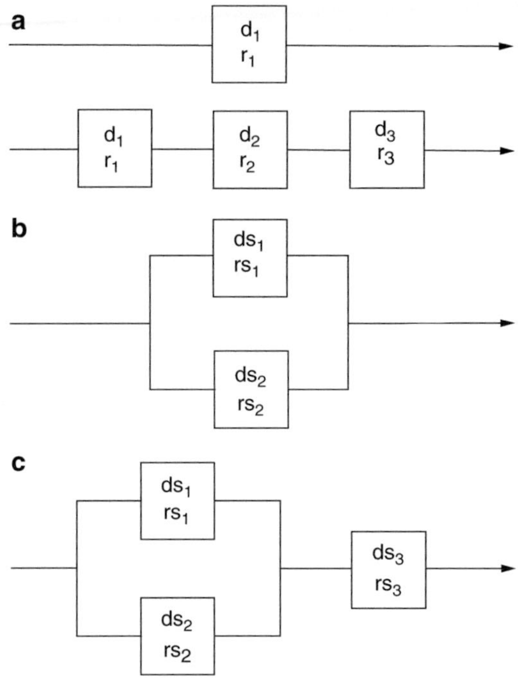

Fig. 12.2 (**a**) System with components in series. (**b**) Redundant system with components in parallel. (**c**) System with components in series with redundant system

Reliability of components is frequently given as failures per year of operating time. Using these numbers, the MTBF can be calculated using

$$d_1 = \text{MTBF}(\text{in hours})$$
$$d_1 = 8760\,\text{h}/(\text{failures}/\text{year}).$$

The failure rate (λ) is given by

$$\lambda = 1/\text{MTBF}$$

and the component reliability (R_1) for one year is given by

$$R_1 = e^{-(\lambda)(8760)}.$$

For a single component, the availability (A) is given as the total operating time over the total time, or

$$A = (\text{MTBF})/(\text{MTBF} + \text{MTTR}).$$

12.3.1.2 Systems with Components in Series

For the system shown with three different components in series (see Fig. 12.2b)

$$d = \text{MTBF}\left(\text{in hours}\right)$$
$$r = \text{MTTR}\left(\text{in hours}\right).$$

The characteristics of each individual component can be calculated using

$$\lambda_i = 1/d_i$$
$$A_i = d_i/\left(d_i + r_i\right)$$
$$R_i = e^{-\lambda_i t}$$
$$R_i = e^{-\lambda_i (8760)} \text{ for one year.}$$

Using these, the combined failure rate (failures per year) becomes

$$\lambda_s = \lambda_1 + \lambda_2 + \lambda_3$$

or

$$\lambda = R_1 * R_2 * R_3.$$

Reliability of the system for one year is

$$R_s = e^{-\lambda_s (8760)}.$$

System availability is

$$A_s = A_1 * A_2 * A_3.$$

Probability of failure during one year is

$$P_s = \left(1 - R_s\right) * 100.$$

MTTR in hours is

$$\text{MTTR} = \text{MTBF}\left[\left(1/A_s\right) - 1\right].$$

where MTBF = $1/\lambda_s$

Forced unavailable time (FUT) is

$$\text{FUT} = \left(1 - A_s\right).$$

12.3.1.3 Systems with Redundant Components

1. *Parallel (redundant) systems*: Reliability can be dramatically increased by installing a parallel (redundant) system. The simplest of these is a system that operates satisfactorily if either one of two parallel components functions (see Fig. 12.2c).

   The reliability for such a system for one year can be calculated using

$$R_s = \left[1-(1-R_{s1})(1-R_{s2})\right].$$

The combined failure rate is

$$\lambda_s = \ln(1/R_s).$$

The system availability is

$$A_s = \left[1-(1-A_{s1})(1-A_{s2})\right]$$

and the probability of failure during one year is

$$P_s = (1-R_s)*100.$$

2. *Parallel (redundant) systems with maintenance provisions*: If the components can be repaired, the reliability of the systems described above also becomes a function of the time required to repair the system.

Using constant failure rate for two identical units, the steady-state availability is

$$A = \mu / \lambda + \mu$$

where the repair rate (μ) is [(MTTR) – 1] and the MTBF is

$$\text{MTBF} = \mu / \left(2\lambda^2\right).$$

For two different components in parallel, the MTBF is

$$\text{MTBF} = (\lambda_a + \mu_b)(\lambda_b + \mu_a) + \lambda_a(\lambda_a + \mu_b) + (\lambda_b + \mu_a) / \lambda_a \lambda_b (\lambda_a + \lambda_b + \mu_a + \mu_b)$$

and the steady-state availability for n blocks is

$$A = 1 - \prod_{i=1}^{n} (\lambda_i / \lambda_i + \mu_i)$$

or for two parallel components

$$A = 1 - \left[(\lambda_1 / (\lambda_1 + \mu_1))(\lambda_2 / (\lambda_2 + \mu_2))\right].$$

12.3.1.4 Common Cause

Redundancy calculations frequently lead to reliability numbers that are outside the realm of reason, for example, one failure in a thousand years. In reality, even redundancy of components still leaves a chance that the parallel system will fail from a common mode. Examples of this include common electrical connections, common control wiring, or the chilled water piping. One suggested method to account for common failure modes is to use an additional component in series with the redundant system as shown in Fig. 12.2c. A reliability of between 0.01 and 0.1 times the reliability of the components in the individual redundant components seems to provide results that are comparable with actual experiences.

12.4 Risk Analysis

Risk analysis is very important in considering what are called the failure modes. Determine first what is a failure and, second, the consequence of the failure. Most of the time it is more important to know all the failure possibilities than to have an in-depth knowledge of some and miss others. This first step is to find out what could happen, not how often or how severe.

The use of an event tree is most helpful. Particular care must be given to find "common mode" causes. Common mode failures are those single events that can cause several units to fall. A main power switch is a typical common mode failure because it may disconnect several units where an individual circuit fuse will cut out only an individual unit.

Risk is defined differently by an engineer than a by layman. Table 12.1 indicates salient differences between the two.

The primary goal of any building system reliability analysis is to reduce the probability of accidents and the attending human, economic, and environmental losses. The human losses include death, injury, and sickness or disability, and the economic losses include production, revenue, or service shutdowns; loss of capital equipment; legal costs; and regulatory agency fines. Environmental losses are air and water pollution and other environmental degradations such as odor, vibration, and noise.

Failures occur when an initiating event is followed by system failures. The three types of basic failure causes are:

1. Humans: operator error, design error, and maintenance error
2. Hardware failure: power source component failure, leakage of toxic refrigerants from a valve, loss of motor lubrication, incorrect sensor measurement, controller communication failure, etc.
3. Environmental disaster: earthquake, storm, flood, tornado, lightning, etc.

Disastrous failures are frequently caused by a combination of failure events, that is, a hardware failure plus human error and/or environmental faults. Typical policies to minimize these accidents include:

1. Equipment redundancies

Table 12.1 Risks and difference in technical treatment and layman treatment

Approach	Technical treatment	Layman treatment
1. Risk acceptance	Emphasizes sound decision-making, because of finite resources for risk reduction and impracticability of achieving zero risk; checks the tendency to ignore nondollar costs in such tradeoffs and provides net benefit to system; neglects indirect and certain long-term benefits without explicit decision criteria and structured analyses	Greater tendency to judge risk in absolute terms; ignores risks of no-action alternatives to rejected technology; gives greater weight to nondollar costs; emphasizes personal rather than system benefits; includes both qualitative and quantitative benefits but tends to neglect indirect and long-term benefits; tends to distort equity considerations in favor of personal interest
2. Risk assessment methods	Quantitative, experimental, computational	Qualitative, intuitive, impressionistic
	Risk = consequence × probability	Incomplete rationale
	Fault trees/event trees	Emotional input to value judgments
	Statistical calculation	Personal experience/memory
	Engineering test equipment and simulators	Media accounts
		Political exchange
3. Basis of information	Established institutions and through qualification of experts	Nonestablishment sources; limited ability to judge qualifications or nonpractical engineering knowledge
4. Risk attribute evaluation	Provides objective, conservative assessment; considers broad range of high and low estimates; and gives equal weight to diverse views over treatment of incommensurables and discount rate	Tends to exaggerate or ignore risk; gives greater weight to catastrophic and immediate issue except for overall exposure to failure; gives greater weight to dreaded risk

2. Energy transportation media and source backups
3. Inspection and maintenance
4. Safety systems such as sprinklers, fire walls, and relief valves
5. Fail-safe and fail-soft design

12.5 Failures and Relationships

A system consists of hardware, materials, and plant personnel, is surrounded by its physical and social environment, and suffers from aging (wear-out or random failure). Accidents are caused by one or a set of physical components generating failure events. Each physical component in an integrated system is related to the other components in a specific manner. The system environment, in principle, includes the entire world outside the plant, so an appropriate boundary for the environment is necessary to prevent the initiating-event and event-tree analyses from diverging.

A primary probability risk analysis objective is to identify the causal relationships between human, hardware, and environmental events that result in failures or accidents and to find ways of ameliorating their impact by building system redesign and upgrades. The causal relationships can be developed by event and fault trees, which are analyzed both qualitatively and quantitatively. After the combination of the basic failure events that lead to losses is identified, the building system can be improved and losses reduced.

12.6 Fault Trees

The fault tree is structured such that an undesired system failure event appears at the top and is linked to more basic failure events by event statements and logic gates. The fault-tree analysis is restricted only to the identification of the system and component causes that lead to one particular top event. System specification requires a careful delineation of component initial conditions. All components that have more than one operating state generate initial conditions. For example, if the initial quantity of fluid in a tank is unspecified, the event "tank is full" is one initial condition, while "tank is empty" is another.

The fault-tree structure provides the following:

1. Reveals system weakest links
2. Points out the segments of the building system important to the system failure of interest
3. Provides a graphic aid
4. Provides options for qualitative and quantitative system reliability analysis
5. Allows concentration on one particular system failure at a time
6. Provides an insight into system behavior

Fault-Tree Construction In large fault trees, mistakes are difficult to find, and the logic is difficult to follow or obscured. For large systems, it is advised to divide the system into logical boundaries. The construction of fault trees is perhaps as much of an art as a science. Fault-tree structures are not unique; no two analysts construct identical fault trees (although the trees should be equivalent in the sense that they yield the same cut set or combination of causes).

To find and visualize causal relations by fault trees, gate symbols and event symbols are used as main building blocks. Gate symbols include AND, OR, inhibit, priority AND, exclusive OR, and voting. Event symbols are rectangle, circle, diamond, house, and triangle. The symbols for the gates are listed in Fig. 12.3. A gate may have one or more input events but only one output event.

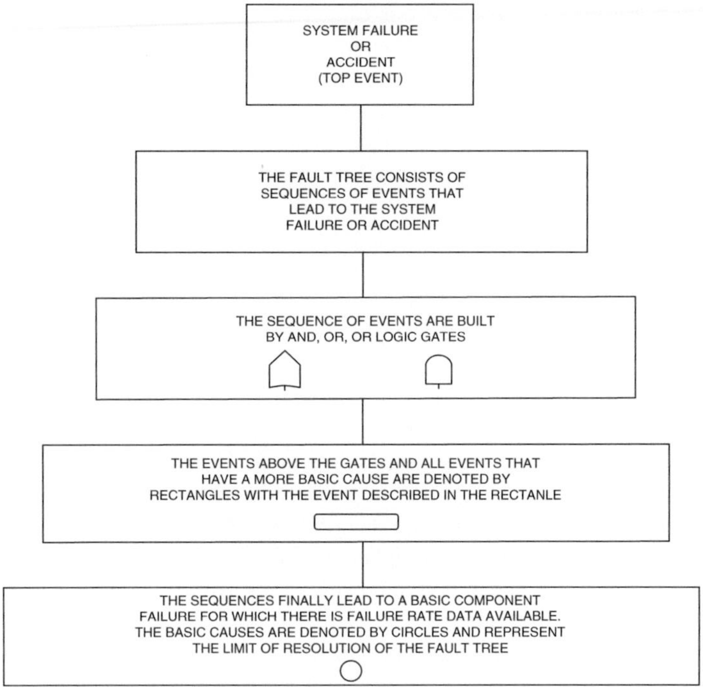

Fig. 12.3 Fundamental fault-tree structure (Copyright© 1996. IEEE. All Rights Reserved)

Gate Symbols

Connect events according to their causal relationships. The *AND gate* output event occurs if all input events occur simultaneously, and the *OR gate* output event happens if any one of the input events occurs. Examples of OR and AND gates are shown in Fig. 12.4. For example, consider a fuel pumping system shown in Fig. 12.5; the system event "pressure tank overflows" happens when two events, "pump motor runs too long" and "overflow protection device does not operate," occur simultaneously. The latter event happens when either one of the two events, "high level cutoff protective device does not exist" or "device is not suitable," occurs. Showing these events as rectangles implies they are system states. If the event "overflow device malfunction," for example, were a basic cause, it would be circled and become a basic hardware failure event. The causal relationship expressed by an AND gate or OR gate is deterministic because the occurrence of the output event is controlled by the input events.

	GATE SYMBOL	GATE NAME	CAUSAL RELATION
1	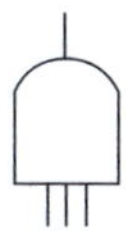	AND GATE	OUTPUT EVENT OCCURS IF ALL INPUT EVENTS OCCUR SIMULTANEOUSLY
2		OR GATE	OUTPUT EVENT OCCURS IF ANY ONE OF THE INPUT EVENTS OCCURS
3	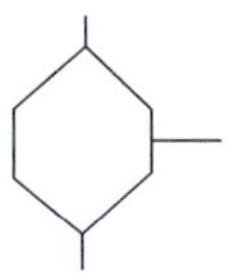	INHIBIT GATE	INPUT PRODUCES OUTPUT WHEN CONDITIONAL EVENT OCCURS
4	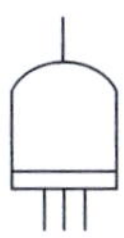	PRIORITY AND GATE	OUTPUT EVENT OCCURS IF ALL INPUT EVENTS OCCUR IN THE ORDER FROM LEFT TO RIGHT
5		EXCLUSIVE OR GATE	OUTPUT EVENT OCCURS IF ONE, BUT NOT BOTH, OF THE INPUT EVENTS OCCUR
6		m-OUT-OF-n GATE <VOTING OR SAMPLE GATE>	OUTPUT EVENT OCCURS IF m-OUT-OF-n INPUT EVENTS OCCUR.

Fig. 12.4 Gate symbols (Copyright© 1996. IEEE. All Rights Reserved)

There are causal relationships that are not deterministic. Consider the two events "overflow occurs" and "protective device fail to operate." The causal relationship here is probabilistic, not deterministic, because a protective device misoperation does not always result in overflow. The *hexagonal inhibit gate* is used to represent a probabilistic causal relation. The event at the bottom of the inhibit gate is an input event, whereas the event to the side of the gate is a conditional event. The output event occurs if both the input event and the conditional event occur. In other words, the input event causes the output event with the (usually constant, time independent) probability of occurrence of the conditional event. In contrast to the probability of

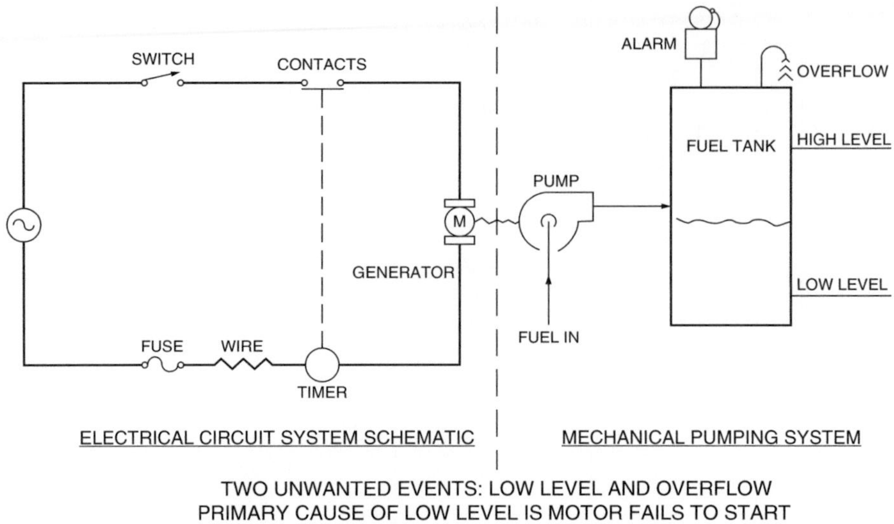

Fig. 12.5 Fuel pumping system: electrical/mechanical schematic diagram

equipment failure, which is usually time dependent, the inhibit gate frequently appears when an event occurs with a probability according to a demand.

Another gate logically equivalent to an AND gate is the *priority AND gate.* It causes the output if input events occur in a specific order. The occurrence of the input events in a different order does not cause the output event. Consider, for example, a system that has a principal power supply and a standby power supply. The standby power supply is switched into operation by an automatic switch when the principal power supply fails. Power is unavailable in the system if the switch controller fails first and then the principal unit fails. The failure of the switch controller after the failure of the principal unit does not yield a loss of power because the standby unit has been switched correctly.

Similarly *exclusive OR gates* describe a situation where the output event occurs if either one, not both, of the two input events occurs. Consider a system powered by two generators. A partial loss of power can be represented by the exclusive OR gate. In the case of multiple generators, a *voting gate* based on the occurrence of *m-out-of-n* input events represents a partial loss. This situation can be expressed by the m-out-of-n gate.

To examine various special fault-tree cases by forcing some events to occur and other events not to occur, a *house* symbol is used. When we turn on the house event, the fault tree presumes the occurrence of the event and vice verse when we turn it off. *Triangles* are used to cross-reference two identical parts of the causal relationships. The two triangles have the same identification number. The transfer-out triangle has a line to its side from a gate, whereas the transfer-in triangle has a line from its apex to another gate.

See Fig. 12.7 to construct a fault tree for the integrated system shown in Fig. 12.5. This tree construction uses component failure characteristics shown in Fig. 12.8.

Event Symbols Describe occurrence (see Fig. 12.6). In the fault-tree schematics, a *rectangle* usually denotes an undesirable system event, resulting from basic failures through logic gates. The *circle* designates a basic component failure that represents the lowest level of a fault tree. Circles usually represent events (components) for which failure rate (likelihood of occurrence) data are available. "Pump fails to start," "pump fails to run," or "pump is out for maintenance" are examples of basic component failures represented by a circle. *Diamonds* are used to symbolize undeveloped events, in the sense that a detailed analysis into the basic failures is not carried out because of lack of information, money, or time. "Failure due to sabotage" is an example of an undeveloped event. Such events are removed frequently prior to a quantitative analysis. They are included initially because a fault tree is a communication tool and their presence serves as a reminder of the depth and bounds of the analysis. Most secondary failures are diamond events.

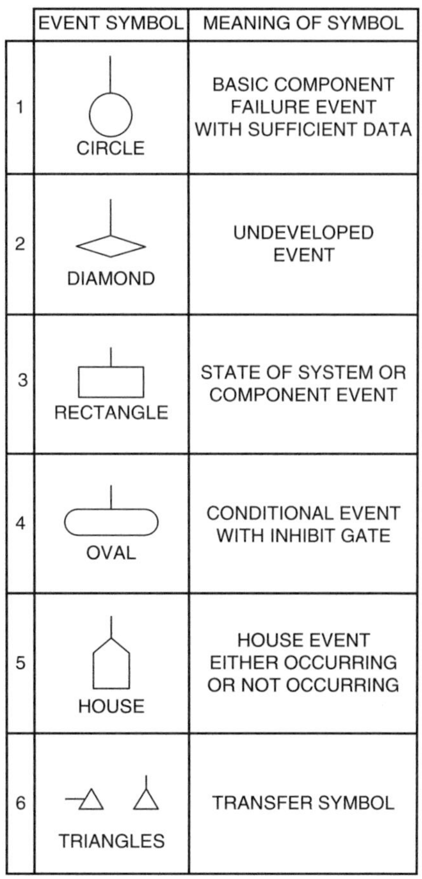

Fig. 12.6 Event symbols (Copyright© 1996. IEEE. All Rights Reserved)

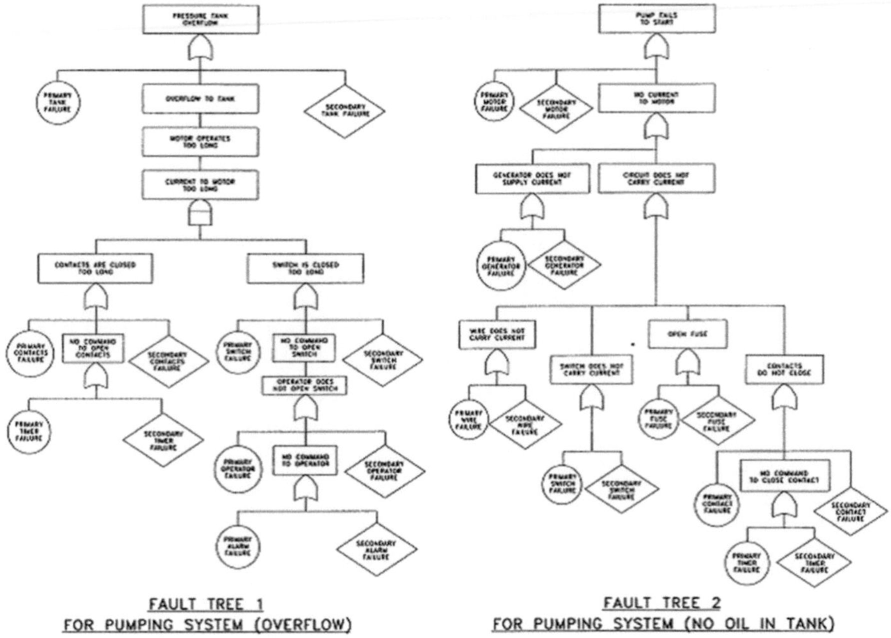

Fig. 12.7 Fuel pumping system—fault trees

12.7 Intelligent Building Automation Reliability

The term fault-tolerant intelligence computing can be defined as "the ability to execute specified algorithms correctly regardless of hardware errors and program errors." The understanding of fault tolerance can be helped first by understanding building intelligent system reliability, data protection, and maintainability needs. Faults in intelligence systems can be defined as "the deviation of one or more logic variables in the computer hardware from their design-specified values." A logic value for a digital computer is either a zero or a one. A fault is the appearance of an incorrect value such as a logic gate "stuck on zero" or "stuck on one." The fault causes an "error" if it in turn produces an incorrect operation of correctly functioning logic elements. Therefore, the term "fault" is restricted only to the actual hardware that fails.

Faults can be classified in several ways. Their most important characteristic is a function of their duration. They can be either permanent (solid) or transient (intermittent). Permanent faults are caused by solid failures of components. They are easier to diagnose, but usually require the use of more drastic correction techniques than do transient faults. Transient faults cause 80–90 % of faults in most systems. Transient faults or intermittents can be defined as random failures that prevent the proper operation of a unit for only a short period of time, not long enough to be tested and to become permanent. Permanent fault-tolerant techniques must then be used for system recovery.

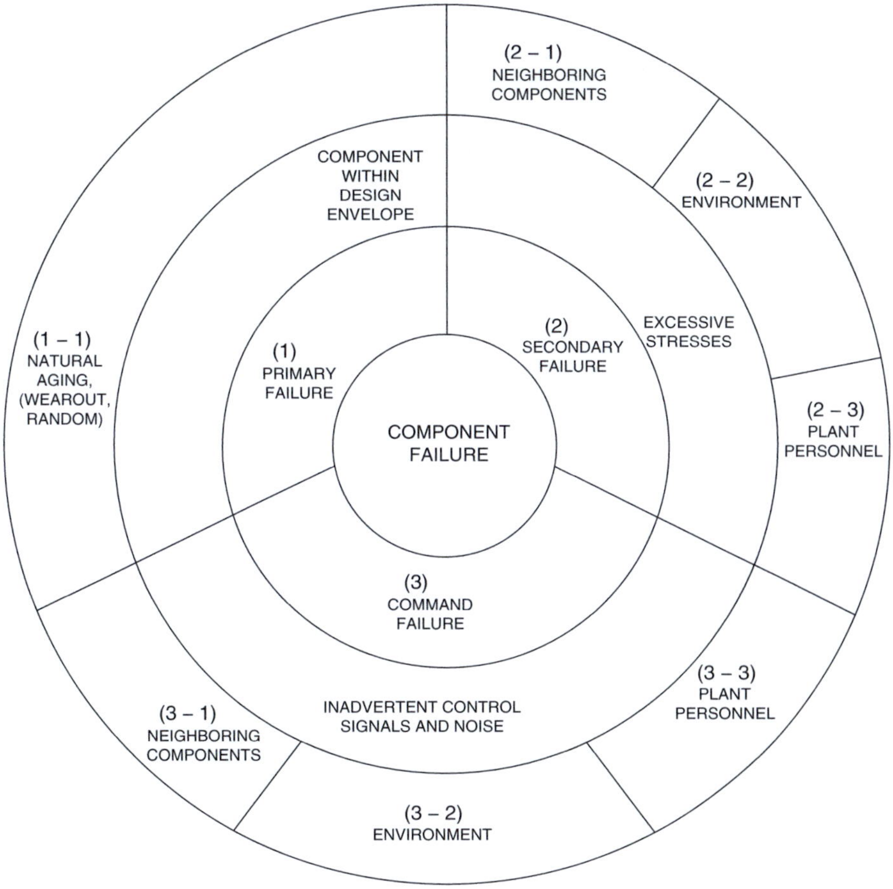

Fig. 12.8 Component failure characteristics (Copyright © 1996. IEEE. All Rights Reserved)

12.7.1 Computing Reliability

The goal of system reliability or of fault-tolerant computing is either to prevent or allow recovery from faults and continue correct system operation. This also includes immunity to software faults induced into the system. To achieve high reliability, it is essential that component reliability be as high as possible. As the complexity of computer systems increases, almost any level of guaranteed reliability of individual elements becomes insufficient to provide a satisfactory probability of successful task completion. Therefore, successful fault-tolerant computers must use a judicious selection of protective redundancy to help meet the reliability requirements. The three redundancy techniques are:

1. Hardware redundancy
2. Software redundancy
3. Time redundancy

These three techniques cover all methods of fault tolerance. Hardware redundancy can be defined as any circuitry in the system not necessary for normal computer operation should no faults occur. Software redundancy, similarly, is additional program instructions present solely to handle faults. Any retrial of instructions is also known as time redundancy.

12.7.2 Hardware Redundancy

This is the set of all of those hardware components that need to be introduced into the system to provide fault tolerance with respect to operational faults. These components would be superfluous should no faults occur, and their removal would not diminish the computing power of the system in the absence of faults.

In achieving hardware fault tolerance, it is clear that one should use the most reliable components available. However, increasing component reliability has only a small impact on increasing system reliability. Therefore, it is more important to be able to recover from failures than to prevent them. Redundant techniques allow recovery and thus are very important in achieving fault-tolerant systems. The techniques used in achieving hardware redundancy can be divided into two categories: static or masking redundancy and dynamic redundancy.

Static techniques are equally effective in handling both transient and permanent failures. Many different techniques of static redundancy can be applied. The simplest or lowest level of complexity is by a massive replication of the individual components of the system. For example, four diodes connected as two parallel pairs that are themselves connected in series will not fail if any one diode fails "open" or "short." Logical gates in similar arrangements can also guard against single faults and even some multiple faults for largely replicated systems.

More sophisticated systems use replication at higher levels of complexity to mask failures. Masking is virtually instantaneous and automatic. It can be defined as any computer error correction method that is transparent to the user and often to the software. Redundant components serve to mask the effect of hardware failures of other components. Instead of using a mere massive replication of components configured in fault-tolerant arrangements, identical nonredundant computer sections or modules can be replicated and their outputs voted upon. Examples are triple modular redundancy (TMR) and more massive N-modular redundancy (NMR) where N can stand for any odd number of modules.

In addition to component replication, coding can be used to mask faults as well as to detect them. With the use of some codes, data that have been garbled, that is, bits changed owing to hardware errors, can sometimes be recovered instantaneously with the use of redundant hardware. Dynamic recovery methods are, however, better able to handle many of these faults.

Higher levels of fault tolerance can be achieved more easily through dynamic redundancy, implemented through the dual actions of fault detection and recovery. This often requires software help in conjunction with hardware redundancy. Many dynamic methods are extensions of static techniques.

More effective use of massive redundancy in components often results from dynamic control. Redundant modules or spares can have a better fault tolerance when left unpowered until needed, because they will not degrade while awaiting use. This technique—standby redundancy—can use dynamic voting techniques to achieve a high degree of fault tolerance. The union of dynamic and standby redundancy is referred to as hybrid redundancy. Hybrid redundancy requires additional hardware to detect and switch out faulty modules and switch in good spares.

Error detecting and error-correcting codes can be used dynamically to achieve fault tolerance in a computing system. Coding refers to the addition and rearrangement of the bits in a binary word that contains information. The strategy of coding is to add a minimum number of check bits (the additional bits) to the message in such a way that a known degree of error detection or correction is achieved. Error detection and correction is accomplished by comparing the new word, which should be unchanged after transmission, storage, or processing, with a set of allowable configurations of bits. Discrepancies thus discovered signal the existence of a fault that then can be corrected if enough of the original information remains intact—that is, the original binary word can be reconstructed provided a set number of bits in the coded word have not changed. Encoding and decoding words with the use of redundant hardware can be very effective in detecting errors. Often, through hardware or software algorithms, incorrect data also can be reconstructed. Otherwise, the detected errors can be handled by module replacement and software recovery actions. The actions taken are dependent on the extent of the fault and of the recovery mechanisms available to the computing system.

12.7.3 Software Redundancy

Redundant software plays a major role in most fault-tolerant computers. Even computers that recover from failures mainly by means of hardware use software to control their recovery and decision-making processes. The level of software used depends upon recovery system design, and recovery design is dependent upon the expected type of error or malfunction. Different schemes have been found to be appropriate for the handling of different errors. Some types of error correction are most efficiently accomplished with hardware. Others need only software, but most use a mixture of the two.

For a functional system, that is, one without hardware design faults, errors can be classified into two varieties:

1. Software design errors
2. Hardware malfunctions

The first category can be corrected mainly by means of software. It is extremely difficult for hardware to be designed to correct for programmer's errors. Software methods, though, are often used to correct hardware faults, especially transient ones. There are several software errors that computers may be designed to detect, including the use of illegal instructions (i.e., instructions that do not exist), the use of privileged instructions without authorization, and address violations (i.e., reading or writing into locations that are beyond those of usable memory). These limits can often be set physically on the hardware. Computers capable of error detection cause interrupts, which route the program to specific locations in memory. The programmer, knowing these locations, can add his own code to branch to his specific subroutines which can handle each error specifically as he sees fit.

There are several methods of software recovery from software errors. As mentioned before, parallel programming where alternative methods are used to determine a correct solution can be used when an incorrect solution can be determined. Some less sophisticated systems print out diagnostics so that the user can correct the program off-machine. This should only be a last resort for a fault-tolerant machine. Nevertheless, a computer should always keep a log of all errors incurred, memory size permitting.

Preventive measures used with software methods refer mainly to the use of redundant storage. Hardware failures often result in a garbling or a loss of data or instructions read from memory. If hardware techniques such as coding cannot recover the correct bit pattern, those words will become permanently lost. Therefore, it is important to at least duplicate all necessary program and data storage so that it can be retrieved in case one copy is destroyed. In addition, special measures should be taken so that critical programs such as error recovery programs are placed in nonvolatile storage, that is, read-only memory. Critical data also should be placed in nondestructive readout memories, such as plated-wire memories.

The second task of the software in fault tolerance is to be able to detect and diagnose errors. Software error detection techniques often can be used to detect hardware faults that are transient. This is important since "a relatively large number of malfunctions are intermittent in nature rather than solid failures." Time-redundant processes, that is, repeated trials, shall be used for their recovery.

Software error detection techniques do not locate the sources of the errors. Therefore, diagnostic test programs are often used to locate the module or modules responsible. These programs usually test the extent of the faults at the time of failure or do periodic tests to determine malfunctions before they manifest themselves as errors during program execution. Almost every computer system uses some form of diagnostic routines to locate faults. In a fault-tolerant system, the system itself initiates these tests and interprets their results, rather than requiring the outside insertion of test programs by operators.

Part II
Integrating Nature and Understanding Behavior

Chapter 13
Integration of Mother Nature into Smart Buildings

Helen Santiago Fink and Ingrid Kaltenegger

Come forth into the light of things, let nature be your teacher. —*William Wordsworth*

13.1 Biophilia

13.1.1 Introduction: What Is Biophilia?

Running in a park, cuddling with a pet and enjoying the views of the mountains or the seaside are common behaviors that define human beings' affinity for nature. It is this that constitutes the concept of biophilia[1]—human beings' inherent relationship (or love from "philia") with the natural environment and other livings. Yet, biophilia goes beyond seemingly superficial interests to denote a human dependency on the natural environment for biological and physical existence. Human beings are part of the global biosphere and source essential resources to live and build our homes, businesses, and communities from nature. We are physically, economically, and socially intertwined with the natural environment for the air we breathe, raw materials we procure, and interactions we seek with other living species.

In many current societies, biophilia and the appreciation of nature's extensive services lay dormant. Our technological revolution, introspective design, and disjointed projects in our built environment and cities coupled with weak environmental educational curriculums account for much of this disconnect and apathy toward nature—yet at significant socioeconomic costs. We are trending today toward techno-/video-philia. Yet, technology can help bridge the gap between humans and nature by bringing aspects of the natural environment to where it may not be physically or economically possible. Through simulations and sensory applications,

[1] Biophilia was popularized by Harvard biologist Edward O. Wilson in his book Biophilia (1984) and elaborated thereafter in the The Biophilia Hypothesis (1993) with Yale's social ecologist, Stephen R. Kellert.

A. Ahuja, *Integration of Nature and Technology for Smart Cities*,
DOI 10.1007/978-3-319-25715-0_13

nature can be brought into inaccessible and closed spaces and contribute to environmental justice and greater understanding of the invaluable services the natural world provides to the human species.

Technological biophilia is a new term being given to the simulations of nature (by technology) to induce the cognitive, sensory, emotional, and physiological reactions humans experience in the actual natural environment, although not to the same degree. Technologies such as sky ceilings and digital cinema windows can replicate the sky and ocean environments, including their sounds and motions, to trigger relaxation and stress relief. Luminous virtual windows allow the perception of daylighting and can be customized to create any natural vista while creating an illusion of depth and space. Such technologies can form part of biophilic design to enhance the built environment through greater connectivity of interior and exterior spaces and between humans and nature.

Accepting that technology has now become integral in our lives, it behooves us all the more now to learn to live in harmony with the natural environment. We need to decouple (or de-link) our use of natural resources from the negative environmental impact underway. We need to better work with the finite supply of earth's resources and protect the vulnerability of the very elements such as water and air that we need to survive. Technology is playing a valuable role in this learning process and necessary societal change through the use of smart meters, environmental management systems, monitoring sensors, closed loop systems, and the like. Social media can alert us of new ideas and foster collective impact of human behaviors toward environmental stewardship and sustainable living. Technology's information flows and biocomputations can better integrate biophilic design into our built environments and help awaken biophilia among society to foster a sense of environmental consciousness. In doing so, we can mobilize individuals and communities to coexist with nature in a manner that makes use of our technological advances to promote human well-being, economic productivity, innovation, and smart sustainable lifestyles.

The argument for biophilia in this book is premised on its capacities for a better quality of life for all. We know the natural environment reduces stress, enhances physiological levels, heightens mental capacities, and accelerates curative and restorative properties.[2] Health care establishments have been pioneers in incorporating nature into their facilities to exploit the beneficial impact on the human body and its effects in reducing physical and emotional stress and ultimately medical costs. Nature is a fountain of health and offers the individual and communities an array of services and benefits if only the symbiotic relationship among all living things was more widely understood and integrated into our living, work, educational, administrative, and leisure spaces.

---

[2] Studies by Kaplan (1995), Kuo and Sullivan (1998, 2001), and Frumkin (2003) as well as Roger Ulrich's seminal study "*View through a Window May Influence Recovery from Surgery*" (Science (1984)), in which hospital patients covering from gallbladder surgery experienced reduced curative times and required less pain medication on account of simple views of trees from their room windows, underscore the value of nature on the human being.

It is time to arouse biophilia in the individual and society to stimulate environmental commitment and behavioral change to preserve the essence of life—nature itself. Technology can be a valuable tool for this purpose and facilitate information dissemination, as well as to monitor, report, and verify (MRV) biophysical conditions, consumption patterns, and human impact on natural resources. Technology is also a powerful educator and can connect numerous areas of knowledge to promote a multidisciplinary approach to learning and inspire creativity and innovation using the natural world as a reference. The Biophilia Educational Program[3] developed by the song artist, Bjork, incorporates technology, music, and the natural sciences through the use of apps/applications to communicate and integrate the concepts of nature into educational school curriculums. This integration and cooperation among disciplines better prepares us to tackle local and global challenges including air pollution, climate change, water scarcity, food security, and sustainable development.

13.1.2 Biophilic Design

Biophilic design is considered an ethic and process, rather than a technique that holistically integrates many of nature's qualities and functions into our buildings, cities, and lives. Compared to green design that aims at maximizing energy performance in buildings, biophilic design serves to unite humans and the natural world by creating spaces for access and sensory engagement to reap its salubrious benefits as well as optimize the possibilities of using nature as a reference point for shaping our communities and actions. Designing with nature provides the medium to revive and strengthen our biophilic tendencies and catalyze environmental consciousness toward behavioral change for broad environmental stewardship.

Roger Ulrich's studies catalyzed the development of "evidence-based design," whereby quantitative data and baseline measurements contribute to the prioritization of building systems and design elements used in building construction. Hospitals and more recently schools have taken the lead in embracing biophilia and incorporating greenery, views, and/or visual simulations (e.g., sky ceiling and daylighting, see Fig. 13.1) of the natural environment into their spaces to exploit the positive healing benefits for patients and heighten alertness of students. Other environments including offices and prisons have started to recognize the proven value of biophilic features, e.g., daylighting, to enhance mental processes and reduce aggression.

Bringing nature inside is the goal of biophilic design in buildings. And for cities, it means we are surrounded by natural elements, sounds, vegetation, and the like wherever we turn. Biophilic design is equally important in our homes where we spend 80–90 % of our time as well as in our neighborhoods and communities that shaped our norms and influence our actions. The more natural elements, biological materials, and ecological patterns we incorporate into current building design and

---

[3] http://biophiliaeducational.org/.

Fig. 13.1 Sky ceiling and daylighting

public spaces, the greater employee and student productivity levels, enhanced social serenity, and creative human activity.

The pillars of biophilic design comprise (1) nature in the space, whereby elements of the natural environment, e.g., vegetation, water, sounds, are integrated into the built environment and are most effective in generating biophilic benefits; (2) analogues of nature, which refers to the use of materials, patterns/fractals, and illustrations that represent and/or simulate nature and its qualities. Fractals are the designs of nature that offer insight into life's geometry and functionality and offer tremendous opportunity to learn from the efficiency and resiliency of living things; (3) *nature of the space*, which suggests evoking the ambiance of human evolution in the savannah landscapes and creating spaces that offer protection, inspiration, and tranquility to stimulate human creativity, environmental appreciation, and connectivity to other living things.

The following attributes contribute to nature's palette of opportunities for architects, planners, and designers to create invigorating and rewarding environments that reinforce human biophilic predispositions:

- Natural light, e.g., day/sunlight: increases serotonin, a neurotransmitter, in the body that alleviates pain throughout the nervous system; daylighting in office and school buildings demonstrate increased student performance and employee productivity and a lower rate of absenteeism, all of which have implications on financial savings and economic benefits.
- Vegetation and greenery: documented to expedite healing and reduce stress and promote employee productivity either by natural or simulated views of trees, parks, and the like as well as reduce energy consumption when applied as green infrastructure (see Fig. 13.2); access to natural environments facilitate recovery from mental fatigue.[4]

[4] Attention Restoration Theory asset exposure to nature heightens mental acuity and focus (Kahn 2008) (Rocky Mountain Institute and Carnegie Mellon University 2004).

Fig. 13.2 (**a**) Example of nexus of interior—exterior natural spaces; (**b**) Example of vertical garden. (**c**) Example of green façade and vertical garden

- Discovery and mystery: nature inspires artistry, imagination, and innovation. Yet, sadly, Earth Institute at Columbia University uncovered among American K-12 students a lack of teaching about the values and benefits of the natural environment and rather a counterproductive focus on the dangers of nature toward society.

Other aspects of biophilic design can include transition and open spaces that provide access to the outdoors and natural elements to create opportunities for nature's sounds, smells, and climatic temperaments to enhance connectively with the human person. The use of seasonal representations and local topographical

Table 13.1 Biophilic design applications

Biophilic design attributes	Building applications	Design spaces
Natural and daylight	Glazing of doors, windows, walls, skylights	All interior rooms, courtyards, patios, terraces, balconies, southern/best site orientation
Vegetation and greenery	Living/green walls, green roofs, interior potted plants, window boxes, greenery simulations, green/blue/terra cotta colors for walls and ceilings	Interior and exterior spaces, gardens, courtyards, patios, terraces, balconies, building facades, surface areas
Discovery and mystery	Atriums, cathedral-like ceilings, interior balconies	Interior and exterior spaces
Seasonal and topographical elements	Porches, balconies, terraces, water walls, natural materials	Open and transition spaces, passages connecting interior and exterior spaces

Source: author

elements, e.g., moving water, is a design feature that helps strengthen our bond to the multitude of local biodiversity. Actual versus simulated symbols or illustrations of nature are both valuable in design, although actually seeing or touching the real thing has greater impact. Recognizing that all settings may not lend themselves to direct access to nature, the use of illustrations and murals of skies, landscapes, and vegetation can produce similar positive effects and economically and structurally facilitate connectivity with the natural environment at most all locations and for a broader public.

Biophilic design benefits from an understanding of the local microclimate to ensure the architecture is in balance with the natural surroundings. Consideration of local resources, materials, and how other living species exist in the particular ecosystem can provide valuable information to guide human development toward a more resilient and cost-effective trajectory. Buildings constructed primarily of glass and steel in hot climates may not be the best choice and normally warrant significant budgets for cooling and mechanical systems. Such climates would reap greater socioeconomic and environmental benefits from the introduction of local vegetation as design complements to induce evapotranspiration for naturally cooling. Similarly, water-scarce regions need to design for local realities and incorporate systems for water harvesting, fog capture, gray water recycling and ecosystem conservation (Table 13.1).

Biophilic design also incorporates the use of fractals or nature's patterns and sequencing of systems that form part of the physical and biological processes of the natural environment. This creates a form of organic architecture used traditionally by indigenous groups and some architects, including Antonio Gaudi, Frank Lloyd Wright, and Frank Gehry, to speak to the inherent biological affinities of the human being (see Fig. 13.3).

Fractals illustrate nature's intricacies and allude to its internal processes and thereby are windows of opportunity for designers and other disciplines to learn from the elegant efficiency of the natural environment. Trees have been a source of inspiration for architects and designers for centuries on account of their strong support

Fig. 13.3 (**a**) Antonio Gaudi's Sagrada Familia Basilica, (**b**) Fractals of Romanesco Broccoli

structure. Antoni Gaudi's design for the Sagrada Familia Church (Fig. 13.3a) invoked the use of trees to create dendriform structural columns to support the crown load of intertwined hyperbolid vaults mimicking the tree branches role in nature. By inclining the columns and applying the rotational variations of tree truck fractals to their design, they were able to support the mechanical traction forces of the structure. Present-day technology allows designers to better visualize and optimize the multiple fractal combinations found in nature (see Fig. 13.3b) to create more resilient and smarter buildings.

People tend to respond more positively to natural forms than to grids and checkerboards and may be why fractals have been used for generations by indigenous cultures to create designs for colloquial attire, living settlements, tools, and ceramics—reinforcing biophilia's presence throughout societies and time.

Current and future societies stand to benefit in multiple ways from embracing nature and a biophilic design ethic for smartening our built environments, enhancing human functionality and well-being and sparking new technological innovation. Close examination of the patterns, forms, and functions found in our ecosystems can allow us to learn from 3.8 billion years of earth's history and encourage replication of nature's many lessons and experiences for better quality of life, sustainable development, and a harmonious equilibrium within the context of our planetary boundaries. Such thinking forms the basis for the growing body of science known as biomimicry spearheaded by Janine Benyus.[5]

13.2 Biomimicry

To learn from nature lies the innovation of the future. Smart buildings and smart cities can constructively address present-day challenges, only if they can harness and utilize essential resources in ways that minimize greenhouse gas emissions and guarantee resiliency of its inhabitants and habitats. How do other living things

---

[5] *Author of 1997 book, Biomimicry: Innovation Inspired by Nature.*

source and optimize water, energy, and food for existence? What adaptive measures do other species undertake to safeguard against disaster? The answers are embedded in our natural environment. How we can recognize and effectively utilize this encyclopedia of biodiversity and knowledge for sustained human development? According to ecologist Stephen Kellert, we destroyed nature at our own peril, but rather how can we inculcate nature into our lives for its wisdom and our human longevity?

Biomimicry offers a new way of perceiving our natural world. The science of biomimicry strives to replicate natural and circular processes to optimize resource use and strengthen resiliency in our societies. Nature's modus operandi or "life's genius" creates life and also contributes to sustaining life of all living species on Earth. Biomimicry views nature as "model, measure, and mentor." By gleaning nature's best practices and intricate capacities of efficiency and multifunctionality, we can constructively improve our technological, industrial, and mechanical systems to sustainably support human societies and the resources we depend upon rather than simply exploit its richness. The fundamental principles of biomimicry encourage designers to mimic natural form, natural processes, and natural ecosystems. Nature's forms offer efficiency in function. The sleek form of the kingfisher's beak guided engineers' improvements of Japan's bullet train to reduce noise levels and decrease energy consumption by 15 % while increasing velocity by 10 %. Nature's processes optimize resources in a closed loop without the use or creation of toxins. The field of green chemistry has embarked on mimicking such processes for industrial and pharmaceutical production as examples resulting in significant economic savings for both businesses and consumers. Nature's processes offer solutions for the reduction of greenhouse emissions and their development are being encouraged by foundations and governments as positive alternatives to the use of geoengineering to address our climate change urgency. As in nature, biomimicry goes beyond form and function to also constructively impact the ecosystem of the product or design being developed. It embraces a life-cycle approach and positively contributes to the other species and elements in the chain of production and life.

The company Calera of California which has mimicked the processes inherent in coral reefs. By producing calcium carbonate from seawater or industrial flue gas and similar waste emissions, it creates cement for a number of building products and simultaneously sequesters CO2 in a productive and safe manner (see Fig. 13.4). Learning how (marine) organisms make use of biomineralization through the production of hardening minerals can advance the collaborative research potential of the earth sciences, biological, and chemistry disciplines to create new products that embody nature's processes with more resource efficiency and resiliency.

Calera's nature-inspired technology could significantly reduce the amount of carbon emissions generated by the concrete and cement industries, which have increased substantially during the past decades on account of increased urbanization and rapid construction in China and other developing countries. Cement production, a necessary binding agent for concrete, accounts for 5–6 % of global human-generated emissions and is growing at 2.5 % annually. Such innovation is

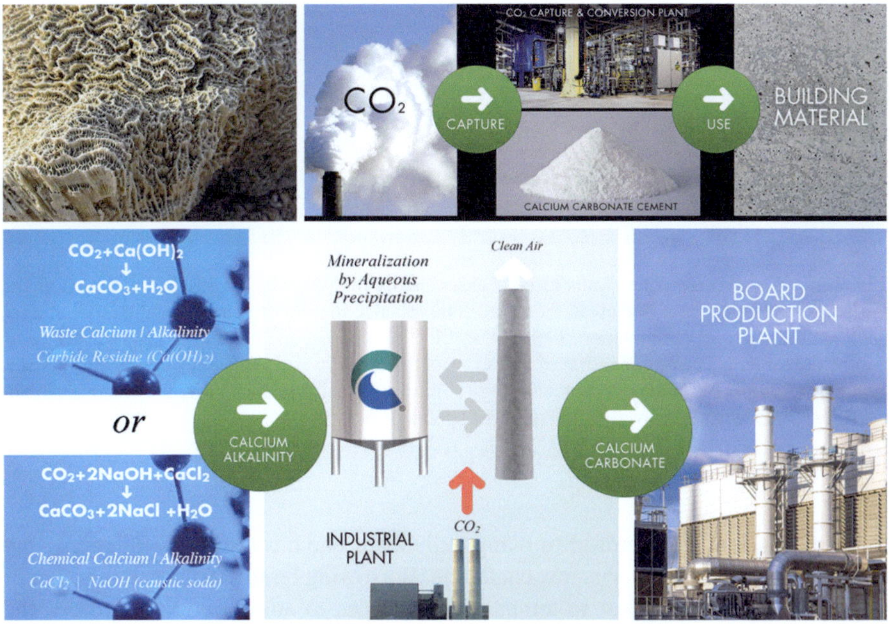

Fig. 13.4 Coral reef exoskeleton produced utilizing CO2 (Source: Calera, http://www.calera.com/—mimics the process of coral reefs with its innovative CO2 sequestering technology)

revolutionary considering concrete is the second most widely consumed human product after water![6]

The lotus leaf properties have been incorporated into building technologies and biomimetic materials for the purposes of repelling water and cleaning surfaces. The plant's micro- and nanostructures contain waxy properties (see Fig. 13.5) that have guided engineers to replicate its rough surface characteristics for paints and coatings as well as glass to benefits from the polarity of water and its cohesive properties to clean surfaces. Self-cleaning glass is commercially available and contains a hydrophilic layer to prevent water adhesion and a photocatalytic layer to oxidize organic matter. Similarly, lotus effect coatings exist for buildings and facades to induce self-cleaning yet have met challenges in durability of its effectiveness once surface damage occurs. Nature's lotus leaf continuously regenerates itself, thereby preserving its super hydrophobic qualities, whereas current human technologies still warrant further development to fully achieve nature's superior capacities.

Applying biomimicry to building architecture can embody structural, aesthetic, and mechanical qualities learned from our natural habitats including our human bodies. The excessive cooling common in many retail and office buildings in the United

[6] Columbia University's Earth Institute, http://blogs.ei.columbia.edu/2012/05/09/emissions-from-the-cement-industry/.

Fig. 13.5 *Right image*: The "lotus effect," associated with Wilhelm Barthlott, a biologist at the University of Bonn, and borrowed from the lotus plant's superhydrophobic qualities is another such naturally inspired innovation. *Left image*: Electron microscope photograph of the surface of a lotus flower leaf. The combination of surface roughness and water-repellent wax crystals gives it superhydrophobic properties. Nanotechnology solutions for self-cleaning, dirt, and water-repellent coatings; http://www.nanowerk.com/spotlight/spotid=19644.php

States is not only a discomfort to many individuals and a wasteful use of energy but also often an operational systems challenge. Borrowing lessons from the function of human muscles that adjust to temperature changes to attain optimal comfort, the earlier New York architect firm of Decker Yeadon has developed a prototype of a homeostatic façade system[7] that acclimates itself without mechanical or human programming. Dielectric elastomers prepared with silver electrodes are shaped into ribbons and incorporated into a double-skin glass façade that react to solar radiation by contracting and expanding, thus shading the building or not. Reduced mechanical system costs, energy conservation, lower emissions, and building occupant comfort are among the benefits resulting from this engineered muscle analogue.

The Eastgate Building in Harare, Zimbabwe, is a pioneering example of how nature's other living species can guide building design and development. The architect Mick Pierce applied the principles used by termites to construct their towers to design a passive heating and cooling system for the 55,000 m^2 mixed-use building. The nature inspired design replaced the standard mechanical air-conditioning system generating a savings of $3.5 million dollars and 35 % less energy consumption than comparable structures in Harare by making use of convection currents and the building's concrete mass for thermal storage (see below Fig. 13.6).

The observation of air currents as applied to the principles of convection have provided designers the opportunity to create natural ventilations systems to complement other building technologies.

The Anaheim Regional Transportation Intermodal Center (ARTIC)[8] in southern California uses the principles of convection to create a natural cooling system that conserves 50 % of the energy consumption in 67,000 ft^{2}of space. As a major transportation

[7] http://www.asknature.org/product/0c2046bdbcb82fbcec31c8b4030f6e6b.

[8] http://www.hok.com/design/type/aviation-transportation/anaheim-regional-transit-center-artic/; http://www.articinfo.com/.

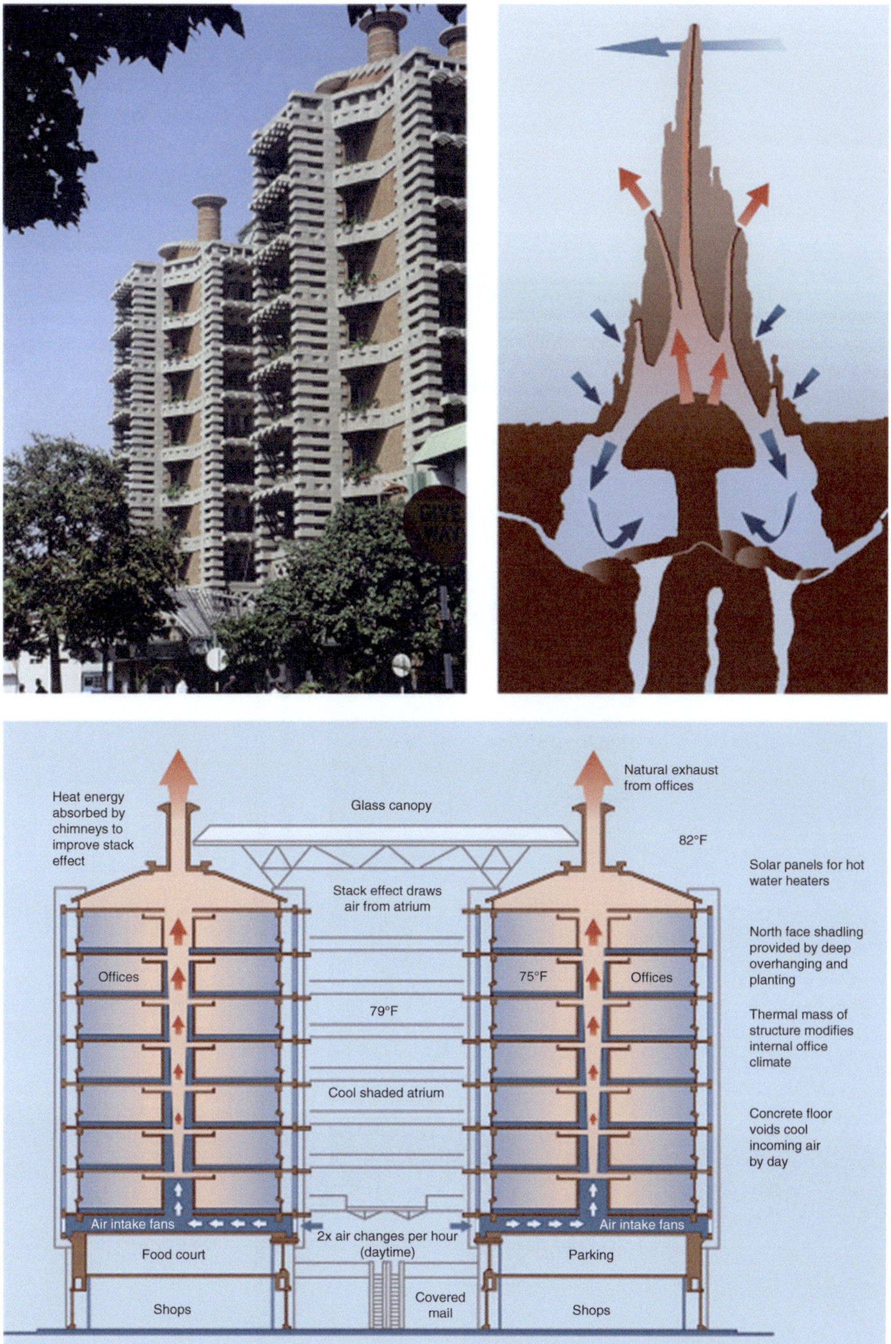

Fig. 13.6 Harare's Eastgate Building (Source: http://ehp.niehs.nih.gov/wp-content/uploads/2013/01/ehp.121-a18.g004.png)

Fig. 13.7 National Oceanic and Atmospheric Administration Daniel K. Inouye Regional Center, Pearl Harbor, Hawaii, USA (Source:http://www.hoklife.com/2013/07/16/paul-woolford-on-noaas-new-pacific-regional-center-campus-in-hawaii/)

hub, ARTIC is also representative of the necessary mixed use developments of the future that integrate the built and natural environments optimized by the use of technology to monitor and synchronize operating systems for greater efficiency of resources while creating urban spaces that encourage sustainable human behaviors. The Parkview Building in Beijing requires no air conditioning to cool its 200,000 m^2 but rather lies on an exterior envelope that creates an insulting microclimate adaptable to the day and seasons benefiting again from the principles of air convection and passive measures for natural ventilation.

The principles of biomimicry are embodied in the National Oceanic and Atmospheric Administration (NOAA) Regional Center located at Ford Island in Honolulu[9] and the site of Pearl Harbor (see Fig. 13.7). Designers took guidance from the local microclimate to create a space that optimizes the benefits of daylighting, seawater cooling, and gray water recycling. Inspired by the native trees and function of the leaves to reflect, transmit, and internally scatter light, the designers were able to illuminate a space of 350,000 ft^2 system with minimal electrical lighting by using a grid of apertures and reflectors to distribute the daylight as needed. The project's application of life principles contributed to the reduction of lighting loads of more than 50 % and energy conservation of 33 %.

China's building boom has inspired local creativity and growing talent among designers to look to the experiences of nature to address the country's growing pains. The Absolute World Towers, designed by Beijing MAD Architects, break the traditional skyscraper liner design by embodying an organic and even curvaceous form that inspires human interest and captured the local real estate residential market. The design evokes motion and piques human imagination symbolic of the simplicity and elegance of nature (see Fig. 13.8) and biomimicry's fundamental principles. The creativity of the design in turn pushed the envelopes of engineering and building trades to innovate practices to bring the design to structural life. The Towers' unique aesthetic and emotion-inspiring design now punctuate the municipal skyline and are giving definition to the surrounding urban space.

As our world becomes more populated, adding 2.4 billion by 2050 to reach 9.6 billion, and more urbanized at 70 % by 2050, the built environment will become an even more dominating feature of human evolution. We stand to benefit substantially from understanding how nature and other living species address similar challenges. The design of cities and living habitats will need to evolve and transform within future resource constraints. New technologies that allow buildings to breathe and react to movements of the sun[10] can provide creative means of natural ventilation and maximizing solar gains. Such technologies will lessen the urban heat island effect and mitigate air-conditioning needs in our buildings—all in a manner reflective of environmentally inspired architecture. With rapid urbanization, the growing demand for energy, water, food, and other fundamental life-sustaining resources

---

[9] http://www.hoklife.com/2013/07/16/paul-woolford-on-noaas-new-pacific-regional-center-campus-in-hawaii/.

[10] Doris Sung, Ted Talk 2012, http://www.ted.com/talks/doris_kim_sung_metal_that_breathes/transcript?language=en.

Fig. 13.8 Absolute World Towers cantilevered balconies

will only be met if and when individuals and societies learn to respect the natural environment and start embracing nature and its principles to optimize its indispensible human benefits in the immediate present and the long term.

13.3 Economics of Biophilia[11]

In recent years, architects, planners, and designers have started to incorporate biophilic attributes into building design moving toward a more holistic and comprehensive understanding of what constitutes a smart building. In 2012, the International Federation of Consulting Engineers (FIDIC) published a Sustainability Logbook as guidance for the profession to address a number of key criteria in the urban and built environment, including issues of biodiversity. However, the practice of preserving natural habitats, maintaining ecological corridors, and incorporating local vegetation and natural sounds is far from mainstream and still lacks broad acceptance by the multifaceted building industry. Despite the existing body of evidence in favor of nature's optimizing human benefits, the reality of our world is largely dictated by economics, and those of the near term, without considering the opportunity and imbued costs of present-day actions and investments. A broader economic perspective is required to provide a more informed position on the monetary value of biophilic design in our cities and built environment.

Considering the substantial employee cost of a company/organization's budget of over 80 % (see Fig. 13.9), investing in biophilic design is a cost-effective strategy

[11] Terrapin Bright Green, LLC (2012).

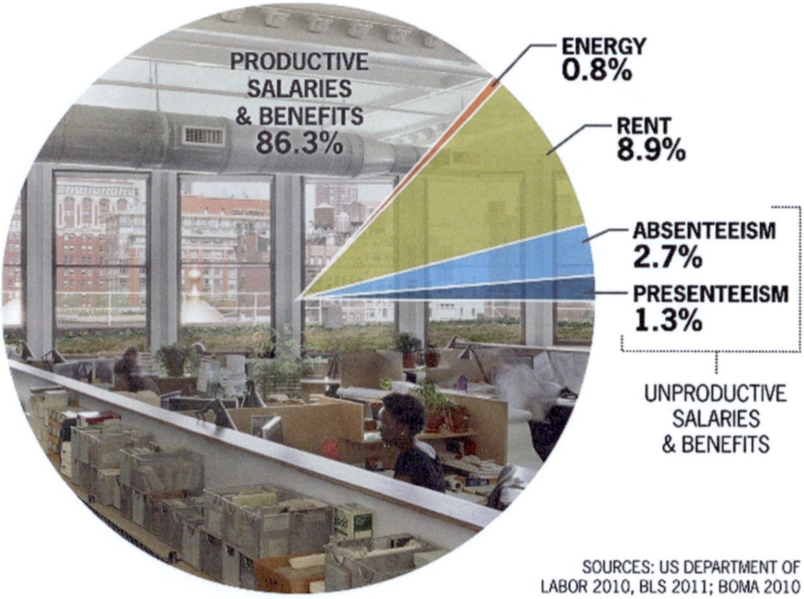

Fig. 13.9 Company employee costs

to optimize worker productivity and enhance the office ambiance. Estimated benefits of over $3.6 million per year for middle-size employers have been documented[12] as a result of increased productivity and reduced absenteeism associated with reduced stress levels. Medical facilities and hospitals are already aware of the economic benefits evidenced by reduced recovery times after surgical procedures and overall lower costs on account of faster restorative capacities induced by a biophilic environment. The biophilic benefits for schools and learning environments as well as societies at large are significant. The incorporation of daylighting in school buildings heightens cognitive capacities resulting in students' tests scores improvements between 7 and 26 %, and broad macroeconomic benefits from reduced school drop rates and better skilled labor forces.[13] Public facilities including social housing and prisons also benefit through reduced levels of aggression, domestic violence, and number of felonies. The real estate market has traditionally benefited from higher property prices for those with accessibility to parks, lakes, and other ecosystems. More recently, studies indicate higher consumer sales (40 % increase) in retail buildings with daylighting and natural elements.

[12] Terrapin Bright Green, LLC (2012), The Economics of Biophilia.

[13] The Alliance for Excellent Education (2007) estimates that if the students who had dropped out of the class of 2007 had graduated high school, the national economy would have benefited, from an additional $329 billion in income over their lifetimes.

13.4 The Natural Way Forward

The solutions to many of our present- and future-day (urban/life) challenges already exist. They are both technological and biophysical in essence benefiting from the efficient, innovative, and resilient properties of each domain. The science of biomimicry offers many opportunities yet to be discovered, inasmuch as individuals embrace the ingenuity of the natural environment through awaken biophilia for its answers. Clearly, biophilia on its own will not solicit the behavioral change toward smarter (and sustainable) lifestyles necessary of individuals and societies alike. The appreciation (or love) of nature can be catalyzed in the design of our buildings and cities but must be complemented as urgently in our education curriculums from onset as well as in the development of technologies, adoption of government policies, availability of financial incentives, promotion of information and awareness campaigns, and cultivation of new social and community norms.[14]

Quantitative evidence now exists to make the business case for employing biophilic design in our buildings and cities to enhance human health, address environmental urgencies, and invigorate economic productivity. The challenge lies in overcoming business as usual practices within the traditionally fragmented building industry and addressing the split incentive-like obstacles among investors to pursue a truly "smart" trajectory. Only then can we expect to reap the social and economic payoffs of optimization in and of our built environment. Real estate developers and investors among the wider population stand to profit from buildings that incorporate passive design, green infrastructure, renewable energy sources, and connectivity to urban services and natural elements. Within the global climate change agenda, the building sector remains a "low-hanging fruit" for mitigating greenhouse gas emissions, while its multiplier potential (cobenefits) for advancing socioeconomic change among societies in live, work, learn, and leisure environments is equally as valuable and mutually reinforcing. Policies such as the European Buildings Directive is one example of normative action that aims to catalyze market forces to provide energy efficient, low carbon, and zero-energy (new) buildings. We need to maximize the potential of our buildings as primary human habitats to enlighten individuals and transform (societal) behaviors toward smarter sustainable lifestyles. Biophilia offers us a blanket of assurance to integrate the environmental, social, technological, aesthetic, and economic for the benefit of all.

13.5 Green Areas and Roofs, Green Facades and Walls, Urban Gardening, and Green Corridors

Not only bringing nature inside, as it is the goal of biophilic design in buildings, is important but also the nature that is outside of the building matters. Integrating nature into the city can have many facets: from parks and green corridors to green

[14] See research article "Promoting behavioral change towards lower energy consumption in the building sector" (Santiago Fink (2011)).

Fig. 13.10 Climate protection for everyone: the city of Vienna provides funding for green facades, which greatly contribute to a good microclimate (Source: https://www.tuwien.ac.at/aktuelles/news_detail/article/9650/)

roofs and green facades, to vertical farms and urban gardening. This chapter tries to shed light on those different facets.

13.5.1 Green Areas in Cities

What is considered a "green space" or "green area" in a city? When we are talking about a smart space, we are talking about planned naturally green spaces in a city; this can be parks, gardens, and squares but also green roofs and facades. It has to be more than some streets that are lined by trees but also these contribute to the greening of a city.

Statistics[15] show that people prefer to live in cities where they can breathe, where they can see trees and plants and flowers. A good example is Vienna as one of the greenest cities in the world due to continuous planting of new avenue trees and municipal funding for projects to green up courtyards and facades (Fig. 13.10).

[15] http://www.skyscrapercity.com/showthread.php?t=1660203.

Fig. 13.11 Highline New York (Source: http://www.dezeen.com/2013/06/04/stephen-burks-the-high-line-new-york/)

Cities that are ranked among the world's most livable cities like Sydney (46 %) and Stockholm (30 %) have a higher percentage of green space and consume lowest energy to manage storm drain to sewage plants compared to other cities.

Other nonscientific or technological advantages of a green space, integrating nature into the city, is that nature in the city makes it more beautiful and attractive for people (and investors!). Who prefers to live in a city, gray and full of buildings, with no green spots and almost no animals (Figs. 13.11 and 13.12)?

Benefits that go hand in hand with a more beautiful environment are improved air quality (capturing airborne pollutants and noxious gases) and a moderation of the urban heat island effect (cooling down the air temperature). This helps with reducing energy and maintenance costs of air handlers, air filters, purifiers, pumps, and compressors.

13.5.2 Filtering (fine) Particulate Matters and the Greenhouse Gas Effect

Among air pollutants, particulate matter (PM) is considered to be the most serious threat to human health. Particulate matter, also known as particle pollution is composed out of several "ingredients," such as acids, organic chemicals, and particles from dust or soil (see also: http://www.epa.gov/pm/). There is a differentiation between "inhalable coarse particles" (which are larger than 2.5 μm and smaller than 10 μm in diagram) that can be found near roads or industry sites and "fine particles,"

Fig. 13.12 Promenade Plantee in Paris (Source: http://europeantrips.org/promenade-plantee-the-first-elevated-park-in-the-world.html)

being smaller than 2.5 μm in diameter and mostly being formed when gases from power plant and cars react in the air.

Plants help reduce the levels of $PM_{2.5}$, but the effect is hardly measurable. It could be shown that biogenic particles, which are released into the air by trees, can react as nuclei for particle matters, but mostly for particle bigger than PM_{10}.

The graph in Fig. 13.13 shows the sizes of different particles:

13.5.3 Cooling Effects

Several studies,[16] conducted in different areas of the world, show the effect of green areas on microclimate and temperature difference from up to 5–6 °C less temperature in green spaces compared to sealed areas. Vegetation can cool down cities through its evaporation and therefore contributes to reducing the urban heat island (UHI) effect and energy required to cool the skin of buildings due to shades and surroundings.

Correlations between the shape of a park and the temperature outside of the park could be shown.[17] Most important seem to be the canopy of the trees and the height of the trees, but also the size of the parks, their shape, and the composition (trees, bushes, grass, etc.) play an important role. The bigger the park, the bigger the range of the cooling effect; it can be as much as 300 m from the edge of the green space but is also depending on the building structure surrounding the park.

---

[16] Bencheikh and Rehid (2012); Huang et al. (2008); Feyisa et al. (2014).

[17] Feyisa et al. (2014).

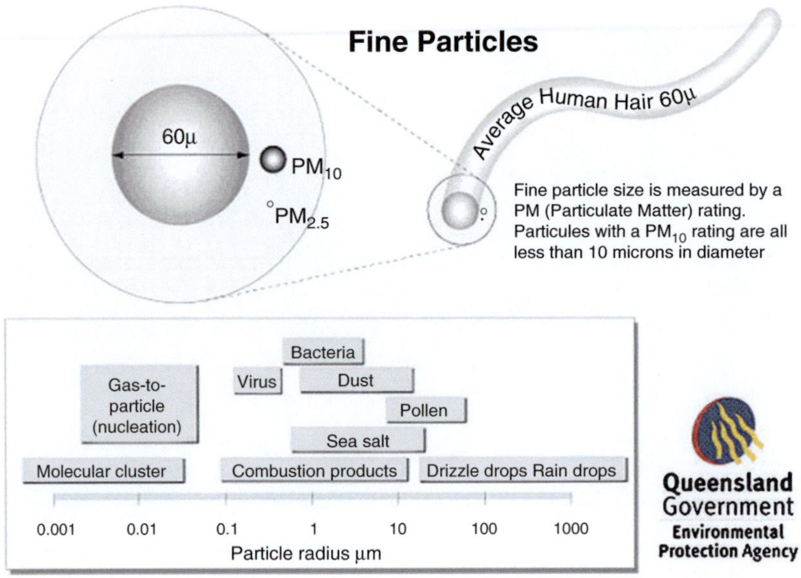

Fig. 13.13 Size of different fine particles (Source: QLD EPA at http://www.heggies.com/extranet/willoughbyAQ/ParticulateMatter.aspx)

A big influence is also given by the different times at day and night: whereas at daytime the biggest cooling effect is given by (high) trees with a wide canopy, they allow ventilation by night but only low air exchange with the atmosphere. Grassy underground heats up at day and cools down more at night but with only little airflow.

13.5.4 *Improved Health and Well-Being and Safety*

Green in a city can also contribute to more well-being of people, be it due to the better air quality, the cooler environment, the noise reduction through vegetation and recreation areas, but also by the potential to serve as community hubs, increasing social cohesion, a sense of community, and greater public safety. Numerous studies have shown that, in general, open green spaces that allow for visibility increase the feeling of personal safety as the person can better identify potential risks.[18]

These studies also reveal that in contrary to what crime specialists often suggest (to remove as much vegetation as possible), vegetation is in some way linked to lower levels of crime in residential neighborhoods. The reasons are twofold: a

[18] https://depts.washington.edu/hhwb/Thm_Crime.html.

greater sense of safety on the one hand and less aggressive behavior on the other side. Trees and other vegetation make a dull, barren space—a pleasant place where people like to be, feel comfortable, and foster informal surveillance that prevents crime.

So when we stroll through a city, which types of green can we see? The following pages try to give an overview on the different types of green (besides parks) that can be present in a city.

13.6 Green Roofs

First of all, let us have a look at the definition of a green roof: a green roof is a roof that is partially or totally covered with different kinds of vegetation, growing on a special medium. Despite its name, a green roof also shows many other colors than green: especially succulents change their leaf color throughout the year and appear in red, orange, and yellow. And flowers show their blossoms in all possible colors.

Two main types of green roofs are distinguished: extensive and intensive, but, of course, also combinations are possible. The difference lies in the thickness of layers and the plant depths.

An extensive roof (also referred to as "eco-roof" or "low-profile") has thinner and less numbers of layers, so therefore, it is much lighter, is less expensive, and requires very low maintenance. Extensive green roofs are the choice if there is limited or no human access and an ecologic cover of the roof is desired. Plants that grow best on such a roof are "Alpine-type" plants like sedums and other succulents, flowering herbs, and certain grasses and mosses as they tolerate drought, wind, frost, and heat. These plants are low growing with general maximum plant heights of 16–24". Extensive green roofs can be constructed on slopes up to 30°. Maintenance is relatively low but still from time to time the plants, the substrates, and the layers should be checked. Even if the right choice of plant has been made, watering might be necessary from time to time, so a water source should be nearby. In most cases, an irrigation system is advisable but also ponds or rainwater collectors on roofs can serve as a water source (Fig. 13.14).

An intensive roof (also referred to as "high profile") has a much broader variety of plants, as the growing media varies up to 15 feet or more. This type of green roofs is meant to serve as recreation areas and provide an interaction area between people and nature. Therefore, these roofs also have to be flat and need much more maintenance as other green roofs. An intensive green roof is more like a real garden and needs also watering and more care for the plants, which also includes the installations of an irrigation system.

Fig. 13.14 Extensive green roof (Source: http://www.zinco.ca/extensive.html)

13.6.1 Green Roof Construction

In general, a green roof consists of several components; Fig. 13.15 shows them all:

A green roof consists of several layers on top of the roof itself: the supporting structure (made of wood, steel, or other material), covered by a vapor control layer, some insulation, a waterproof membrane, and a protection mat which prevents roots from penetrating the lower construction. The drainage layer protects the plants from rotting by draining the water surplus. It also stores the surplus water to give it back to the plants in dry periods. A filter layer prevents the drainage layer from fine particles from the growing medium. On top of the growing medium (or substrate), the vegetation is planted, depending on the type of the green roof, climate, and preferences of the green roof owner.

13.6.1.1 The Growing Medium

A regular garden soil is in most cases too heavy for roofs, so a special substrate is used: a typical mix is to use 1/3 clean topsoil, 1/3 compost, 1/3 perlite, or other inorganic material. Nevertheless, there is no universal substrate; it has to be adjusted to the local climate and the plants on the roofs. In recent times, substrates containing recycling materials such as crushed porcelain are being tested to see if they can substitute or complement the usually available green roof substrates (Fig. 13.16).

Fig. 13.15 (**a**) Intensive green roof (Source: http://www.nbcnews.com/id/24056306/ns/business-going_green/t/green-roofs-popping-big-cities/#.Vbh3j7kw_Gg; Photo courtesy of Green Roofs for Healthy Cities). (**b**) Structure of a green roof (Source: https://greenerheights.wordpress.com)

13.6.2 Green Roof Plants

As a green roof is (almost) like a garden on the ground, the same factors have to be taken into consideration for choosing the appropriate plants: local climate, sun exposure, water requirements, and personal preferences. The difference to a traditional garden is the depth of the growing medium on the roof, which depends on how much weight the roof can carry.

Fig. 13.16 Recycled porcelain from MSU's Campus (Source: http://www.hrt.msu.edu/greenroof/research-projects/substrates.html)

On an extensive roof, the growing medium is quite shallow and inorganic, so some kind of succulents would be the best choice. These plants are quite frugal and need only low maintenance, which is ideal for extensive roofs that are not directly meant to be accessed a lot by people.

If the growing medium is deeper (up to 10 cm), one can choose between more plants: several types of grasses, herbs, and wildflowers are able to grow under these conditions. Herbs also have the nice side effect, besides looking nice and providing some ingredients for you meals: it adds nice fresh smell to the surrounding buildings!

In general, the more the thickness of the growing medium grows, the taller the plants can be. Special forms of green roofs are "container gardens," where plants are maintained in pots. This is a form of a "roof garden." One example, which became very popular, exists in Copenhagen, Denmark, the "tag tomat." This is an open-source project, which started on the roof of a garbage shelter in 2011 with a few self-watering containers. The Danish word "tag" means "roof" in English, so it means literarily "roof tomatoes." The goal behind this project was not to green roofs but to make things grow, using ones hands for that, and make social communities grow http://www.tagtomat.dk/short-summary-in-english/ (Fig. 13.17).

13.6.2.1 Smart Maintenance of Plants on Smart Green Roofs

Especially if we are talking about an intensive green roof, an irrigation system is necessary. On smart green roofs, this can be powered by solar energy. Another option is to collect rainwater in cisterns to irrigate the roof in dry spell.

The plants might also need some support in form of fertilizers but also pesticide to protect them against enemies. Pest find their way up to green roofs, no matter how high up they might be. Nevertheless, smart pest control maintains a healthy environment. For fertilizers, it should be used as economically and ecologically as possible. For both chemicals, it is necessary that they do not go into the runoff and are contained in the structure of the roof.

Fig. 13.17 TagTomat ("Roof Tomato"), Denmark (Source: http://www.tagtomat.dk/taghaven/)

Examples for green roofs can be found in the international green roof and green wall database: http://www.greenroofs.com/projects/.

More Sources:

http://www.greenroofguide.co.uk/biodiversity-and-planting/

http://greengarage.ca/greenroofs/plants.php

13.6.3 Benefits of Green Roofs

Green roofs serve several purposes for a building:

- They have an insulation effect which helps to save on energy in the building for heating as well as for cooling, leading to significant cost savings and an increasing comfort inside the building. A green roof protects the building from direct solar heat especially in the summer, whereas in winter, it reduces heat loss through its added insulation. Just covering a (black) roof, one of the biggest sources of heat in a city during summer, helps to a great extent to cool the building and lower the temperature of the air around the building, thus mitigating the urban heat island effect (Fig. 13.18).
- The greater insulation offered by green roofs can reduce the amount of energy needed to moderate the temperature of a building. For example, research published by the National Research Council of Canada found that an extensive green roof reduced the daily energy demand for air-conditioning in the summer by over 75 % (Liu 2003, http://www.csustentavel.com/wp-content/uploads/2013/10/termica-gr-canada.pdf).
- Therefore, green roofs can also reduce the demand on power plants, and potentially decrease the amount of CO_2 and other polluting by-products being released into the air. As other green space on the ground, also plants on a green roof can filter particles out of the air.

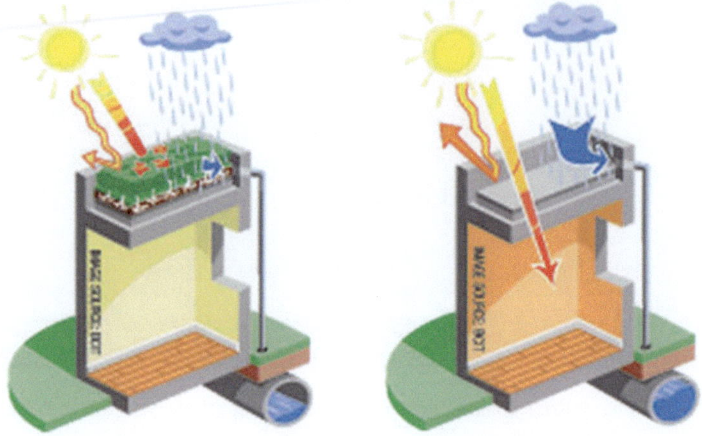

Fig. 13.18 Comparison between green roof and traditional roofs (Source: http://commons.bcit.ca/greenroof/faq/why-green-roofs-benefits/)

- Green roofs can absorb rainwater, which is stored in the substrate and taken up by the plants from where parts of it are returned into the atmosphere by transpiration and evaporation. Research shows that in summer, depending on the plants and depth of growing medium, green roofs retain 70–90 % of the precipitation that falls on them; in winter, they retain between 25–40 %. For example, a grass roof with a 4–20 cm (1.6–7.9 in.) layer of growing medium can hold 10–15 cm (3.9–5.9 in.) of water (http://www.greenroofs.org/index.php/about/greenroof-benefits). But green roofs not only retain the rainwater (and therefore reduce the volume of the runoff) but also delays the time at which it gets into the sewer system, taking stress out of the storm water infrastructure. The plants on the roof also act as natural filters so that the water running off the roof is cleaner than the runoff from a conventional roof.
- Last but not least, they provide a bit of nature for wildlife and citizens by creating habitat for plants and animals and increasing well-being of humans and beautifying the environment. Green roofs act as a stepping stone habitat for migrating species; they can link species together that would otherwise be fragmented (see also green corridors) (Fig. 13.19).

Blumau in Austria is a very special place: it is not only a spa but a piece of art, combining the countryside with the creative architecture of Friedensreich Hundertwasser and, on the way, being the first spa in Europe presenting a sustainability report in 2004, where also green roofs contributed to eco diversity (Fig. 13.20).

Examples for green roofs are many around the world. Germany is the country with the most green roofs in the world and the most advanced knowledge in this technology, and green roof research centers were founded in several German cities like Hannover and Berlin.

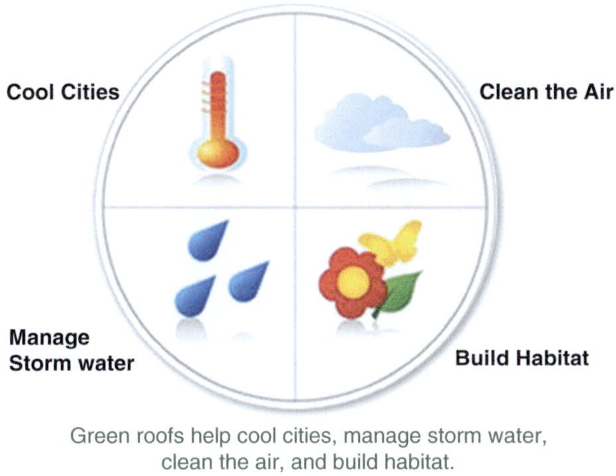

Fig. 13.19 Benefits of green roofs (Source: http://www.greenroofstoday.co.uk/environmental-benefits-of-green-roof-systems.html)

Fig. 13.20 Bad Blumau, Austria (http://curious-places.blogspot.co.at/2011/02/rogner-bad-blumau-bad-blumau-austria.html)

Fig. 13.21 Green wall in Mexico City (Source: https://en.wikipedia.org/wiki/Green_wall)

Switzerland owns one of Europe's oldest green roofs at Wollishofen in Zurich, which was built already back in 1914. Sweden claimed to have the first green roof botanical garden which was set up in Augustenburg, Malmö, in 1999 and the International Green Roof Institute opened in April 2001.

13.7 Green Facades and Green Walls

The differentiation between green facades and green walls is not always clear. The design and the types of plants used on vertical surfaces expand and so the systems overlap and become harder to define. Green walls normally signify that plants are kept in containers rather than having plants that climb up a wall. They are also known as "living walls," "bio-walls," or "vertical gardens" and can also be found inside buildings (Fig. 13.21)!

13.7.1 Green Facades Plants

For creating green facades, plants can be rooted in soil beds at the base of the building and/or in elevated planters. Green facades can be attached to existing walls or be built as freestanding structures and offer many benefits as cooling the air temperature, filtering the air from particulate matters and CO_2, providing shade and

Fig. 13.22 Self-clinging plant attaches with adhesive suckers (Source: http://www.growinggreen-guide.org/technical-guide/design-and-planning/plant-selection/construction-and-installation-of-green-facades/)

protection from the outside and also pleasing the eyes of residents and passersby. Medical studies indicate that these make people more relaxed and mentally healthy.

The plants that grow on green facades mostly have two modes of climbing up a wall:

Self-clinging plants have adhesive suckers and form a self-supporting vegetation layer on a solid wall or surface. Due to their characteristics, self-clinging plants leave traces on the wall they are climbing which is only a problem if the plant has to be removed (Fig. 13.22).

Another method of climbing is by twining and tendrils. These plants need a support system where they can climb up, such as a cable or a trellis, which have to be attached to the surface (Fig. 13.23).

Some plants attach to structures by twining stems or by hooking and clasping tendrils (Image Leanne Hanrahan) (Fig. 13.24)

Green facades, as well as green roofs, can (potentially) reduce energy consumption while shading the outside of buildings and cooling the wall surface by evaporation of the plants. Research shows[19] that green facades reduce wall surface temperatures by as much as 25 °F (14 °C) compared with exposed wall surfaces. But requirement to plants are high: In the deep urban canyons with lots of shade, heat, and wind, they have to be quite undemanding and resilient. Plants that are exposed to bright light in elevated and shadeless setting have to be highly tolerant to that.

[19] Tilley et al. (2012).

Fig. 13.23 Supporting climbing structure (Source: http://www.growinggreenguide.org/technical-guide/design-and-planning/plant-selection/construction-and-installation-of-green-facades/; image by Leanne Hanrahan)

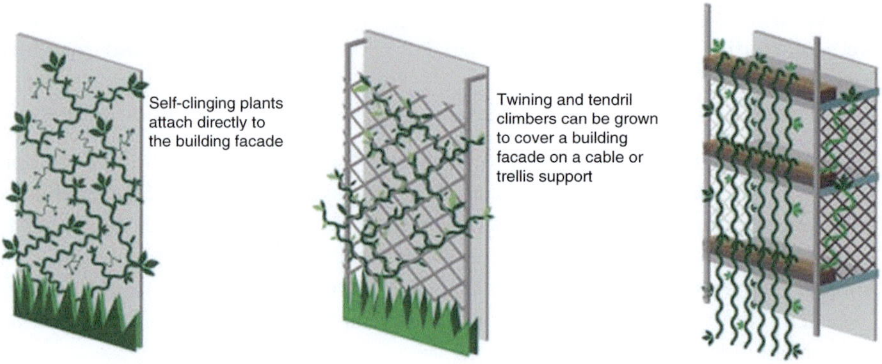

Fig. 13.24 Different climbing plants (Source: http://www.growinggreenguide.org/technical-guide/design-and-planning/plant-selection/construction-and-installation-of-green-facades/)

13.7.2 Shading and Cooling of Plants

Green facades are often installed because they provide an attractive look to a building wall, or they may be used to block out a view or to provide shade for a building. But they can also create a better and cooler microclimate, by directly shading the building and by cooling from the leaves of the plants and their loss of water to the air by evaporation. With their leaves, green facades will also provide some protection of the building surface and some retention of storm water and capture airborne particulate matter and gaseous pollutants.

Fig. 13.25 Bosco Verticale, Milan (Source: http://assets.inhabitat.com/wp-content/blogs.dir/1/files/2011/10/Bosco-Verticale-lead.jpg)

One exceptional example for a green façade, resp. a vertical garden or better to say "vertical wood" is the "Bosco Verticale" in Milan, Italy. "This is a system that optimizes, recuperates, and produces energy" (http://inhabitat.com/bosco-verticale-in-milan-will-be-the-worlds-first-vertical-forest/_bosco-verticale/#ixzz3YVv5a7rk) (Fig. 13.25).

Having plants on every side of the building, including the roof, helps to a great extent to balance the microclimate and to filter particle matters and absorb CO_2 from the air. The vast amount of green also protects the building from radiation, sun and heat, and acoustic pollution and helps to achieve dramatic energy savings. By choosing plants that will lose their leaves in autumn, sunlight will reach the apartments; during wintertime, the plants protect the building from wind, release humidity, and produce oxygen (Fig. 13.26).

On the project website, architect Stefano Boeri notes that if the units were to be constructed unstacked as stand-alone units across a single surface, the project would require 50,000 m^2 of land and 10,000 m^2 of woodland. Bosco Verticale is the first offer in his proposed BioMilano, which envisions a green belt created around the city to incorporate 60 abandoned farms on the outskirts of the city to be revitalized for community use.

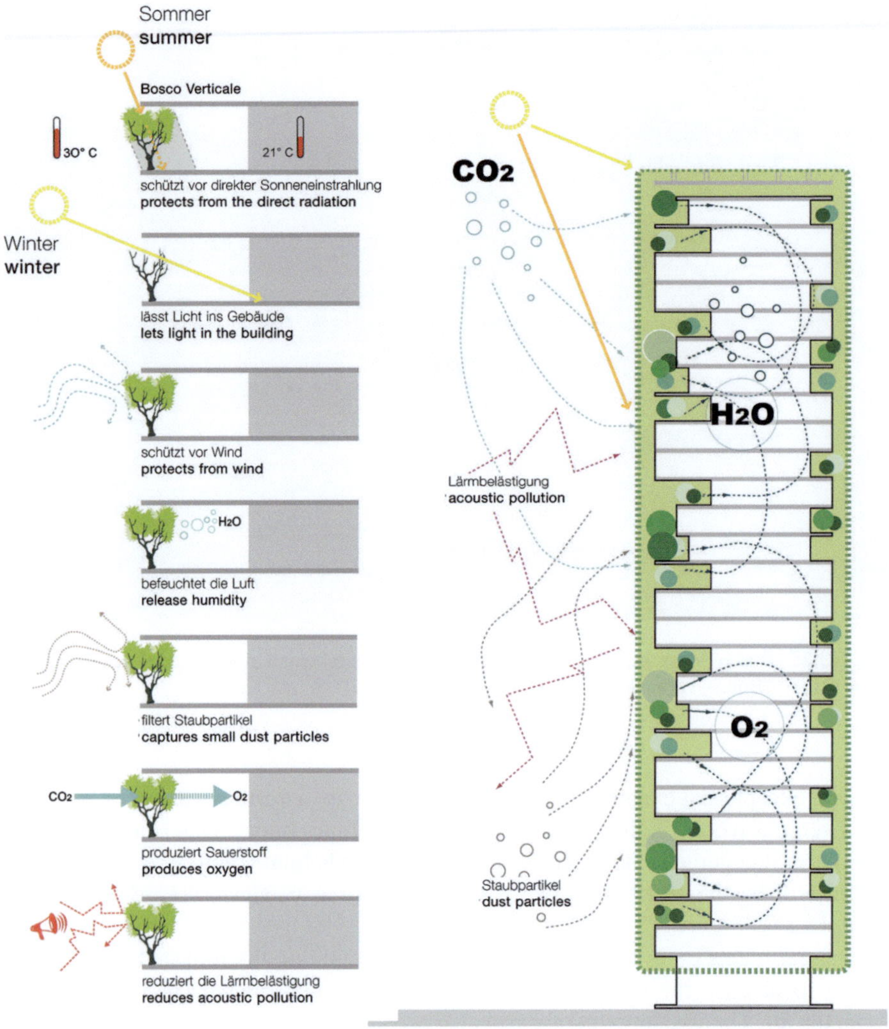

Fig. 13.26 Benefits of a vertical wood (Source: http://www.fassadentechnik.de/news/FT_06_00_02_Bosco_Verticale.php by Boeri Studios)

13.8 Urban Gardening/Farming/Agriculture

All these words stand for growing food in a city or quite populated area, only the intensity is different: "Farming" and "agriculture" have a more commercial touch; plants are grown to be sold and not for sharing or own consumption. According

Fig. 13.27 Urban gardening in Graz, Austria (Source: http://derstandard.at/1363711289560/Grazer-Kraeuter--und-Gemuesemeile-Ernten-darf-jeder)

to the definition of Bailkey and Nasr, urban agriculture is the "the practice of cultivating, processing, and distributing food in or around a village, town, or city,[20] whereas urban agriculture can also involve animal husbandry, aquaculture, agroforestry, urban beekeeping, and horticulture." Urban agriculture and farming can take different forms, from being more of a social movement for sustainable communities to provide more food security, especially in developing countries. Urban "gardening" is more likely a private initiative. For example, in Graz (Austria), a group of people put up containers in parts of the cities and planted herbs, fruits, and vegetables and whoever wants can harvest these plants (Fig. 13.27).

These kinds of projects will not be able to substitute other food systems but can contribute to some extent to give people a glimpse on how to grow vegetables and fruits in cities which most of urban people have unlearned or never known. Besides that, this can also serve as a source of community empowerment. One exceptional example is the small Victorian mill town of Todmorden, West Yorkshire, England. This example shows that it is possible to supply a whole town (with approx. 17,000 inhabitants) through urban gardening. In addition to the supply of food, the project also resulted in a better social cohesion and in a decreased crime rate.

The project was started in 2008 by some very active people around Mary Clear, grandmother of 10 grandchildren, and Pam Warhurst, a former restaurant owner in

[20] Bailkey, M., and J. Nasr. 2000. From Brownfields to Greenfields: Producing Food in North American Cities. Community Food Security News. Fall 1999/Winter 2000.

Fig. 13.28 Todmorten, England (Source: http://foodtank.com/news/2014/01/food-hero-pam-warhurst-and-incredible-edible-transform-todmorden)

town. Since then, the "Incredible Edible ethos" has been taken up by other communities, now more than twenty all over the world.

Until now, 70 big beds have been put up throughout the whole town, filled with strawberries, raspberries, apples, blueberries, beans, potatoes, carrots, salad, and other fruits, vegetables, and herbs (Fig. 13.28).

The beds are situated nearby school, supermarkets, parking lots, and different other places. They all offer the opportunity for the citizens to get fresh food for free at any time. This concept was not only welcomed by the people but is also very much respected and not abused.

13.9 Green Corridors

The examples in the former paragraphs showed all different benefits green spaces in cities could bring to the urban landscape, from cooling and cleaning the air to decreasing flood risk and improving the quality of life. Green spaces have even been shown to reduce crime rates and slow city traffic. More "added value" from plants can even be gained by fostering sustainable urban food growing.

Per definition, a green corridor is "a thin strip of land that provides sufficient habitat to support wildlife, often within an urban environment, thus allowing the movement of wildlife along it. Common green corridors include railway embankments, river banks and roadside grass verges."[21] On these "avenues," consisting of

[21] https://en.wiktionary.org/wiki/green_corridor.

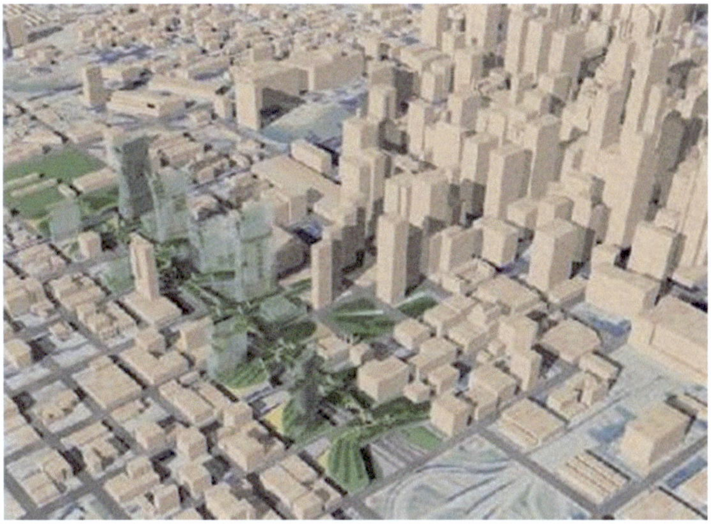

Fig. 13.29 A green corridor through the city (Source: http://easa.antville.org/stories/960542/)

all different green spaces already mentioned, e.g., roofs, facades, walls, parks, etc., animals can travel and plants can propagate (Fig. 13.29).

References

Bailkey M, Nasr J (2000) From brownfields to greenfields: producing food in North American Cities. Community Food Security News. Fall 1999/Winter 2000

Bencheikh H, Rehid A (2012) The effects of green spaces (palme trees) on the microclimate in aride zones, case study: Ghardaia, Algeria. Energy Procedia 18:10–20

Feyisa GL et al (2014) Efficiency of parks in mitigating urban heat island effect: an example from Addis Abeba. Landsc Urban Plan 123:87–95

Fink HS (2011) Promoting behavioral change towards lower energy consumption in the building sector. Innovation Eur J Social Sci Res 24(1-2):7–26

Frumkin H (2003) Healthy places: exploring the evidence. Am J Public Health 93(9):1451–1456

Huang L et al (2008) A fieldwork on the diurnal changes of urban microclimate in four types of ground cover and urban heat island of Nanjing, China. Build Environ 43:7–17

Kahn PH Jr, Friedman B, Gill B, Hagman J, Severson RL, Freier NG et al (2008) A plasma display window? The shifting baseline problem in a technologically-mediated natural world. J Environ Psychol 28:192–199

Kaplan S (1995) Review of the biophilia hypothesis. Environ Behav 27:801–804

Kuo F et al (1998) Transforming inner-city neighborhoods: trees, sense of safety, and preference. Environ Behav 301:28–59

Kuo F, Sullivan W (2001) Aggression and violence in the inner city – effects of environment via mental fatigue. Environ Behav 33(4):543–571

Terrapin Bright Green, LLC (2012) The economics of biophilia

Tilley D, Price J, Matt S, Marrow B (2012) Vegetated walls: thermal and growth properties of structured green facades. Final Report to Green Roofs for Healthy Cities, Toronto (available from author)
Wilson E (1984) Biophilia. Harvard University Press, Cambridge, MA

Suggested Readings

18 Apr 2014. Web. 17 Mar 2015. http://www.washingtonpost.com/business/case-in-point-mimicking-nature-to-cut-carbon-dioxide-emissions/2014/04/18/7574f6e6-c4de-11e3-bcec-b71ee10e9bc3_story.html
A lecture by Nataly Gattegno, Future Cities Lab & California College of the Arts. April 29, 2015, HOPES[20]. https://www.youtube.com/watch?v=kv1U1lgUh3I
Architecture that imitates life (2009) http://harvardmagazine.com/2009/09/architecture-imitates-life
Biomimetic Architecture: green building in Zimbabwe modeled after termite mounds. Inhabitat – Sustainable Design Innovation, Eco Architecture, Green Building http://inhabitat.com/building-modelled-on-termites-eastgate-centre-in-zimbabwe/
Biophilia in practice: buildings that connect people with nature (2006) AIA. http://www.aia.org/practicing/groups/kc/AIAS077174?dvid&recspec=AIAS077174
Bulkeley H, Tuts R (2013) Understanding urban vulnerability adaptation and resilience in the context of climate change, Local Environment: The International Journal of Justice and Sustainability, 18:6, 646-662, DOI:10.1080/13549839.2013.788479 http://dx.doi.org/10.1080/13549839.2013.788479
Cement from CO2: a concrete cure for global warming. Scientific American Global RSS. 7 Aug 2008. http://www.scientificamerican.com/article/cement-from-carbon-dioxide
China's Green Buildings. http://www.solaripedia.com/13/150/parkview_green_simply_passive_(beijing,_china).html
Climate matters. State of the planet emissions from the cement industry comments. 1 Jan 2012. Web. 17 Mar 2015. http://blogs.ei.columbia.edu/2012/05/09/emissions-from-the-cement-industry/
Conniff R (2009) Interview with Stephen Kellert: reconnecting with nature through green architecture, environment 360, 3 Sept 2009. http://e360.yale.edu/feature/reconnecting_with_nature_through_green_architecture/2164/
Heschong L, Wright R (2002) Daylighting and human performance: latest findings. ASHRAE J
Howard E (1902) Garden Cities of To-Morrow (London, 1902. Reprinted, edited with a Preface by F. J. Osborn and an Introductory Essay by Lewis Mumford. (London: Faber and Faber, [1946]):50–57, 138–147. http://urbanplanning.library.cornell.edu/DOCS/howard.htm
Indicators for management of urban biodiversity and ecosystem services. In: Elmqvist, Th, Fragkias M, Goodness J, Güneralp B, Marcotullio PJ, McDonald RI, Parnell S, Schewenius M, Sendstad M, Seto KC, Wilkinson C (eds) Urbanization, biodiversity and ecosystem services: challenges and opportunities. A global assessment. Springer, pp 699–718. doi:10.1007/978-94-007-7088-1_32
Iasef Md R, Mario S (2014) Tree-inspired dendriforms and fractal-like branching structures in architecture: a brief historical overview. Frontiers of Architectural Research, Elsevier
Kellert S, Heerwagon J (2008) Biophilic design: the theory, science, and practice of bringing buildings to life. Wiley, Hoboken
Kellert S et al (1993) The biophilia hypothesis. Island Press, Washington, DC
Lorenzi MG, Francaviglia M (2010) Art & mathematics in Antoni Gaudí's architecture: LA Sagrada Famĺia (I). J Appl Math 3(1)
Marberry SO (2010) A conversation with Roger Ulrich', Healthcare Design. http://www.healthcaredesignmagazine.com/article/conversation-roger-ulrich?page=3
Nanotechnology solutions for self-cleaning, dirt and water-repellent coatings. Nanotechnology solutions for self-cleaning, dirt and water-repellent coatings. Web. 17 Mar 2015. http://www.nanowerk.com/spotlight/spotid=19644.php

Nina K, John R, Wei F, Nan Z, China Energy Group Environmental Energy Technologies Division Lawrence Berkeley National Laboratory Qing Ye Shenzhen Institute of Building Research. Comparative Policy Study for Green Buildings in U.S. and China, Apr 2014
Paul Woolford: how biomimicry Inspired the design of NOAA's Daniel K. Inouye Regional Center in Hawaii. http://www.hoklife.com/2013/07/16/paul-woolford-on-noaas-new-pacific-regional-center-campus-in-hawaii/
Restorative biophilic design. Helena Van Vliet Architect LLC. Web. 17 Mar 2015. http://www.helenavanvliet.com/
Rocky Mountain Institute, Carnegie Mellon University (2004) Environmental Building News. EBN 13(10v) https://www2.buildinggreen.com/article/productivity-and-green-buildings
Secretariat of the Convention on Biological Diversity (2012) Cities and Biodiversity Outlook. Montreal, 64 pages
Wilson EO (2002) The future of life. Vintage Publishers
http://assets.inhabitat.com/wp-content/blogs.dir/1/files/2011/10/Bosco-Verticale-lead.jpg
http://biophiliaeducational.org/
http://commons.bcit.ca/greenroof/faq/why-green-roofs-benefits/
http://curious-places.blogspot.co.at/2011/02/rogner-bad-blumau-bad-blumau-austria.html)
http://derstandard.at/1363711289560/Grazer-Kraeuter--und-Gemuesemeile-Ernten-darf-jeder
http://dx.doi.org/10.1080/13549839.2013.788479
http://easa.antville.org/stories/960542/
http://greengarage.ca/greenroofs/plants.php
http://inhabitat.com/bosco-verticale-in-milan-will-be-the-worlds-first-vertical-forest/_bosco-verticale/#ixzz3YVv5a7rk
http://www.bdcnetwork.com/8-grand-green-roofs-and-walls
http://www.csustentavel.com/wp-content/uploads/2013/10/termica-gr-canada.pdf
http://www.designweneed.com/enviromental-advantages-of-green-rooftops/
http://www.epa.gov/pm/
http://www.greenroofguide.co.uk/biodiversity-and-planting/
http://www.greenroofs.com/projects/.
http://www.greenroofs.org/index.php/about/greenroofbenefits
http://www.greenwalls.si/de/izvedeni_projekti_zelene_stene.html
http://www.growinggreenguide.org/technical-guide/design-and-planning/plant-selection/construction-and-installation-of-green-facades/
http://www.heggies.com/extranet/willoughbyAQ/ParticulateMatter.aspx
http://www.hrt.msu.edu/greenroof/research-projects/substrates.html
http://www.ineffableisland.com/2012/08/20-colleges-embracing-green-roof-trend.html
http://www.planningportal.gov.uk/general/glossaryandlinks/glossary/g
http://www.skyscrapercity.com/showthread.php?t=1660203
http://www.tagtomat.dk/short-summary-in-english/
http://www.zinco.ca/extensive.html
https://depts.washington.edu/hhwb/Thm_Crime.html
https://greenerheights.wordpress.com
https://www.wieninternational.at/en/content/environmental-city-vienna-50-green-space-en

Chapter 14
Taking Advantage of Natural Elements Surrounding the Building

Anil Ahuja

14.1 Geothermal Energy

What is geothermal energy? Ten miles beneath the earth's surface, an almost limitless source of energy waits to be mined. It's not oil, coal, or natural gas, but the energy from the rocky, hot interior of 420 °C, or 800 °F, or more. This intense heat, known as geothermal energy, turns underground water into the steam that erupts from geysers. These natural pockets of steam, within a mile of the surface, are too rare to provide large amounts of natural geothermal energy worldwide.

Through plate motion and vulcanism, some of this energy is concentrated at high temperature near the surface of the earth. Energy is also transferred from the deeper parts of the crust to the earth's surface by conduction and by convection in regions where geological conditions and the presence of water permit. Geothermal energy as applied to building systems is the thermal energy within the earth's crust. Calculations show that the earth, originating from a completely molten state, would have cooled and become completely solid many thousands of years ago without an energy input in addition to that of the sun. It is believed that the ultimate source of geothermal energy is radioactive decay within the earth.

Because of variation in volcanic activity, radioactive decay, rock conductivities, and fluid circulation, different regions have different heat flows (through the crust to the surface), as well as different temperatures at a particular depth. The normal increase of temperature with depth (i.e., the normal geothermal gradient) is about 13.7 °F per 1000 ft of depth, with gradients of about 5–27 °F per 1000 ft being common. The areas that have higher temperature gradients and/or higher-than-average heat flow rates constitute the most interesting and viable economic resources. However, areas with normal gradients may be valuable resources if certain geological features are present.

A. Ahuja, *Integration of Nature and Technology for Smart Cities*,
DOI 10.1007/978-3-319-25715-0_14

14.1.1 Introduction to Geothermal Systems

Geothermal systems that produce essentially dry steam are vapor dominated. While these systems are valuable resources, they are rare. Hot water (fluid-dominated) systems are used more common than vapor-dominated systems and can be produced either as hot water or as a two-phase mixture of steam and hot water, depending on the pressure maintained on the production system. If the pressure in the production casing or in the formation around the casing is reduced below the saturation pressure at that temperature, some of the fluid will flash, and a two-phase fluid will result. If the pressure is maintained above the saturation pressure, the fluid will remain single phase. In fluid-dominated systems, both dissolved gases and dissolved solids are significant. The quality of the fluid varies from site to site, from a few hundred parts per million (ppm) to over 300,000 ppm dissolved solids.

The following characteristics have a major influence on selection of a geothermal system. Because the costs of geothermal systems are primarily front-end capital costs, annual operating costs are relatively low.

- Depth of resource
- Distance between resource location and application site
- Well flow rate
- Resource temperature
- Temperature drop
- Load size
- Load factor
- Composition of fluid
- Ease of disposal
- Resource life

14.1.2 Applications in Building Systems

The primary applications for the direct use of geothermal energy are space heating, sanitary water heating, and space cooling (using the absorption process). While geothermal space and sanitary water heating are widespread, space cooling is rare. Where space heating is accomplished, sanitary water heating or at least preheating is almost universally accomplished. Domestic water heating in a district space heating system is beneficial because it increases the overall size of the energy load, the energy demand density, and the load factor.

Whenever possible, the domestic hot water load should be placed in series with the space heating load to reduce system flow rates. In many cases, buildings that derive their heat from geothermal fluids operate their heating systems at less than conventional supply water temperatures because of low resource temperatures and the use of heat exchangers to isolate the fluids from the building loop (see Fig. 14.1 for a central plant groundwater system arrangement with heat exchangers). In geothermal systems,

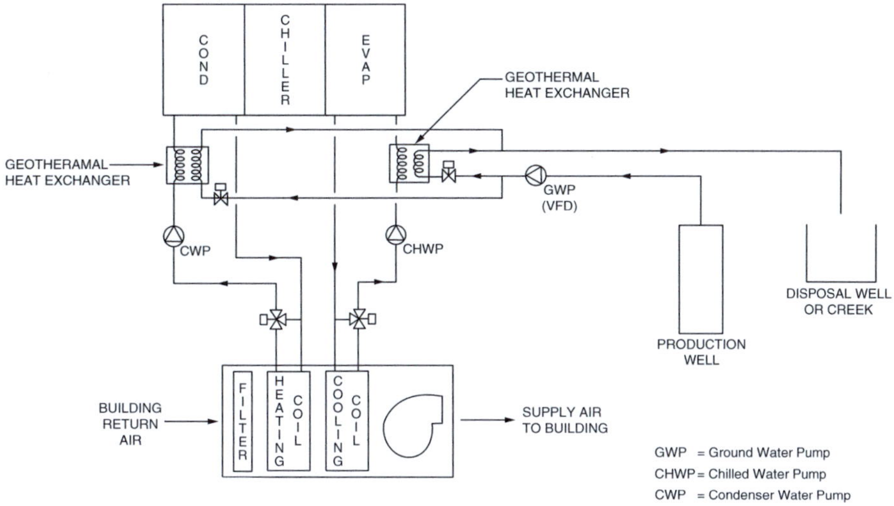

Fig. 14.1 Central plant groundwater/geothermal system

it is frequently advisable to design geothermal heat exchange surface larger than normal surface requirement. The selection of equipment to accommodate these considerations can enhance the feasibility of using the geothermal source.

14.1.3 Components of Geothermal Systems

The primary equipment used in geothermal systems includes pumps, heat exchangers, and piping. While some aspects of these components are unique to geothermal applications, many of them are of routine design. However, the great variability and general aggressiveness of the geothermal fluid necessitate limiting corrosion and scale buildup rather than relying on system cleanup. Corrosion and scaling can be limited through (1) proper system and equipment design or (2) treatment of the geothermal fluid, which is generally precluded by cost and environmental regulations relating to disposal.

The terminal equipment used in geothermal systems is the same as that used in nongeothermal heating systems. However, certain types of equipment are better suited to geothermal design than others. Finned coil, forced-air systems are generally the most capable of functioning under the low-temperature/high-temperature difference conditions. One or two additional rows of coil depth compensate for lower supply water temperatures. While increased delta-T affects coil circuiting, it improves controllability. This type of system should be capable of using supply water temperatures as low as 100–120 °F (37.7–48.8 °C).

Radiant floor panel systems are able to use very low water temperatures, particularly in industrial applications with little or no floor covering. The availability of new nonmetallic piping has renewed the popularity of this type of system. In industrial settings, with a bare floor and relatively low space temperature requirements, average water temperatures could conceivably be as low as 95 °F (35 °C).

Heat pump systems take advantage of the lowest-temperature geothermal resources. Loop heat pump systems operate with water temperatures in the 25–100 °F range and central station heat pump plants (supplying four-pipe systems) in the 45 °F range. Baseboard convectors and similar equipment are the least capable of operating at low supply water temperature. At 150 °F (65.5 °C) average water temperatures, derating factors for this type of equipment are on the order of 0.43–0.45. As a result, the quantity of equipment required to meet the design load is generally uneconomical. This type of equipment can be operated at low temperatures from the geothermal source to provide base-load heating capacity. Peak load can be supplied by a conventional boiler.

14.2 Thermocline Lakes

Lakes are stratified bodies of water in which temperature declines with the depth. Heat is transferred to lakes by three primary modes: radiation from the sun, convective heat transfer from the surrounding air (when the air temperature is greater than the water temperature), and conduction from the ground.

14.2.1 Introduction

Heating of lakes is primarily accomplished by solar radiation, which can exceed 300 Btu/h per square foot of lake area, but it occurs primarily in the upper portion of the lake unless the lake is very clear. About 40 % of the solar radiation is absorbed at the surface. Convective heat transfer to the lake occurs when the lake surface temperature is lower than the air temperature. Wind speed increases the rate at which heat is transferred to the lake, but maximum heat gain by convection is usually only 10–20 % of maximum solar heat gain. The conduction gain from the ground is even less than convection gain.

Likewise, cooling of lakes is accomplished primarily by back radiation, which typically occurs at night when the sky is cool, triggering evaporative heat transfer at the surface. Convective cooling or heating in warmer months contributes only a small percentage of the total because of the relatively small temperature differences between the air and the lake surface temperature. The last major mode of heat transfer, conduction to the ground, does not play a major role in lake cooling.

In the winter, the coldest water is at the surface. It tends to remain at the surface and freeze. The bottom of a deep lake stays 5–10 °F warmer than the surface. This condition is referred to as winter stagnation. The maximum density of water occurs at 39.2 °F (4 °C), not at the freezing point of 32 °F (0 °C). As spring approaches, surface water warms until the temperature approaches the maximum density point of 39.2 °F. Later in the spring, as the water temperatures rise above 45 °F (7.2 °C), the winter lake stratification becomes unstable, and circulation loops begin to develop from top to bottom. This condition is called spring overturn. This pattern continues throughout the summer. The upper portion of the lake remains relatively warm, with evaporation cooling the lake and solar radiation warming it.

The lower portion (hypolimnion) of the lake remains cold because most radiation is absorbed in the upper zone, circulation loops do not penetrate to the lower zone, and conduction to the ground is quite small. The result is that in deeper lakes with small or medium inflows, the upper zone is 70–90 °F (32.2 °C), the lower zone is 40–55 °F, and the intermediate zone (thermocline) has a sharp change in temperature within a small change in depth. This condition is referred to as summer stagnation.

As fall begins, the water surface begins to cool through back radiation and evaporation. With the approach of winter, the upper portion begins to cool toward the freezing point, and the lower levels approach the maximum density temperature of 39.2 °F. Many lakes do exhibit near-ideal temperature profiles. However, a variety of circumstances can disrupt the profile. These circumstances include (1) high-inflow/-outflow rates, (2) insufficient depth for stratification, (3) level fluctuations, (4) wind, and (5) lack of enough cold weather to establish sufficient amounts of cold water for summer stratification.

14.2.2 Application and Components in Building Systems

Thermal stratification of water often causes large quantities of cold water to remain undisturbed near the bottom of deep lakes. This water is cold enough to adequately cool buildings by simply being circulated through heat exchangers. A heat pump is not needed for cooling, and energy use is substantially reduced. Heating can be provided by a separate source or with heat pumps in the heating mode. Prechilling or supplemental total cooling are also permitted when water temperatures are between 50 and 60 °F.

The heat pumps are used to transfer heat to or from the air in the building. To put these heat transfer rates in perspective, consider a 1 acre (43,560 ft^2) lake that is used in connection with a 10-ton (120,000 Btu/h) heat pump. In the cooling mode, the unit will reject approximately 150,000 Btu/h (44 kw) to the lake. This is 3.4 Btu/h-ft^2, or approximately 1 % of the maximum heat gain from solar radiation in the summer. In the winter, a 10-ton heat pump would absorb only about 90,000 Btu/h (26.3 kw), or 2.1 Btu/h-ft^2, from the lake.

In a closed-loop system, water-to-air heat pumps are linked to a submerged coil. Heat is exchanged to (cooling mode) or from (heating mode) the lake by the fluid (usually a water–antifreeze mixture) circulating inside the coil. In an open-loop system,

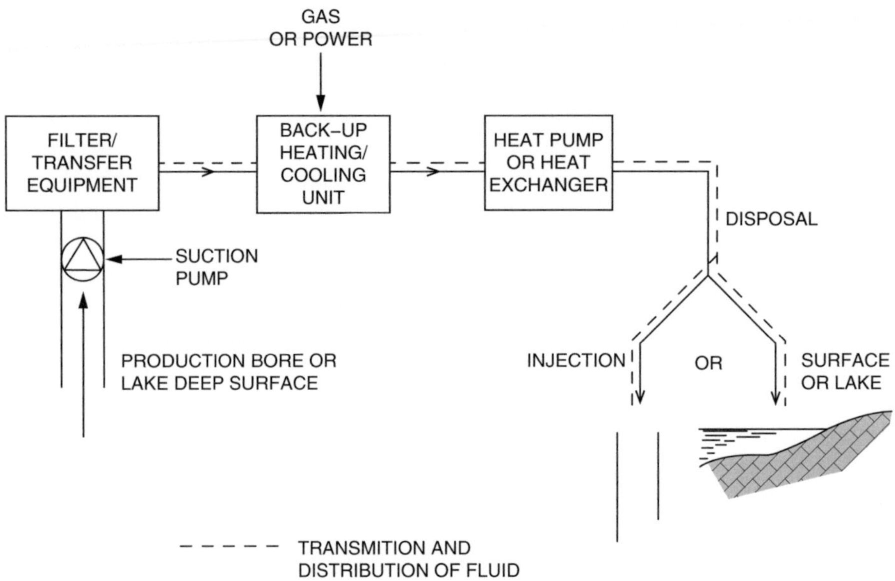

Fig. 14.2 Basic geothermal/thermocline lake direct system

water is pumped from the lake through a heat exchanger and returned to the lake some distance from the point at which it was removed. The pump can be located either slightly above or submerged below the lake water level. For heat pump operation in the heating mode, this type of system is restricted to warmer climates; water temperatures must remain above 42 °F (see Fig. 14.2 for a basic direct use lake system).

Chapter 15
Reduce, Reuse, Recycle

Kavin Singh Chawla and Vasudevan (Raj) Rajaram

We can live sustainably only if we reduce, recycle, and reuse all the things we have been accustomed to in our cities. We are running out of quality water in our cities, and we have become vulnerable to droughts and floods. Our groundwater resource quality and quantity are being adversely affected as we use more groundwater without sufficient recharge of the aquifers. Solid waste in our cities is being sent to landfills further away from our cities, increasing cost of transportation and taking up valuable greenfields. Unless we treat solid and liquid waste as a resource, our quality of life will be adversely impacted in the future. This chapter will focus on managing solid and liquid waste in our buildings and cities.

Reduce is defined as changing our behavior to minimize consumption to only what we need instead of what we want. In other words, it is how we conserve our resources—water, energy, and other products that we use in our lives. We can reduce by monitoring our use and ensuring that we use tools to conserve our resources.

Reuse is defined as using recycled resources instead of using raw materials to produce new products. For example, we can reuse the gray water from our houses to water our plants and gardens. Recycled cans, bottles, and paper are reused, saving precious natural resources.

Recycle is defined as the process of converting used products into new, useful products. For example, we can recycle toilet water to drinking water by treating it and making it safe. We started recycling paper, aluminum, and glass through organized programs in all communities in the United States in the 1980s and 1990s.

15.1 How Do We Manage Solid Wastes?

Cities have been introducing solid waste recycling programs over the last few decades, but the quantity of solid waste diverted from landfills is still around 25%. This is because we have not made an attempt to recycle our food wastes.

A. Ahuja, *Integration of Nature and Technology for Smart Cities*,
DOI 10.1007/978-3-319-25715-0_15

Massachusetts and Oregon have passed laws to divert all food waste from landfills, and this has led to companies developing innovative ways of managing food waste. The types of solid wastes generated in a building or a community include the following:

- Paper and cardboard
- Aluminum and other metals
- Plastic
- Glass
- Electronic waste
- Food and other organic waste

The key to managing these wastes efficiently is to segregate them at source so that contamination of different waste streams is avoided. Ways of managing these waste streams are discussed in the rest of this chapter, with emphasis on food and other organic waste. If each of these waste streams is recycled for reuse, it will create wealth for the community while minimizing the use of natural resources. With some planning, the use of the materials can be reduced, but irrespective of the amount of reduction, recycling and reuse should be emphasized through policies instituted in buildings and cities.

Decentralization of our waste management and managing waste as close to the source as possible is an essential step to turn waste into wealth. By segregating waste through the use of categorized recycling bins (plastics, metal, glass, food, and paper), waste can be sold to recyclers at high prices due to the lack of contamination from other wastes. For example, if recycling bins were installed in a building, the building could sell the plastic, metal, glass, and paper waste to create a new revenue stream and use their food and other organic waste to convert to compost at the building level and biogas at the community level.

Recycling and reuse systems for paper, cardboard, metals, plastics, and glass are well established and markets are already well developed. However, recycling of electrical and electronic waste (e-waste) is in its infancy in the United States. Europe has cradle to grave recycling systems for all manufactured products. E-waste includes a wide range of electrical and electronic devices such as computers, cellular phones, personal stereos, televisions, including large household appliances such as refrigerators and air conditioners. E-wastes contain over 1000 different substances, many of which are toxic and potentially hazardous to our health and our surrounding environment especially when they are not managed appropriately.

15.2 How Do We Manage E-waste?

As the use of computers and appliances is increasing, inevitably e-waste quantities will increase and have to be managed effectively. Potentially hazardous substances in such appliances can be seen on the Table 15.1 below.

Table 15.1 Potential hazardous materials in e-waste

Component	Possible hazardous content
Metal	–
Motor/compressor	–
Cooling	Ozone depletion substance (ODS)
Plastics	Phthalate plasticizer, BFR
Insulation	Insulation ODS in foam, asbestos, refractory, ceramic fiber
Glass	–
CRT	Lead, antimony, mercury, phosphorus
LCD	Mercury
Rubber	Phthalate plasticizer, BFR
Wiring/electrical	Phthalate plasticizer, lead, BFR
Concrete	–
Transformer	–
Circuit board	Lead, beryllium, antimony, BFR
Fluorescent lamp	Mercury, phosphorus, flame retardants
Incandescent lamp	–
Heating element	–
Thermostat	Mercury
BFR containing plastic	BFRs
Batteries	Lead, mercury, cadmium, lithium
CFC, HCFC, HFC, HC	ODS
External electric cables	BFR, plasticizers
Electrolyte capacitors	Glycol

Source: Vasudevan et al. (2016)

15.3 Treating E-waste

Environmentally sound e-waste treatment (EWT) technologies can be identified at three levels. The first level includes decontamination, dismantling, and segregation. The second level includes shredding and special treatment processes like electro-magnetic separation, eddy current separation, cathode ray tube (CRT) breaking and treatment, and density separation using water. The third-level treatment includes recovery of metals and disposal of hazardous e-waste fractions such as mercury, plastics with flame retardants, capacitors, CFCs, lead, and other items.

The third level is where you can see how waste can turn into wealth. It is carried out mainly to recover ferrous and nonferrous metals, plastics, and other items of economic value, which would ideally be carried out at different integrated facilities. The input, output, and unit operations at the third-level treatment are described in the Table 15.2 below.

Table 15.2 Input/output and unit operations for third-level treatment of e-waste

S. no.	Input/WEEE residues	Treatment/recycling technique	Output
1.	Sorted plastic	Recycling	Plastic product
2.	Plastic mixture	Energy recovery/incineration	Energy recovery
3.	Plastic mixture with FR	Incineration	Energy recovery
4.	CRT	Breaking/recycling	Glass cullet
5.	Lead smelting	Secondary lead smelter	Lead
6.	Ferrous metal scrap	Secondary steel/iron recycling	Iron
7.	Nonferrous metal scrap	Secondary copper and aluminum smelting	Copper/aluminum
8.	Precious metals	Au/Ag separation (refining)	Gold/silver/platinum and palladium
9.	Batteries (lead acid and Li ion)	Lead recovery, smelting, remelting, and separation	Lead
10.	CFC	Recovery/reuse and incineration	CFC/energy recovery
11.	Oil	Recovery/reuse and incineration	
12.	Capacitors	Incineration	Energy recovery
13.	Mercury	Separation and distillation	Mercury

Source: Vasudevan et al. (2016)

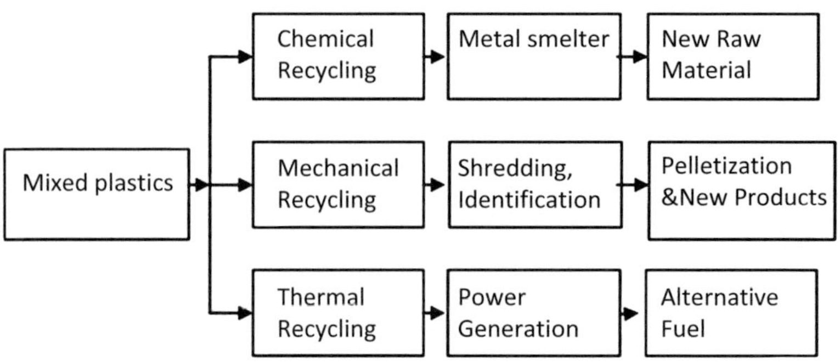

Fig. 15.1 Recycling e-waste plastic parts

Chemical, thermal, and mechanical recycling plays a big part in achieving the outputs desired. Those processes can be understood from the Fig. 15.1 below.

The figure shows how the different types of recycling for plastics can be used to generate a sellable output, truly turning waste to wealth. Processing e-waste is essential in building a smart city, managing the replacement of old technologies, and ensuring that waste is treated in an environmentally safe manner.

As the vision of the "smart city" becomes more and more reachable, emerging technologies are being introduced to the market that help treat other kinds of waste (besides e-waste) in truly intelligent ways that turn it into a valuable, marketable resource.

15.4 Converting MSW into Charcoal Briquettes

An innovative design to all municipal solid waste (MSW) has been developed in India and turns all waste into sellable commodities. This process converts the waste into usable charcoal briquettes and into construction bricks.

The Neway patented process uses no sorting, burning, and land refilling, and at the end of the process, there are no residues created. The end result is a high-calorific-value charcoal powder as well as building material. The briquettes are essentially a type of green coal that is made from municipal waste that can be used as a biofuel for any furnace or boiler.

The first step in the conversion of MSW into charcoal powder is to separate ferrous and nonferrous materials from the raw MSW. A magnetic separator is used for segregation. Next, a trommel screen is used to segregate the combustible MSW from noncombustible MSW. After the waste has been separated, the moisture in the combustible waste is squeezed and removed to the maximum possible extent and the wastewater is treated before disposal. The squeezed combustible waste (called refuse-derived fuel [RDF]) is then shredded into 10-mm sizes. After this, the RDF goes through 2 rounds of pulverization. Pulverizing the waste turns RDF from 10mm to 2–3mm, significantly increasing the surface area. As the moisture is removed, the calorific value of the waste increases. The pulverized combustible waste RDF with low moisture content is fed into the patented reactor, and with very low power consumption, it is converted into charcoal powder with a high caloric value. The entire process can be seen in the Fig. 15.2 below.

Clearly, the conversion of waste to wealth is possible, but the ultimate goal in waste management is to create a system that will create coproducts from the inputs and make sure there is no additional waste sent to a landfill. The inert waste, comprising mainly of construction debris and street sweepings, is processed into bricks for use by the construction industry.

As mentioned earlier, decentralization of waste treatment to a level where each building manages their own waste is a key step for building smart cities. Collecting different kinds of waste using categorized recycling bins will allow the community to sell the waste to recyclers who could use innovative technologies to process it and turn it into useful products. Organic waste conversion to bioenergy can produce the power required to operate the city.

15.5 Processing Organic Waste

Both wet and dry organic waste can be utilized, but it is the wet organic waste that smart buildings can use to generate electricity for their use. In order to do this, anaerobic digestion is used. After categorized bins have been installed in a facility, uncontaminated inorganic waste is sold to recyclers and the organic waste is stored in large bins. The waste can be sold to a waste-processing company or the building

Fig. 15.2 Patented technology for converting municipal solid waste to useful products

can install an anaerobic biodigestor. The following flowcharts demonstrate the potential flow of waste that can create revenue streams (Figs. 15.3, 15.4, and 15.5).

From the above charts, it is obvious that we can convert every waste into a revenue stream (highlighted through oval-shaped elements) or reuse it in the system (hexagon-shaped elements).

15.6 Dry Organic Waste Composting

Processing dry organic waste through decentralized systems will benefit both the community and the building or facility utilizing the system. As landfill laws are becoming tighter for disposing food waste, innovative strategies for dealing with these kinds of waste are emerging.

For smart buildings, a simple dry composting unit should be installed. There are several dry composting systems available, and a suitable system can be installed by

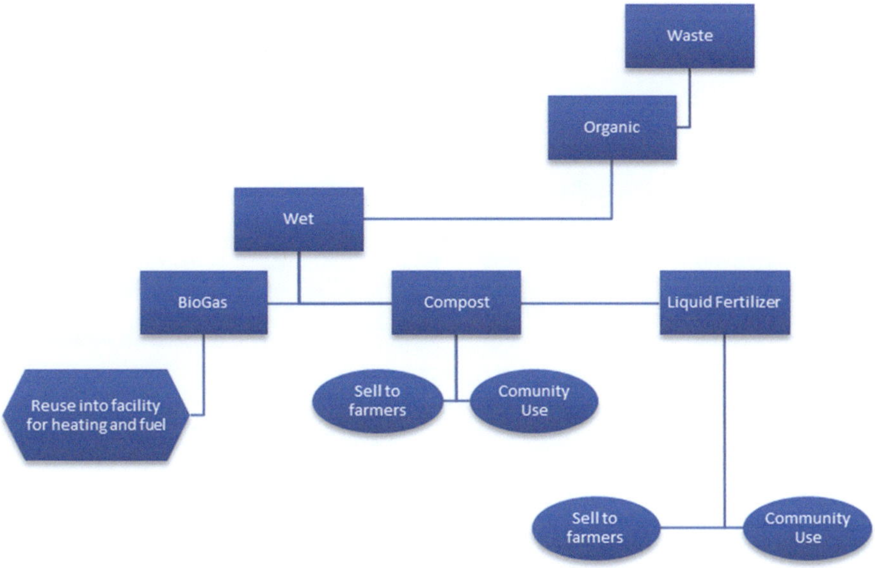

Fig. 15.3 Wet organic waste flow

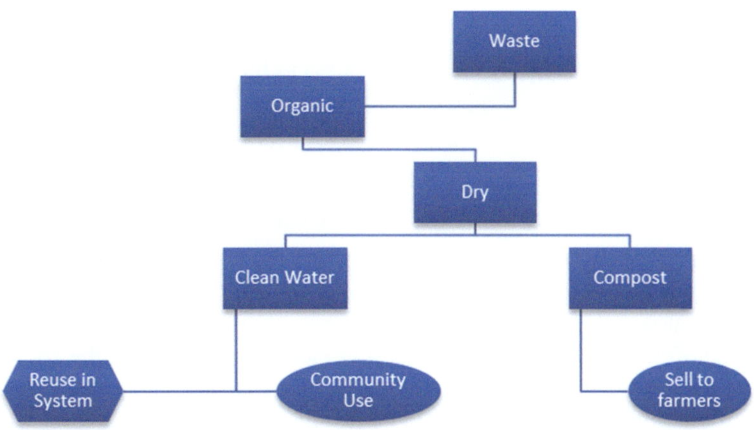

Fig. 15.4 Dry organic waste flow

the building manager (Griffith-Onnen et al. 2013). There are top-loading or front-loading machines which accelerate organic waste decomposition with aeration, moisture management, and various levels of heat combined with microorganisms. The main advantage of these machines is the reduction in waste volume and the turnaround time to create compost from organic waste.

By processing dry waste and producing mulch, compost, or any coproducts, waste to wealth is truly achieved in an environmentally sound manner. Running

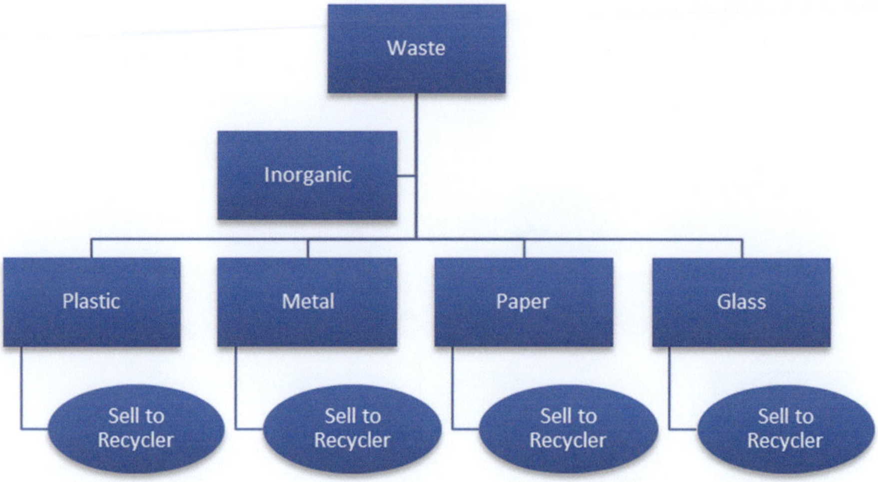

Fig. 15.5 Inorganic waste flow

such a system in a building results in allowing the management to convert their waste into a useful product and creating a revenue stream, truly a "smart" solution to the waste issue.

Processing dry waste is only half the concern for dealing with organic waste; wet organic waste can be extremely useful to process as biogas as well as to create compost and liquid fertilizer.

15.7 Anaerobic Digestion in Communities

From Fig. 15.3 (above), it is clear that the three main achievable byproducts of processing wet organic waste effectively are biogas, compost, and liquid fertilizer; of these, biogas can be converted into heat and electricity for use in the community. Compost and liquid waste can be sold to urban gardens and landscape companies.

Anaerobic digestion systems should be installed at a community level where all organic waste not converted into dry compost is stored for processing. An anaerobic system works quite elegantly, where food and other wet organic waste is fed into a digestion tank. Temperatures and pressure, oxygen levels, and moisture levels are optimized to speed up the decomposition process. From the tank, three products are recovered as mentioned above.

A simple anaerobic biodigestion system provided by the American Biogas Council is shown below (Fig. 15.6).

The anaerobic digestion process consists of the following steps:

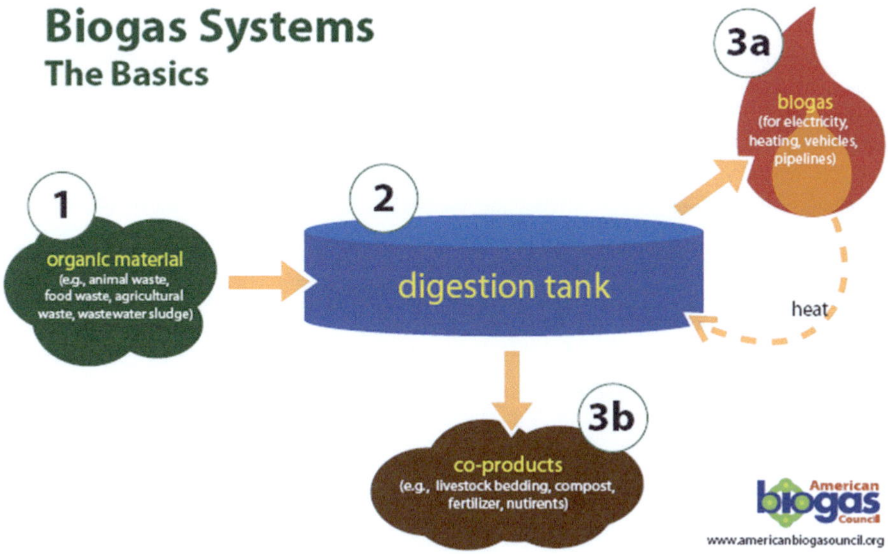

Fig. 15.6 Basic biogas system (Source: American Biogas Council 2014)

1. Load food waste into grinding unit.
2. Mechanical grinding.
3. Ground waste enters processing vessel and is stirred.
4. A small quantity of air and water is introduced into the vessel.
5. Micronutrient, if needed, is added.
6. Biogas is withdrawn and stored.
7. Settle time to recover solids and liquid nutrients.
8. Composting of solid residue and storage of liquid nutrients.

The advantages of the anaerobic system are in three areas: environmental, economic, and energy. On the environmental side, greenhouse gases are minimized in two ways: first, by lower carbon dioxide emissions as food waste hauling by diesel-burning trucks is avoided and, second, by prevention of emissions of methane as a result of food rotting in landfills (Table 15.3).

When food waste decomposes aerobically (i.e., in an environment where oxygen is present), it is considered to be "biogenic," or a part of the natural cycle in which carbon returns to the ecosystem that originally produced it. Other environmentally positive effects of wet system include expected decreases in the use of plastic garbage bags, the absence of chemical rodent- and pest-control measures, and alleviation of odors associated with rotting food. An additional ecological benefit is the prevention of leachate from landfills entering groundwater (Fig. 15.7).

Oft-cited operational advantages include streamlined workflow, the ability to measure environmental performance, and the fact that wet systems may be incorporated into larger sustainability programs.

Table 15.3 Benefits of anaerobic digestion of biomass

Environmental benefits	Energy benefits	Economic benefits
Elimination of malodorous compounds	Production of high-quality fuel	Transformation of waste liabilities into profit centers
Reduction in pathogens	Generation of surplus electricity as well as heat	Value to negative-value feedstock
Production of sanitized compost	Reduction reliance on energy imports	Reduced water consumption
Less dependence on inorganic fertilizers	Promotion of decentralized, distributed power systems	Increased self-sufficiency
Decrease in GHGs emission	Reduced dependence on fossil fuels	
Promotion of carbon sequestration		
Beneficial reuse of recycled water		
Production of groundwater and surface water resources		

Source: Zafar (2014)

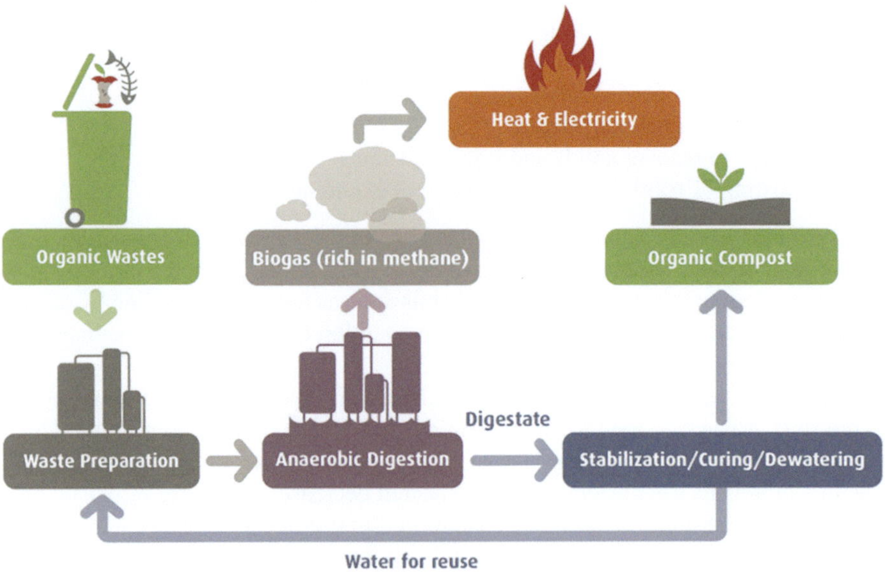

Fig. 15.7 Waste to biogas flowchart (Source: Ortiz 2014)

15.8 Who Are These Systems For?

The sustainable disposal of food waste is a challenge for all levels of society, from the single apartment dweller to the largest food-processing plant. The wet systems discussed here, however, are generally suitable for larger-scale waste generators, given that the machinery is set up to handle more than one ton of organic waste per week, much more than a household produces. Therefore, vendors of wet systems typically market their products to institutional customers such as colleges, country clubs, hospitals, hotels, mall food courts, military bases, prisons, restaurants, stadiums, and supermarkets. If these institutions optimized these systems, a decentralized procedure for treating a large percent of a city's waste will be created and it will reap its benefits to society.

Institutions interested in employing on-site food waste-processing systems will need to consider a number of factors in making their decision. Many of these factors will be specific to the institution: siting needs will vary, and pricing and system configurations are flexible and in some cases negotiable.

On-site food waste management systems are alternatives to landfill disposal and facilitate compliance with the proposed food waste bans in landfills that are being implemented in the United States, Europe, India, and other countries. Generally, those who have employed these technologies have had success at reducing the volume and weight of food waste generated. These solutions are a valuable component of an institutional food waste management program, although companies planning on implementing them need to check local ordinances prior to purchasing a system.

A successful example of a company that has installed an on-site anaerobic digestion system is The Plant in Chicago, Illinois. The Plant is a 4-story building that has several businesses in its premises related to fish harvesting, hydroponic farming, brewing, and baking. All the organic waste produced from these businesses and surrounding organic waste generators is converted to biogas, liquid fertilizer, and compost. By having a digester in their facility, they are able to produce all their electrical needs while producing useful by-products for agriculture and horticulture. A diagram of their system is provided below (Fig. 15.8).

All waste from their hydroponics and aquaculture operations as well as waste from their brewery is fed into an anaerobic digester that produces biogas that triggers a turbine generator and powers the entire facility. If more and more facilities in a city employ integrated systems like these, a "smart" solution to urban waste management will result.

Clearly, managing organic and inorganic solid waste in a zero-waste system is possible and only through innovative technologies can a city become "smart." However, the issue of liquid waste is still a concern and processing it into wealth is also possible.

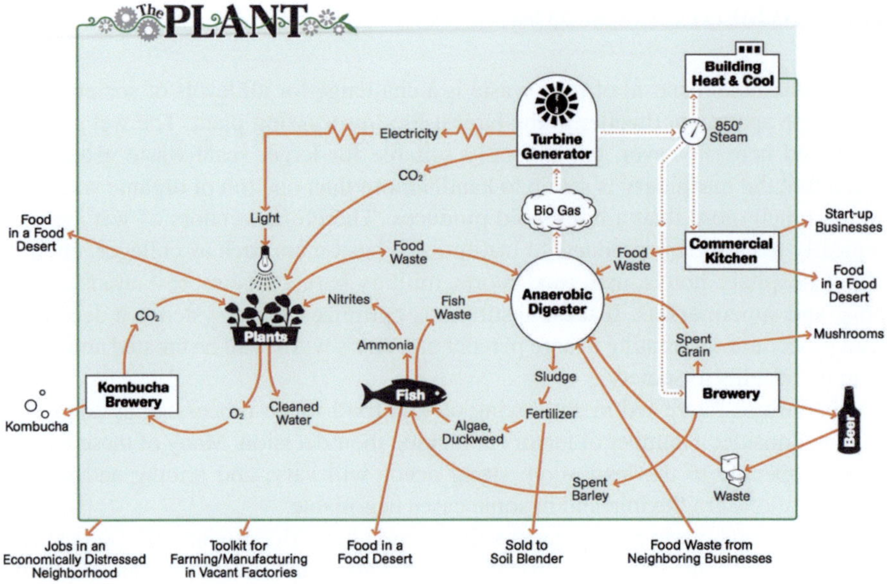

Fig. 15.8 Zero waste system at The Plant, Chicago (Source: The Plant 2014)

15.9 Liquid Waste Management

Liquid waste treatment must be thought of as a cycle, rather than typical treatment plants that discharge treated water into our waterbodies from where they are pumped to provide clean water to the city. Like solid waste, liquid waste can be turned into wealth and can create new revenue streams in the process.

Our homes, offices, and other institutions produce gray water (all wastewater except water from toilets) and black water (from toilets). Gray water can easily be treated where generated and used in urban gardens or used directly in our toilets for flushing. The black water is treated in decentralized treatment lagoons or in large-scale treatment plants. The sewage sludge from these treatment plants is being used to generate energy for use in the treatment plant. The East Bay Municipal Utility District is the only facility in the United States which is using the organic waste generated in San Francisco and combining it with the sludge from their treatment plant to produce energy in excess of their needs. These kinds of systems can be installed in all cities around the world and it will be done in smart cities of the future.

The city of Los Angeles is treating black water and sending it to a filtration plant to convert into drinking water. The system saves a lot of energy since the treated water does not have to be pumped from our water bodies to a filtration plant for reuse as drinking water. Smart cities in the future will locate their water treatment plants next to the wastewater treatment plant and recycle all black water into drinking water. As societies face increasing pressure to provide water for agriculture,

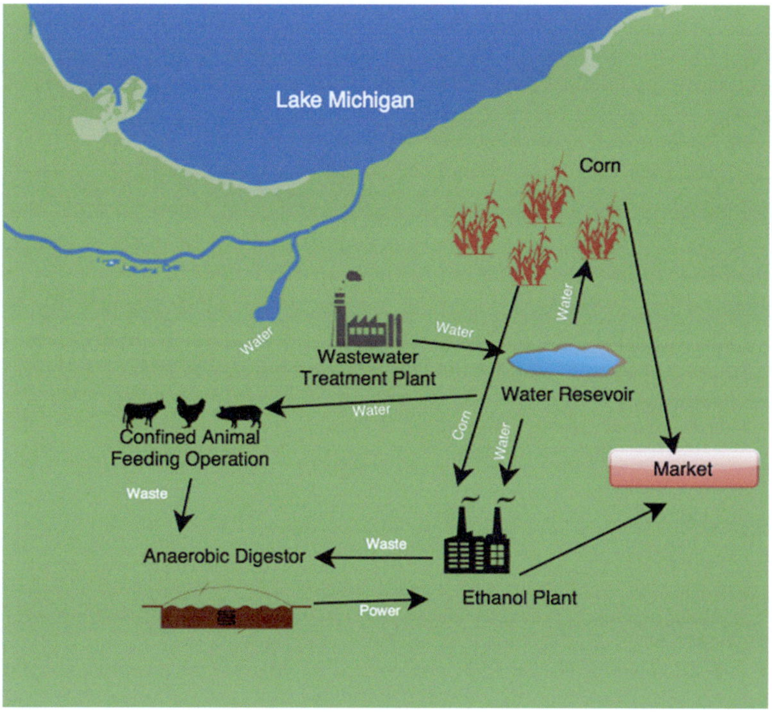

Fig. 15.9 Proposed water treatment in Hammond, IN

industry, and domestic use, they will utilize improved gray water and black water treatment and reuse systems to cope with the increasing demand of growing populations.

Wastewater treatment facilities are facing the need to reduce nitrates and phosphates in the treated water to minimize eutrophication in receiving bodies of water. There is a dead zone in the Gulf of Mexico from the nitrates and phosphates discharged into our rivers from treatment plants. Instead of building expensive denitrification plants at our old wastewater treatment plants, we can devise ways of using the treated water for agriculture (the major user of water in society). An example of the wastewater industry that can work in concert with the agricultural industry to create a zero waste system in treating liquid waste is shown below (Fig. 15.9).

Hammond, Indiana, treats liquid waste and dumps it into the Grand Calumet River, a stream that leads into Lake Michigan, which is the main drinking water source for several states in the United States. From the diagram, you can see how the treated liquid waste does not have to go into our drinking water; instead it can be used in other industries. Once liquid waste is processed, it is piped into water reservoirs to store the treated water. From there, the stored water can be used in corn fields without additional fertilizer as the high nitrates in the water act as a fertilizer for the crops. Also, the water reservoir can send water to an ethanol plant that can process the corn and sell ethanol to the market. The water can also be used for a potential concentrated animal feeding operation (CAFO). From there, the waste produced from the ethanol plant (corn scraps) as well as the waste from the CAFO can be sent to an anaerobic digester that produces bioenergy to power the ethanol plant. All the waste in such a system is recycled and becomes a valuable resource (Report to US EPA by Center for Transformation of Waste Technology 2011).

Thinking about systems as a whole and implementing them in a decentralized manner is the way to manage our waste in "smart cities." Our waste is valuable and we have the technology to treat it innovatively so that waste can be turned to wealth. We cannot afford to manage wastes in a resource-intensive manner. We have to manage wastes using innovative systems that are working in many parts of the world. Smart cities have to adopt the strategies mentioned here for solid and liquid waste management.

References

American Biogas Council (2014) Basics of anaerobic digestion

Center for transformation of waste technology (2011) Hammond, Indiana, water reuse project, report to the Great Lakes National Program Office, Chicago

Issac G-O, Zak P, Jennifer W (2013) On-site systems for processing food waste, a report to the massachusetts department of environmental protection. Northeastern University, Boston, 54 pp

Ortiz J (2014) Puerto Rico's anaerobic digester slated for 2015. Switch Road- Sustainability is on the Way, DW Focus

The Plant Chicago (2014) About us

Vasudevan R, Siddiqui F, Agrawal S, Khan E (2016) Solid and liquid waste management – waste to wealth. To be Published by Prentice Hall India

Zafar S (2014) Anaerobic digestion of livestock manure. Biopact, Mongabay

Chapter 16
The Human Body and Building Synchronization

Anil Ahuja

16.1 Introduction

The body organs are integrated to produce organ systems. Unlike most building systems, body systems are highly organized. To maintain this order, energy must be fed into the system. All living cells depend on the release and use of energy for the maintenance and operation of the living state, just like building systems. However, no building machine or body system can produce energy out of nothing; it can only change energy from one form to another.

Although all of life depends on the sun, only photosynthesizing creatures are capable of trapping light energy directly and, by using simple raw materials, synthesizing organic compounds; these organisms are called *producers*. In discussions of energy, the term "producer" may sometimes be misleading. Producers do not, indeed they cannot, generate energy out of nothing; rather, they convert light energy into chemical energy. Producers are, in fact, converters or transducers of energy, taking the radiant energy of the sun and chemical elements from the earth and transforming these into useful, energy-rich organic molecules. Because photosynthesizing organisms synthesize their own food, these producers are called "self-feeders."

Laws of energy transformations such as that of thermodynamics apply equally to buildings and living beings. The human body is an open thermodynamic system, that is, it must continually be supplied with energy to maintain itself. Energy supplied to the body in the form of food cannot be utilized with 100 % efficiency for cellular work, and at every energy transformation, there is a reduction in the usable energy of the system—usually in the form of heat. This is similar to heat dissipation in buildings, which is part of the energy unavailable for work to operate the system.

The body's energy storage involves breaking the energy-rich chemical bonds of foods through oxidation and trapping a portion of released energy in the chemical bonds of high-energy phosphate compounds called ATP (adenosine triphosphate). This storage is similar to electrochemical storage in buildings. ATP can be likened

A. Ahuja, *Integration of Nature and Technology for Smart Cities*,
DOI 10.1007/978-3-319-25715-0_16

to the “charged” form of the storage battery. Just as a battery runs down and becomes discharged when the chemicals in the electrodes no longer donate electrons, so too there is a “discharged” form of ATP called ADP (adenosine diphosphate). The link between ATP and ADP releases a great deal of energy, like the electron flow link between cathode and anode in a battery. Discharged ATP needs recharging, like batteries in buildings.

The human body is capable of using a variety of foods as sources of energy, roughly equivalent to building systems that could operate on gas, coal, wood, or fuel. The building system lacks such versatility, but humans do not. This is largely achieved because ATP is the single mediator (like currency in commercial systems) between the energy suppliers and the energy users, unlike buildings that have series of mediators that provide a direct usable form of energy such as electricity, steam, or domesticized gas. In the body ATP provides the driving force for most cellular activities and mechanical, chemical, osmotic, and electrical work, whereas in buildings the driving force takes many forms such as chilled water, hot water, compressed air, power, steam, and gas. For this reason, a human cell is about 40 % efficient as an energy-trapping machine; by comparison modern power plants are only 30 % efficient, with the overall system, taking into account source loss, 15–20 % efficient.

16.2 The Body-Building Environment

Our bodies strive to maintain, at all costs, a nearly constant core temperature for our vital organs. This most protected zone takes thermal precedence over the less vital zone of our extremities, such as the arms and legs; next down in priority are our fingers and toes. The most variable thermal zone of all is our skin surface. Similarly, buildings are frequently thermally zoned, and users (paralleling human blood flow) can retreat from—or advance to—the less protected zones as conditions demand.

To maintain the cores of our bodies within a narrow temperature range, we are always generating bodily heat and need to lose this internally produced heat to our environment. The rate at which we produce heat changes frequently, as does the environment’s ability to accept or reject heat. To regulate our bodily heat loss, we have available three common layers between our body cores and our environment: the first skin, our own; the second skin, clothing; and the third skin, a building envelope.

Once the blood and water get our surplus heat from the organ system to the skin surface, we have four ways to pass it to the environment: *convection* (air molecules pass by our surface, absorbing heat), *conduction* (we touch cooler surfaces, and heat is transferred), *radiation* (when our skin surface is hotter than other surfaces “seen” but not touched, heat is radiated to these cooler surfaces), and *evaporation* (a liquid can evaporate only by removal of large amounts of heat from the surface it is leaving). The amount of heat we lose by each of these four methods depends on the interaction of our metabolism, our clothing, and our environment. As air and surface temperature approach our own body temperature, we lose the options of convection, conduction, and radiation. Evaporation is essential, so access to dry, moving air is greatly appreciated.

As air and surface temperature fall, evaporation drops while convection, conduction, and particularly radiation increase. Usually clothing acts as an insulating layer and is particularly effective at retarding radiation, convection, and conduction. As air and surface temperatures fall well below our own, we adjust the second skin. Our second skin is just as likely as our third skin to be dominated by considerations of style more than of thermal regulation; we cannot always count on clothing—or buildings with no environmental systems—to increase our comfort.

A positive definition of comfort is "a feeling of well-being." However, the more common experience of thermal comfort is a lack of discomfort—or being unconscious of how you are losing heat to your environment. The interaction between comfort and those environmental factors can be generally summarized in (Fig. 16.1). The "comfort zone" represents combinations of air temperature and relative humidity that most often produce comfort. Building systems can provide comfort zones inside buildings.

If our second skin does not provide comfort conditions, then our first skin responds by internal activity. (See Fig. 16.2) to draw a parallel between the body and building air conditioning systems. When we are too cold, we begin to increase heat production through shivering and nondirected increases in metabolic rate. Shivering, incidentally, is uncoordinated muscular activity with no other clear purpose than generation of heat. For short periods, shivering can raise resting heat production as much as fivefold; for periods as long as an hour, bouts of shivering can give a two- or threefold increase. Piloerection ("goose bumps")—fluffing up the fur—is a bad joke in humans, but of great use to more ordinarily furry mammals to conserve body heat.

The opposite occurs when we are too hot; blood flow toward the skin surface increases (vasodilation), and the sweat glands greatly increase their secretion of water and salt to the skin surface. We are especially good at sweating, but not at all effective at panting. This increases heat loss by evaporation. The primary sensor in these thermoregulatory feedback loops is located at the base of the brain.

The presence of temperature sensors in the skin is obvious to us all, since we quite literally feel their activity; but do they play a role in thermoregulation? The answer is slightly peculiar. The sensors turn out to be much more important in the response to cold than in the response to heat. When a person is exposed to a heat load, skin temperature changes in a direction opposite that of sweating rate—the more you sweat (really sweat, evaporating the water), the cooler is your skin. Since sweating cools the skin, almost as soon as sweating starts, the sensors in the skin would send quite an inappropriate message, saying that all is well. That message would arrive long before any significant cooling of the core could happen, and it would thus turn off the cooling devices prematurely. Why the thermostat is inside the building should now be obvious.

The human body uses its own anticipatory control to conserve energy, like building systems. For instance, if you begin exercising, then the necessary increases in

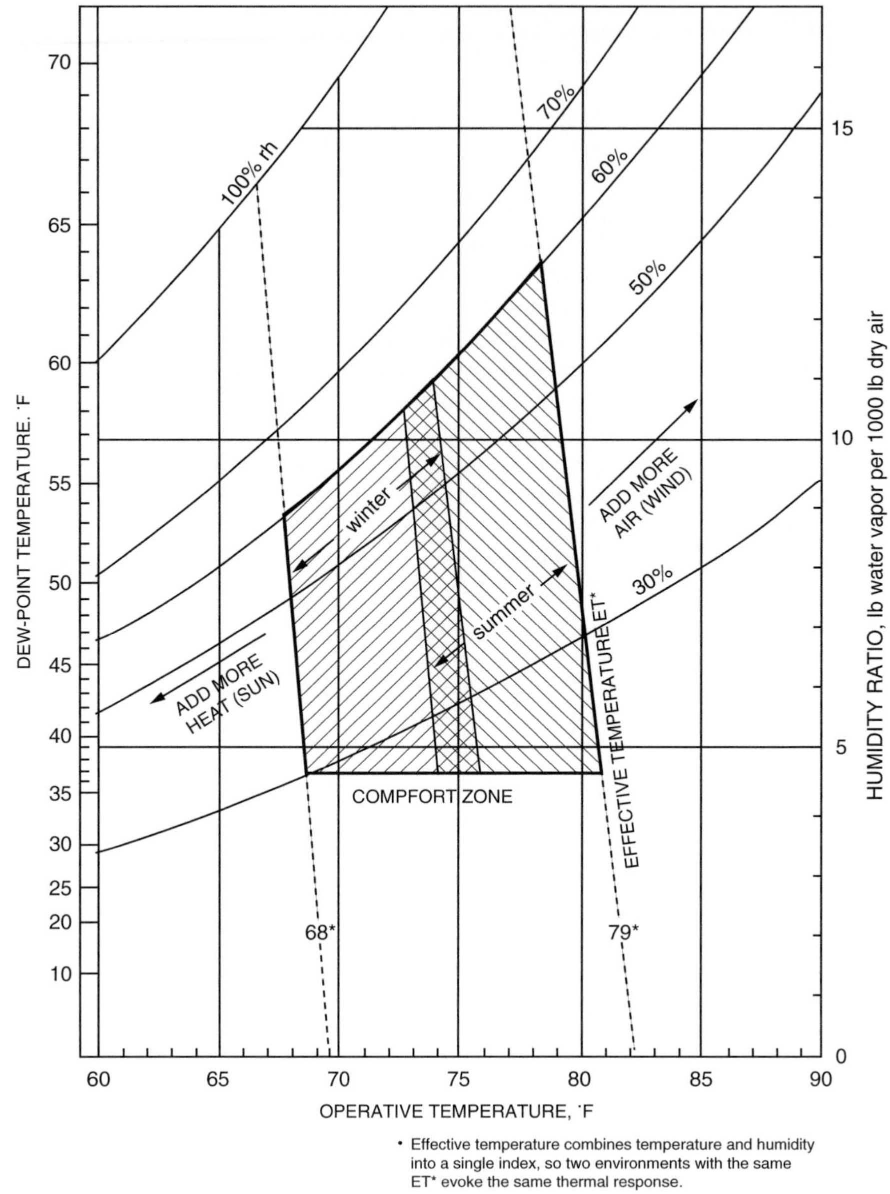

Fig. 16.1 Standard temperature and humidity comfort zone (Reprinted by permission of the American Society of Heating, Refrigerating and Air Conditioning Engineers, Atlanta, Georgia, from the *1993 ASHRAE Handbook—Fundamentals*)

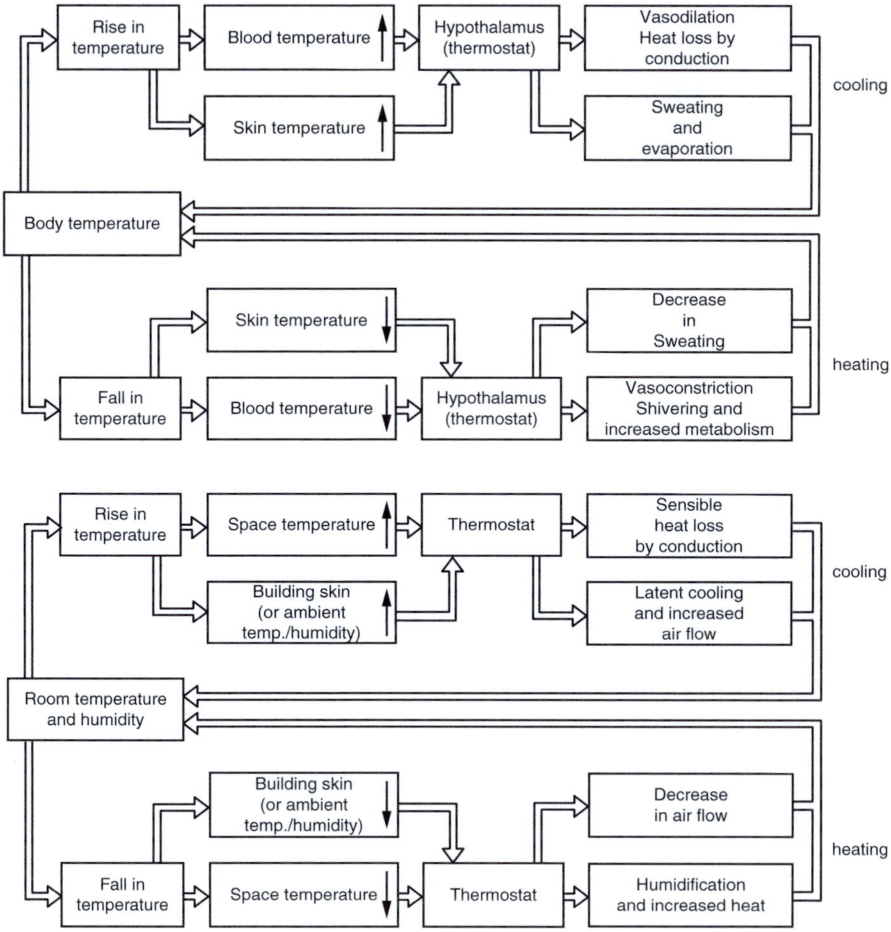

Fig. 16.2 (*Top*) Feedback loops involved in the maintenance of constant body temperature. (*Bottom*) Feedback loops involved in the maintenance of constant room temperature

breathing and circulation start *before* the oxygen and carbon dioxide concentrations in the blood have changed. Both diffusion and circulation take time, and the body anticipates the *lag*. With the increased neural traffic to and from the muscles, the brain, of course, knows that you are up to something. The system therefore has an additional input, one that we might call a "change" detector as opposed to the "error" detecting sensors within the loops. Anticipatory control is always secondary, even though it goes into action earlier, and its role is reduced when the feedback loops get properly aroused.

The longer a feedback system takes either to respond or to recognize that it has responded, the more important anticipatory control becomes. If you are thirsty, you drink water. If you kept drinking until the water was actually absorbed through the wall of the stomach, you would have far exceeded that needed to replace the deficit underlying the thirst, and you would then excrete a lot of water. In short, the system would overshoot badly and perhaps oscillate. Anticipatory control by sensing things such as stomach distention compensates for the relatively long lag of the fundamental, nutritional loops. These feedback loops are like PID (proportional, integral, and differential) control loops commonly applied to maintain the indoor environment of buildings within a comfort zone.

16.3 Electrodynamics of Humans and Buildings

To explain this, a popular model is that of the self-exciting dynamo: A conductor rotating in a magnetic field creates an electron flow. If the flow passes through a coil, the coil itself creates a magnetic field that keeps the flow moving as long as the conductor keeps rotating. Well, the earth rotates. The core is conducting but probably more here and less there because of temperature and/or compositional inhomogeneities. Although the ingredients for the creation of a magnetic field are there, the exact mechanism is still not known. While moving through this naturally occurring earth magnetic field, humans generate electric voltage within his body by induction. Human bodies compensate for externally induced voltages by the development of internal systems that generate electrical pulses that are high enough to overcome this interference generated externally.

The electric field is a function of *only* the electric system voltage level. The higher the voltage, the stronger the electric field will be. The unit of measurement of the electric field is volts per meter. The conductor does not have to have current flowing to generate an electric field. The conductor has to be energized to have an electric field present. An electric field is present in an electric toaster that is plugged into a wall receptacle yet not toasting anything, even if the switch is open. The electric fields of the human body are generated by movement of muscles. To evaluate and compare the effect of an external electric field on the human body, the magnitudes of some natural body electric fields are cited.

The heart produces electric fields on the surface of the chest that are on the order of 50 mV/m. In the area of the brain, fields of 1 V/m over an extremely small area are produced, with values of 1 mV/m over an area of a few centimeters. The differential voltage ("radical differential voltage"—chiropractic term) across the spinal cord that appears on the skin surface can be as high as 13.5 mV standing or as low as 5.5 mV sitting. To produce neural stimulation, the tissue fields are in the order of 1–100 V/m. Brain nerve cells typically produce 0.1 V, 2 ms spike trains with repetition rates in the range of 1–100 Hz.

The human cell has two distinct electrical parts. The membrane of the cell acts like an insulator, and any 60-Hz induced current flows around the cell. The individual cells in the human body can have an electrical field at the surface of the cell up to 10 MV/m (10 million V/m). Through Faraday's law, one should calculate the typical voltage induced by a magnetic field across a cell membrane to determine what body can or cannot tolerate. The levels of internal body voltages, currents, and fields are several magnitudes greater than those of externally induced electrical and magnetic fields to immune body from small level electrical jolts.

Hazardous Voltage and Currents The human body is susceptible to direct application of electric current. The human body can withstand electrical abuse, but there is a limit to what the body can withstand. The upper end is electrocution. Teetering on the border is electroshock therapy. Safety clearances and strict code requirements for building systems set the "safe" limit.

To electrocute a person, the electric chair uses a 5-ampere current at 2640 V. Two thousand volts is usually sufficient to stop the heart, and an extra 640 V are added to compensate for persons with large body mass and any additional voltage drop. Five amperes is used as this amount will not burn the human body. The current is applied for two 1-min periods with a 10-s delay between jolts. The time period is necessary to ensure death.

Electroshock therapy uses 350–800 V peak. The time of application varies from 2 ms to 2 s with a period of 0.5–2 s between applications. The amount of power applied ranges from 59 to 100 J. One source equated the power to that of a 40-W lamp. The application of this amount of current to the brain modifies memory and personality.

In treating heart failure, it is not unusual to apply a momentary shock of 50 W/s or 400–500 W/s of direct current at a cycle of 6 s between shocks. This application, which is at the surface of the skin, will usually reset the heart firing cycle to halt ventricular flutter or fibrillation.

To avoid the above, the maximum amount of current that has been established as acceptable and safe for humans is 9 mA. Between 30 and 250 mA is the danger zone, resulting in cessation of breathing. Beginning at 75 mA, the heart can go into fibrillation. *Ground fault circuit interrupters* (GFCIs), which are used and required by the National Electrical Code, are set to trip at 5 mA.

16.3.1 Radiofrequency Waves and the Human Body

Radiofrequency waves are present in all modern buildings. This section points out some of the effects of radio waves on the human body. Just as the body absorbs infrared and light energy, which can affect thermal balance, it can also absorb other longer wavelength electromagnetic radiation. Visible light (wavelengths 0.4–0.7 μm) and infrared (wavelength 0.7–10 μm) are absorbed within 1 mm of the body surface. The heat of the absorbed radiation raises the skin temperature and, if sufficient, is detected by the skin's thermoreceptors, warning the person of the possible thermal danger. With increasing wavelength, the radiation penetrates deeper into the body. The energy can thus be deposited well beneath the skin's thermoreceptors, making the person less able or slower to detect and be warned of the radiation. Physiologically, these longer waves only heat the tissue and, because the heat may

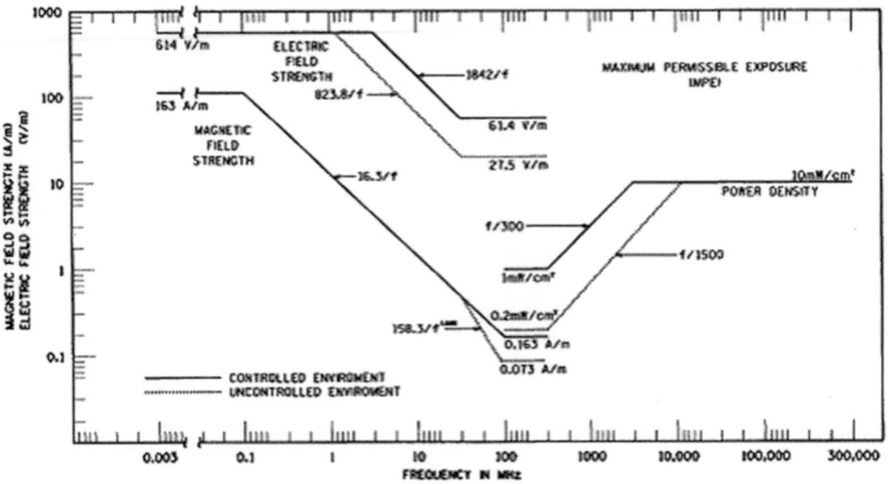

Fig. 16.3 Maximum permissible levels of radiofrequency for human exposure. Note: Maximum permitted power densities are less than half of sensory threshold values (Reprinted by permission of the American Society of Heating, Refrigerating and AirConditioning Engineers, Atlanta, Georgia, from the *1993 ASHRAE Handbook—Fundamentals*)

be deeper and less detectable, the maximum power density of such waves in occupied areas is regulated (see Fig. 16.3).

16.4 Human Body Systems Versus Building Systems

The principal elements of building systems that can be compared with body systems are blood composition (analogous to air quality in building systems), blood pressure (analogous to pumping pressure in buildings), and body temperature control (analogous to building temperature control). These are regulated by adjusting other items such as cardiac (control) output, the resistance (friction) of the vessels (pipes) of the microcirculation, and the relative apportionment of blood (power) to the various organs. These parallels are depicted in a split graphic (Fig. 16.4), half of which shows a building and the other half a human body.

16.4.1 Human Circulatory Versus Building Hydraulic System

The human heart is about the size of a clenched fist. In a life span of 70 years, it beats 2.5 billion times and pumps 40 million gallons of blood. The heart is indispensable to one's very existence. The heart pumps blood, oxygen, and food to every

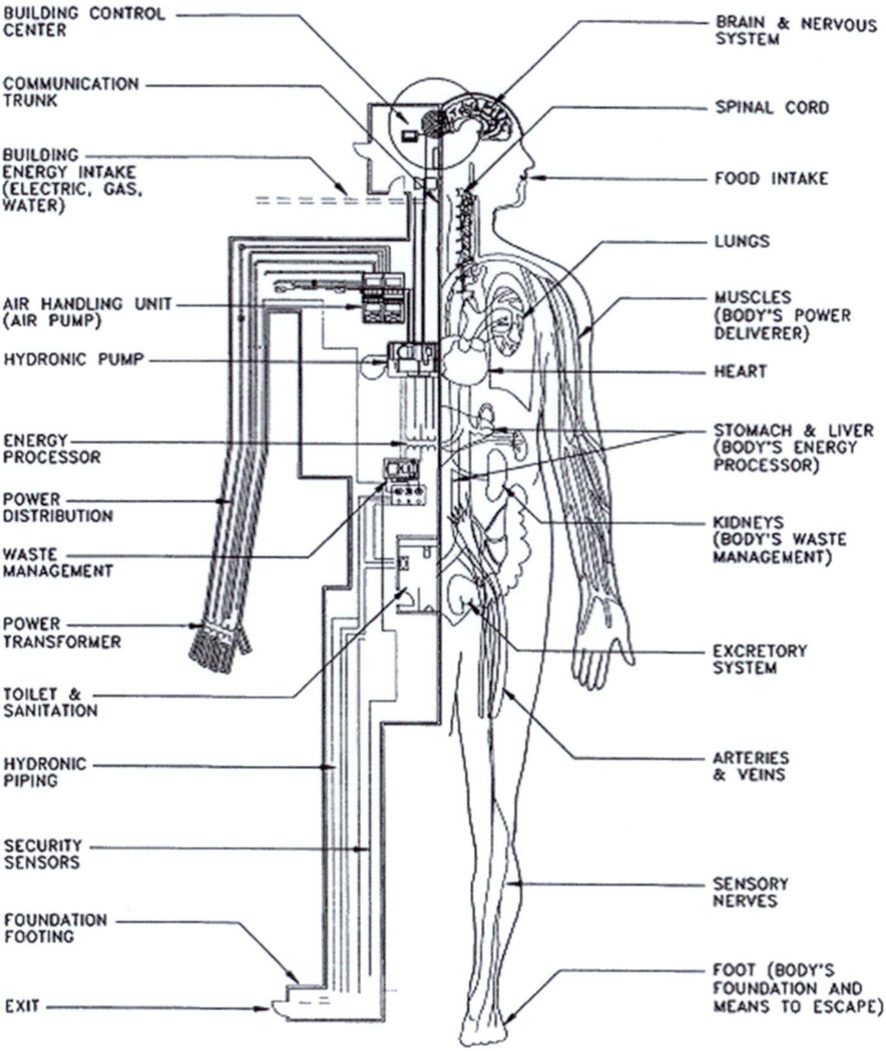

Fig. 16.4 Body-building system integration

cell of the body, and by the return circuit of the bloodstream, cellular wastes and carbon dioxide are removed (see Fig. 16.5) for a comparison between a human circulatory system arrangement and a common arrangement of a building hydraulic system). The human heart has an energy demand, and this varies with its work load, similar to the energy demand of a building hydraulic system. For example, the heart of an obese individual requires more energy to operate than that of a person of more normal girth. It has been estimated that for every pound of fat, there are three extra

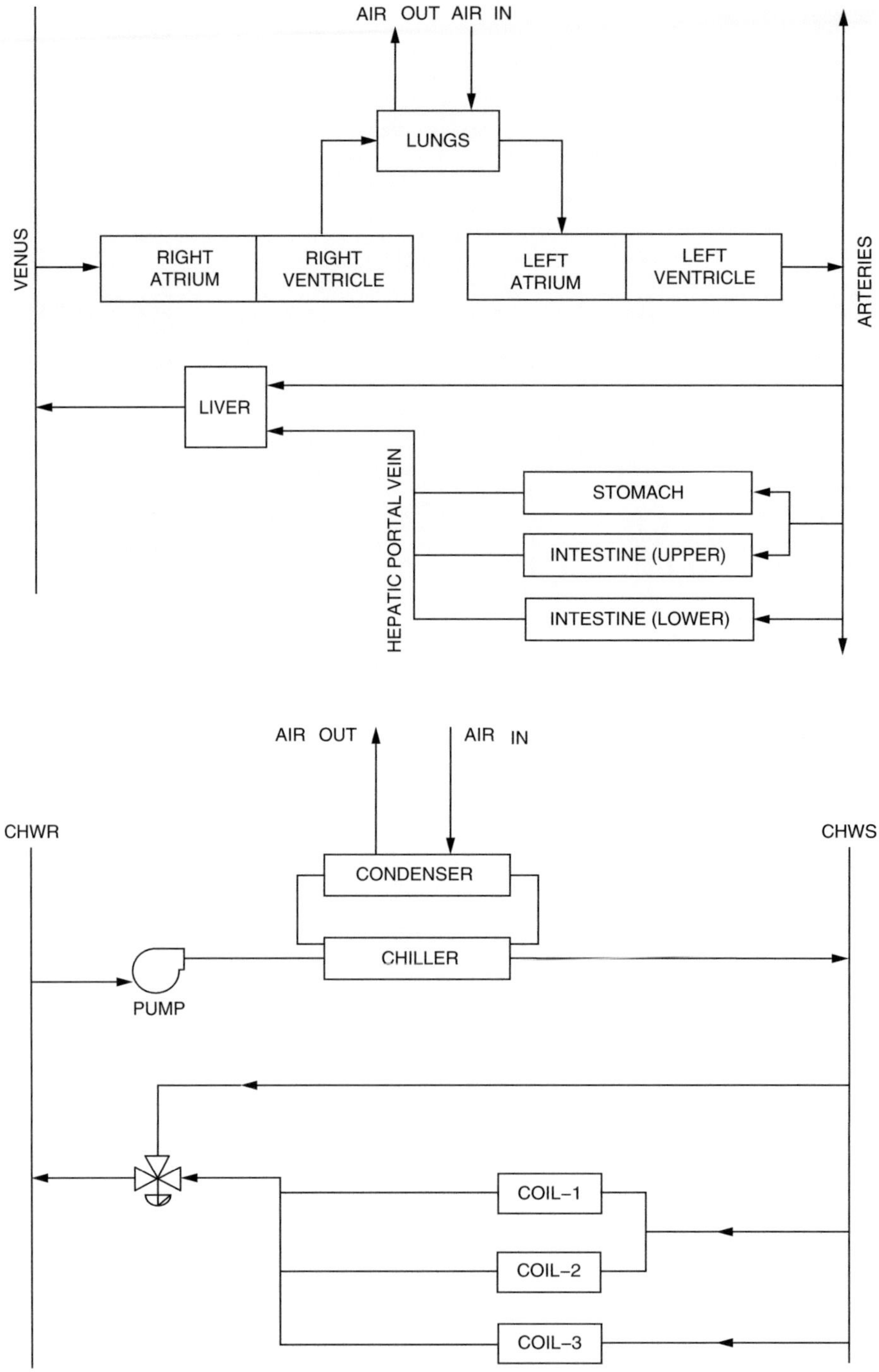

Fig. 16.5 (*Top*) Human circulatory system. The circulatory connections of our liver and functionally adjacent organs, pointing out the way its portal system supplies it with venous blood from the intestines. (*Bottom*) Building chilled water system

miles of blood vessels that require pumping. Similarly large building systems require more pumping and hence more energy.

Both the mechanical integrity of the circulatory system and proper exchange across capillary walls require that the pressure of the circulating fluid be kept within reasonable limits. Average blood pressure is mainly determined by two factors: how hard the heart is pumping and the overall resistance of the microcirculatory vessels. A change of circumstance may lead to an increase in the resistance of some and to a decrease in others and leave still others unchanged. In a building pumping system, a good flow control system works over a range of flows, of temperature, of pressure, of loads, and so forth. The main trick is to put the sensor where control needs to be most effective and where alterations in the controlled variable are least tolerable.

All too many of us regulate our blood pressures at levels too high for our own best interests. In many people with high blood pressure, because the arteries are stiffer, the body pressure sensor is fooled by less pressure on the receptors in the artery walls, so the system will "think" the pressure is lower than it actually is. Thus it will regulate at a higher pressure. Similarly too many buildings operate with oversized pumps ("just to be on the safe side"), and system control is made to regulate flow at a higher pressure.

In the human body, the amount of heat that can be exchanged by conduction across moist interfaces depends on the area exposed; the respiratory surface can be increased tremendously by extensive folding of the surface. Similarly in air handling units, the coil surface area is increased by folding more rows and adding fins. Once the air has passed across the respiratory surface, efficiency of distribution is increased greatly if the respiratory surface is amply supplied with blood. The same applies to flow in coils inside air handling units.

16.4.2 Human Respiratory Versus Building Air System

The exchange of gases between the living organism and its environment is called *respiration.* Respiration in humans involves two related processes. The direct exchange of gases between environment and organism is called *breathing* or *ventilation;* this exchange in the body involves oxygen transfer from the respiratory system to the blood, which carries oxygen to every cell of the body and exchanges it for carbon dioxide. The carbon dioxide is carried back to the respiratory system, where it, in turn, is given up in exchange for more oxygen. In buildings' "respiratory" systems, oxygen is transferred from the atmosphere to the air circulatory system which supplies oxygen to all internal spaces and exchanges it for carbon dioxide. The carbon dioxide is carried back to the air handler where it is exchanged for more fresh air. Although oxygen and carbon dioxide both cross the interface of the lungs, they move independently of each other. They are like total strangers who simultaneously touch the station platform, one boarding and the other leaving the same train. Building air handling units have similar separate chambers (see Fig. 16.6).

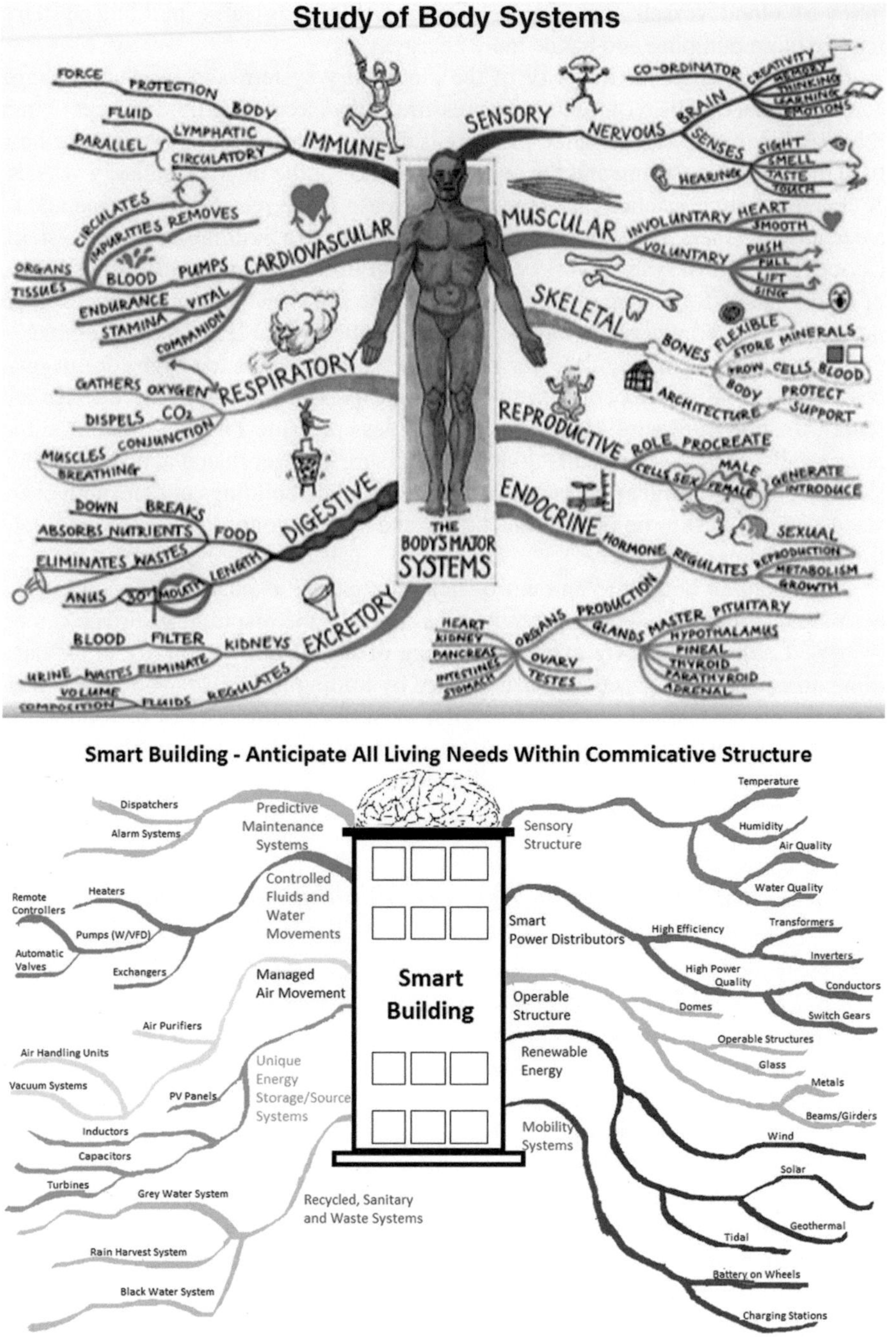

Fig. 16.6 Smart building communicative structure

The human air filtering and conditioning system has unique similarities with building air systems. Air first enters the respiratory tree through the nostrils, where projecting hairs filter out dust and debris. The air then enters the nasal cavity, which is divided by a septum and lined by a ciliated epithelium. The cilia distribute a mucus film that traps debris, and in conveyor-belt fashion, this mucus sheet is moved toward the throat, where it is swallowed or spat out. Beneath the nasal epithelium is a rich network of blood vessels that helps to warm the air as it eddies about in the nasal cavity. Scroll-like turbinate bones and air spaces in the skull (sinuses) form a complex labyrinth of passages that further assist in filtering, warming, and moisturizing the incoming air. Even our tears help moisturize the air we breathe, since tear ducts drain the continuously operating tear glands and empty directly into the nasal chamber.

Warmed, humidified, and scoured of dust and small particles, the air passes from the nasal cavity to the throat (pharynx), where the air and food passages cross. From the pharynx the air enters the larynx (voice box or Adam's apple) through a slitlike opening, the glottis, which is guarded by an elastic flap, the epiglottis. The glottis in

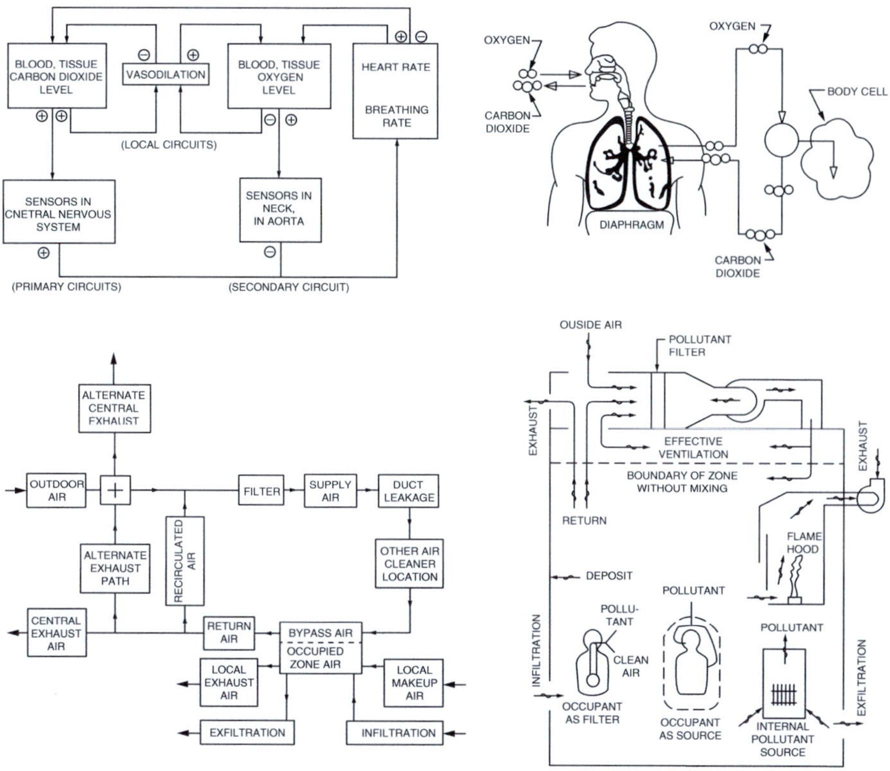

Fig. 16.7 (*Top*) Body control system to control the level of oxygen (and carbon dioxide) in the blood. (*Bottom*) Building IAQ system to control the level of oxygen and other impurities inside the building

human body acts like damper in building air handling unit, the glottis remains open at all times except when we swallow; similarly damper in air handling unit remains open at all times except when air is bypassed.

With each breath, we take in 500 ml of air, but only 350 ml of that air actually reaches the alveoli and the remaining 150 ml is trapped in the upper respiratory passages. Similarly in buildings, all the ventilation air does not reach work area; some air remains trapped in stagnant zones of the building (see Fig. 16.7).

Human beings take 4–10 million breaths a year, equivalent to a fan running at 15 revolutions per minute all the time. The movement of air in the passage from the nose to the lungs, any number of things may go wrong, but if nothing goes wrong, the respiratory system functions reliably. To learn how to increase reliability of building environmental systems, a study of reliable human respiratory systems can provide useful tips. The respiratory center has a built-in safety factor with two feedback loops: one loop is intrinsic in the medulla and the other is extrinsic through the reflex system. This redundant control system is characteristic of many important body functions and is essential for a system that must operate without failure in its lifetime.

How does the respiratory center "know" what the body's oxygen requirements are? At first glance it might seem sensible for the amount of oxygen in the blood to influence the respiratory center directly, so that when the blood oxygen level goes down, ventilation increases to compensate; however, this turns out not to be the case. Increased levels of carbon dioxide in indoor air rather than decreased levels of oxygen stimulate the respiratory center to activity. Although the respiratory centers of the brain monitor carbon dioxide levels in the blood and thus regulate ventilation, there is an independent, or backup, system that registers changes in the amount of oxygen in the blood. The transfer of gases between the blood and other tissues of the body is called *internal respiration* and strictly speaking is a function not of the external respiratory system but of the circulatory system.

16.5 Human Body System Automation Versus Building System Automation

The body receives messages from its own organs and from the external world. We perceive light, sound, odors, pressure, temperature, chemicals, and the like; we think; we move; we have unconscious thoughts and conscious ones. The normal functioning of the body depends both on receipt of stimuli and on production of integrated responses. For the body to perform its activities in coordinated fashion, there must exist a connecting link between stimulus and response, between receptor organ and effector organ; that link must be a system capable of channeling information from one to the other.

As reasoning individuals, we understand why we need a nervous system, but do we reason enough to understand how the nervous system works? The human nervous system is divided into the peripheral nervous system (PNS) and the central nervous system (CNS).

The Peripheral Nervous System: The PNS comprises all of the neurons and nerve fibers outside the CNS and has two subdivisions: the somatic nervous system, with nerves running directly from spinal cord to effector organs (actuators), and the autonomic nervous system (ANS), with nerves running from the spinal cord to effector synapses (ganglion in humans is like an intermediate, N2-level bus in building system automation) outside the CNS and these send signals through nerve fibers to the effector organs.

The CNS consists of the brain and the spinal cord. It serves as a clearinghouse for all nerve impulses, controlling, directing, and integrating all messages within the body. The brain is about the size of a cauliflower and has been described as a "great raveled knot," a "modest bowl of pinkish jelly," or a "messy substance." The CNS is concerned with integrating and coordinating all nervous functions, both voluntary and involuntary. All sensory input arises from the environment (external and internal), and the effector organs (muscles and glands) are the ultimate destination of nerve impulses.

Lying between the CNS and the environment and the CNS and the effector organs is a vast network of nerves and ganglia that constitute the PNS. Hardly any tissue or organ is missed by the complex array of nerves, and thus, through them, the CNS is in continuous contact with nearly every part of the body. But there are other nerve components whose functions ordinarily lie outside our consciousness or con-

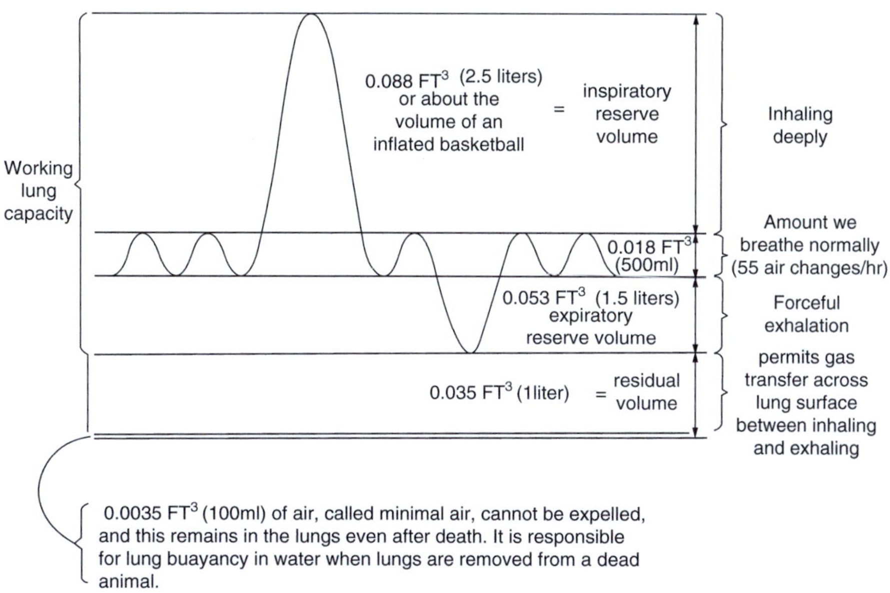

Fig. 16.8 Human body air handling system flow control. The working lung capacity is about 0.0159 ft^3 (4.5 l) as measured by deep breathing in and out, but in actual fact the total lung capacity is 0.194 ft^3 (5.5 l), with 0.03531 ft^3 (1 l) as residual air. To match air changes with breathing rate, a building system needs to provide 55 air changes/hour compared to normal 10–12 air changes/hour

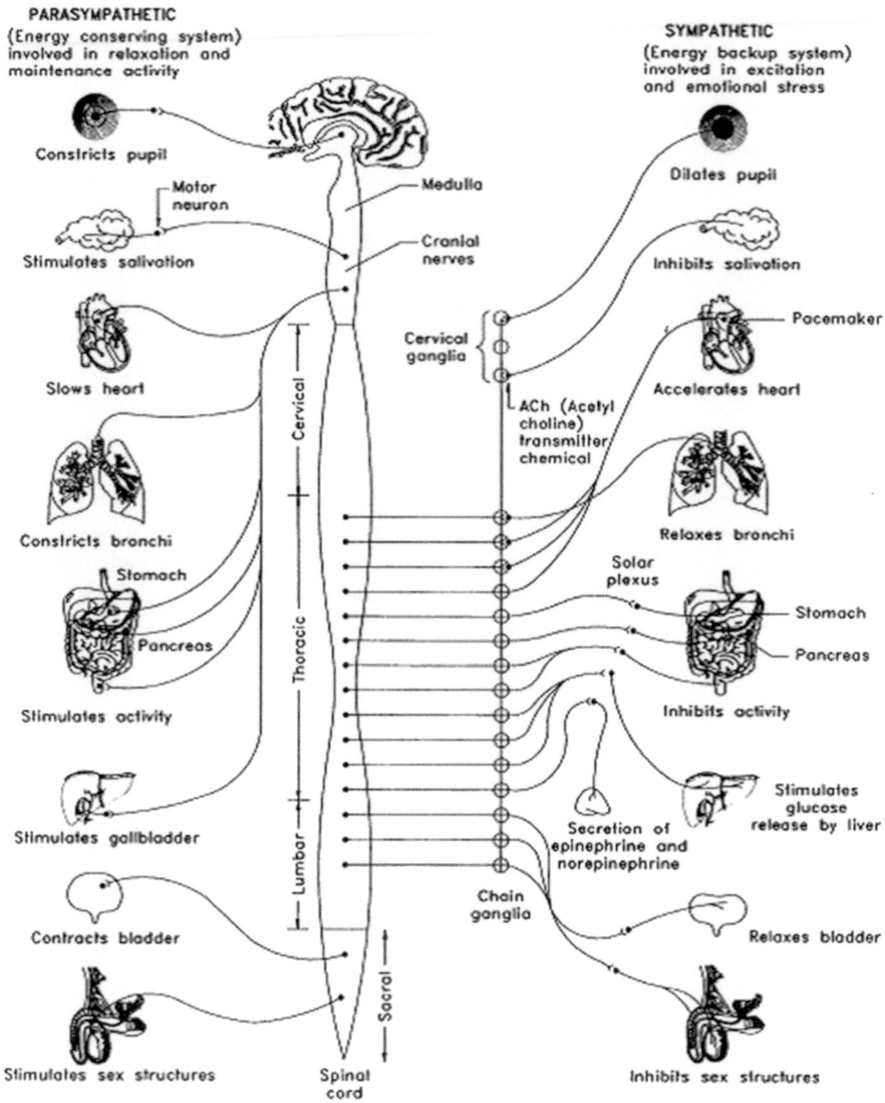

Fig. 16.9 Autonomic nervous system

trol. These components connect with our visceral organs (heart, lungs, kidneys, blood vessels, intestines), and we neither are aware of what these organs are doing nor have voluntary control over them.

Most of the nerves are mixed and so include both sensory and motor components. The motor components of these visceral nerves are extremely interesting and important. They constitute what is known as the *autonomic nervous system.* There are two divisions within the ANS—the sympathetic and the parasympathetic (see Fig. 16.8 and compare with Fig. 16.9).

The sympathetic system mobilizes resources of the body for an emergency, effects that are sometimes called flight-or-fight responses. Thus, in excitation or stressful situations (fear, anger, and the like), the heart beats faster, the stomach muscles relax, the pupils widen, and energy is made available because the brain sympathizes with organs. The *parasympathetic nervous system* is an energy-conserving system; its nerves are involved in relaxation and maintenance activities such as slowing the heart rate, enhancing digestive action, causing blood sugar levels to drop, etc. The condition of a particular organ receiving neurons from the autonomic system is determined by the relative amount of stimulation from the parasympathetic and sympathetic nerves. As a consequence, in the heart, many glands, and smooth muscle, which are innervated by both sympathetic and parasympathetic nerves, a dynamic seesaw balance in organ function is attained.

This is all interesting and rather obvious. But what does it have to do with the building system technology that we look at and deal with every day? Well, let us start with something relatively simple. How does the "intelligent" building control think? Not exactly like humans.

Conventional digital technology is based on bivalence. This means two absolute opposites—yes or no, on or off, 1 or 0, and are or are not. Human logic functions with multivalence, which means that things can be partially so, to some degree, and partially not so, to some degree. For example, we can think of temperature as ranging from cold, to cool, to comfortable, to warm, and to hot without knowing exactly what the temperature is. All building environmental systems operate on the principle of measuring exact temperature and then making a decision to cool or not cool. To take another example from everyday activity, say you are driving and need to make a turn because the road ahead of you is curving. Does your mind analyze the circumstances and immediately advise you the road is curving with a 22° radius, requiring 11° clockwise initial rotation of steering wheel for a duration of 22 s? This may be mathematical logic, but our mind does not function this way to make our bodily systems react. In fact, our mind simply sees the road curving and autonomically makes us turn. All of this happens and we are probably not aware of it. It was sort of fuzzy which is what new technology is trying to achieve and it is called fuzzy logic. Using human perception instead of two-state temperature control building environmental systems can provide a fuzzy–cozy feeling.

As another exercise in logic, consider an apple. A nice, shiny red apple is clearly an apple in every sense of the word. Now take a bite out of the apple. Is it still an apple? Certainly—at least to some degree. Continue to eat the apple. Is it still an apple? At what point is it no longer an apple? In the precise world, an apple is a totally completed thing, and if any element is missing, it ceases to be an apple. In the real world, a partially eaten apple maintains its membership in the community of apples, at least to some degree. Remember, we live in a part-load world where peak load is a special condition. For the sophisticated user or designer who is big on personal comfort and individual control, fuzzy philosophy offers an attractive alternative mode of thinking and analysis. It is an important tool for bridging the gap between the science and art of engineering.

Perhaps age is an even better example. When does a person become an adult? According to law, it is at age 18. If we view this in a precise bivalent way, adulthood assumes yes/no characteristics. But is this reality? Is a person a nonadult up to the very last second before midnight of his 18th birthday, and is he fully an adult in the flash of time after the stroke of midnight? We all know some individuals mature more rapidly than others and that adulthood, even for the most intelligent of individuals, occurs as a result of a more gradual process.

While the technology associated with human logic is of growing and perhaps prime interest to the building system engineering community, in many ways the whole philosophy of fuzzy thinking is of far greater importance. Charles Kettering once noted: "The only difference between a problem and a solution is that people understand the solution." We really do not understand the problem and as a result never get to a real solution. Applying fuzzy logic in this manner allows the designer to transcend building system science and to explore, and perhaps comprehend, the *art* of building systems engineering.

16.5.1 Human Logic Versus Building Control System

How do our bodies integrate action? Two systems in our bodies act as coordinating links between stimulus and response: the endocrine system and the nervous system. The endocrine system regulates the activities of cells by means of hormones. The hormonal system provides for slow communication on a long-term basis. This is similar to time delay control action in building logic control. The nervous system, on the other hand, provides for rapid communication between the various tissues and organs of the body. The nervous system employs electrochemical messages, nerve impulses, that run along specialized nerve pathways receiving and transmitting information to and from various organs. Nerve cells, or neurons, are fundamental units of the nervous system, specialized to conduct electrochemical messages at high speed. Building systems communication protocols are based on neuron chips on the same guidelines popularly known as LON (Local Operating Network), a trademark of EcheLON. Neurons, when bundled together in cablelike form as nerves, are able to regulate the direction in which information flows.

There are three different functional classes of neurons: *Sensory neurons* receive stimuli from the environment and transmit information to the central nervous system (brain and spinal cord). These are comparable to transducers used in building automation systems. *Motor neurons* conduct messages from the brain and spinal cord to the glands and muscles, These are comparable to transmitters used in building systems. *Interneurons* act in an integrative capacity and shuttle signals back and forth between the neurons of various parts of the brain and spinal cord; this is where building systems technology, unlike body systems, is not coming together to share the information. In the body system, a signal travels upward via the spinal cord

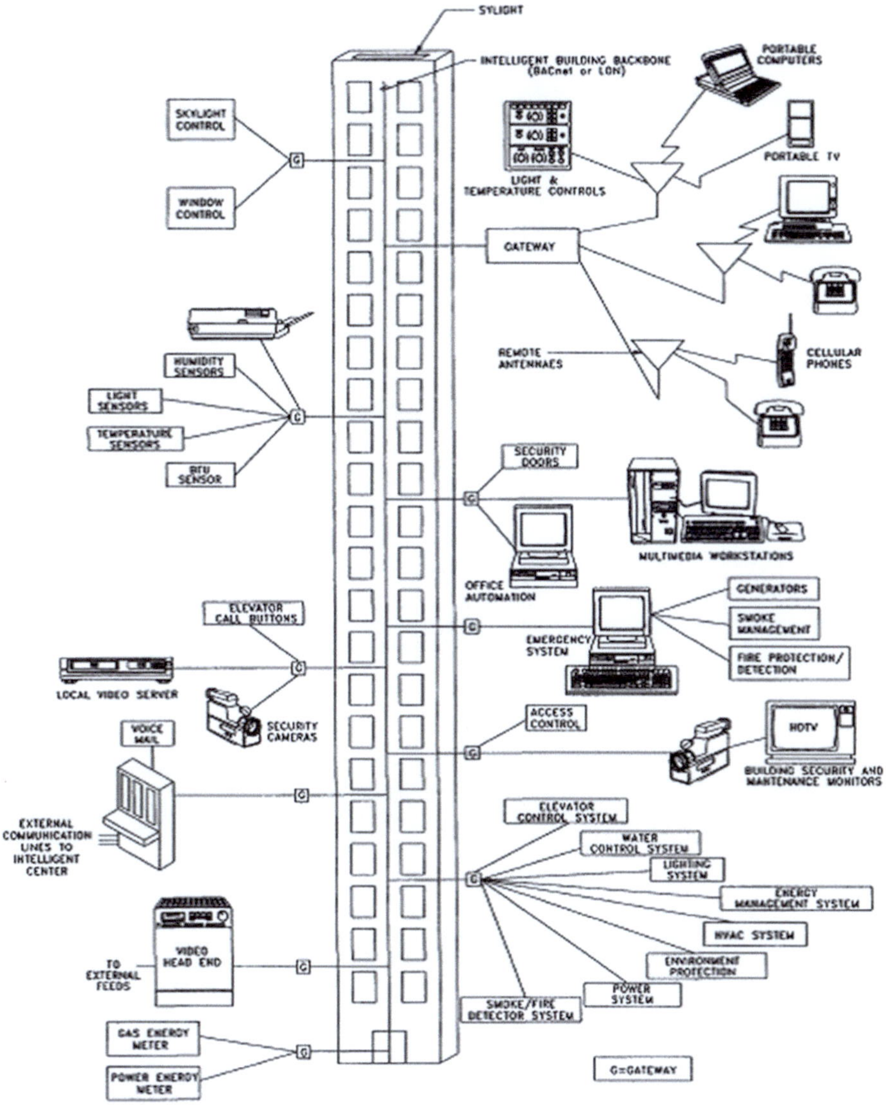

Fig. 16.10 The ideal integrated in-building network

interneurons and through a number of relay centers in the brain before reaching the higher centers (see Figs. 16.10 and 16.11).

For example, let us analyze for a moment a motor that drives an air conditioner or perhaps more appropriately an air handling unit. For simplicity, let us assume that varying the motor speed varies the cooling output. Human logic will integrate this with the feeling of warm, cool, or hot. For example, from temperatures of 65–85 °F, we live in the category of warm, and say from 50 % motor speed to 90 %, our motor

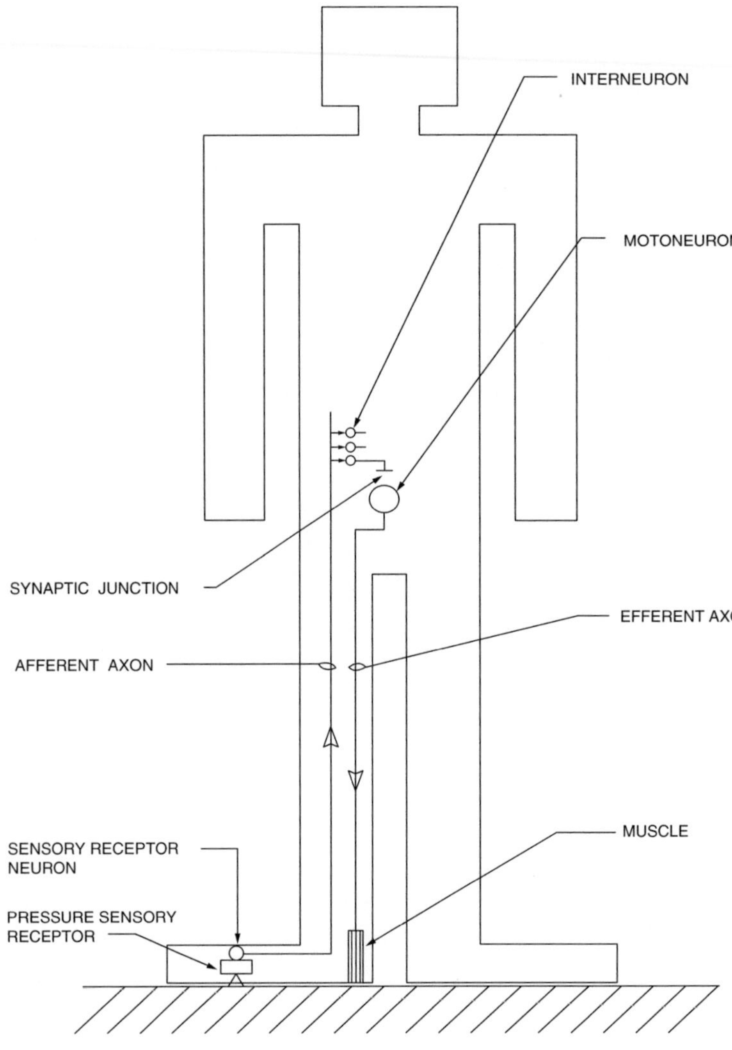

Fig. 16.11 Excitable tissue is called into play when a person steps on a sharp pebble. Illustrated is a reflex arc that follows the excitation sequence: sensory receptor to its neuron to afferent axon to interneuron to motoneuron to efferent axon to muscle. The person starts to jump off the noxious stimulus in 0.025 s

possesses membership in the category of fast. Thus, the human logic rule is *if warm, then fast.* So we have the beginning of a human logic system. Carry the analysis further, *if cold, then stop* motor and *if hot, then go maximum speed.* Notice how the control strategy has developed regions where a human rule applies and we can develop a family of rules that in effect give us a human logic system.

Human philosophy is also capable of assisting in the understanding of part-load conditions. Consider the initial step in an engineering analysis of a building system.

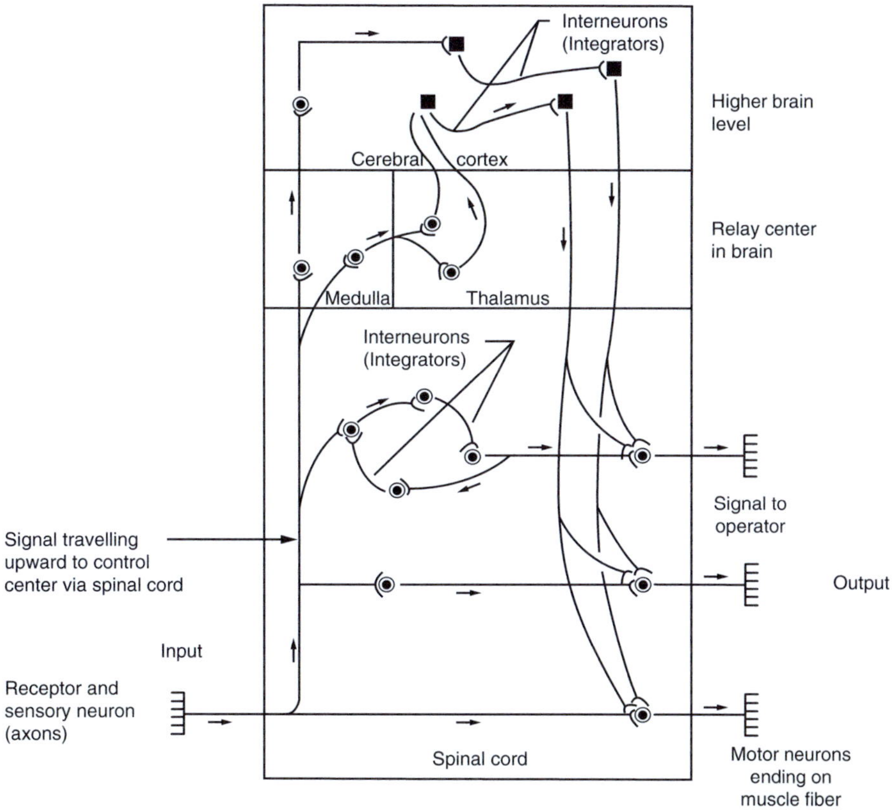

Fig. 16.12 Simplified circuitry of the nervous system. Neurons are arranged into cables consisting of many axons (message receivers) and dendrites (passes signals). Axons are bundled together, to form a multistranded cable, from the nerve fiber or nerves. The collection of axons and dendrites in the brain and spinal cord comprises the information centers

Our prime objective is to determine the peak load and design a system around it. In conventional digital approach, we have two conditions, *peak* and *not peak.* Suffice to say that peak conditions rarely occur. Applying human philosophy to the problem yields another approach, which recognizes that peak work load conditions are a special case similarly Zero work load is also a special situation.

Several observations about human logic (technically called fuzzy logic) are important. First, these systems depend on good input information, so sensing technology comparable to a body system is required. But rather than trying to model the control strategy mathematically with complicated equations, it uses a simpler series of fuzzy rules to develop a fuzzy control system. Since it is more human, it can be used by the less trained, and it is more easily maintainable. Building control systems based on fuzzy logic are especially helpful where nobody knows how to control the system, since all it requires to operate is basic intuition (Fig. 16.12)

Chapter 17
Understanding the Occupant's Behavior

Petra Stieninger

When reading about sustainability in the building sector, buzzwords such as net-zero energy buildings, smart meters, building automation, and internet of things are dominating the discussion. Technologies in this field have been evolving rapidly over the last years and decades, and building systems are getting more and more sophisticated. M2M solutions are facilitating the operations of buildings, the mix of natural and mechanical (or passive and active) solutions for HVAC and lighting are making high-quality indoor comfort more resource efficient, and improved building envelope insulation and the implementation of green roofs support energy conservation for heating in the winter and cooling in the summer.

Technically, a building can be designed for net-zero energy, net-zero water, and net-zero waste, consuming only as much as it produces and reuses itself. High-efficiency appliances, automated building systems, and a well-insulated building envelope can make it happen. However, there is one important factor that can make the operation of an on-paper net-zero energy building unsustainable and wasteful in reality—the occupants of the building.

17.1 Sustainability Is About Changing Behavior

The occupants are the ones who turn on the light, the computer, or the TV. They are the ones who take a shower, water their plants, or clean the dishes. They are the ones who use household or office appliances in order to live their everyday lives. And they are the ones who want to maintain a certain indoor comfort and quality of life. They are the ones who live, work, and play in the building. And by doing this, by living their ordinary, everyday lives, they consume resources; most of the times, without even being aware of it.

When it comes to sustainability and resource efficiency, it always comes down to the user of a certain product, the consumer of a certain service, the person who wants to get from A to B, or, in this case, the occupant of the building.

A. Ahuja, *Integration of Nature and Technology for Smart Cities*,
DOI 10.1007/978-3-319-25715-0_17

The decision of the user results in more or less sustainable or resource-efficient activities. The user decides if they use the high-efficiency energy-saving program of the washing machine or if they use the extra-clean wash program. The user decides if they take the train to work or if they drive in their car. The user decides if they open the window in order to lower the temperature inside the house or if they turn on the air conditioning. In this chapter, most examples will focus on energy efficiency, as this has been my main field of research over the last 6 years; however, it can be applied to other resources and sustainability in general as well.

What Makes Us Consume Energy? What Makes Us Waste Energy? And What Makes Us Save Energy?

We do not behave in unsustainable ways just for the sake of being unsustainable. Thus, we do not consume energy just to consume energy and we would not waste it just to waste it. We are living our everyday lives, sometimes without even knowing how many resources we consume and/or waste every day. The installation of energy-efficient appliances and the insulation of a building may reduce the energy consumption of this particular building from a technical standpoint. It makes the operation of the building technically more sustainable. However, it does not answer the question why we actually need energy input at all in that building. Why do people have to turn on the light and why do they leave it on when they leave their houses? Why do they need air conditioning? And why do they consume almost three times more energy in that same building in the United States than they do in Europe?

We consume energy and other resources in order to satisfy a certain need. It is not the direct need for energy. We turn on the light because we want to read a book, and if the sun is not shining or if our building does not have enough natural light, we need artificial light which requires energy. We need water to keep plants alive that make our building more beautiful and connect us with nature in the built environment. We cook because we are hungry; and in order to heat up the stove, we need energy. We need water to wash our dishes afterwards. We turn up the heat when we are cold, which requires energy. And sometimes we leave on the light when we leave our house just because we forget to turn it off. Sometimes we use inefficient air conditioning to cool our houses instead of just opening a window. We need energy and many other resources in order to produce food, clothes, computers, books, and anything else we need for our everyday lives.

Almost everything we do requires energy of some sort. However, we would never ask for energy. We ask for a warm soup, a chilled glass of wine, a movie, a convenient way to get from A to B, clean clothes, and so on.

Conventional building design focuses on the technical causes of energy consumption. In order to make a building more energy-efficient, engineers and architects design high-efficiency appliances and insulate building envelopes. No one really asks the question why the energy is needed and if there were any alternatives.

The main question they ask is which features of the building need the energy as input and how the input can be reduced while providing the same output.

A technical solution for high energy use for lighting, for example, would be the installation of LED lights instead of conventional light bulbs. The occupant can consume the same amount of artificial light while the needed energy to provide it decreases. The engineer solved the problem of how to provide the same amount of artificial light with less energy input. However, the engineer did not ask the question why the occupants need artificial light at all.

Smart building design does not just install LED lighting. Smart building design asks the question on what the root causes for the energy consumption for lighting in that building are. It is about learning about the occupant's behavior and their daily activities and understanding their particular needs for lighting. Is it because there is not enough natural light in the building or is it because the occupant always leaves the light on when they leave the house? Is it a specific activity of the occupant that needs a lot of light or is it the lack of knowledge on their lighting needs? Are there any alternatives for this specific activity that requires so much light? Are there ways to substitute the activity (root cause of energy consumption) itself or do we have to find a technical solution (technical cause of energy consumption)?

Smart building design focuses on the occupants and their needs. Smart building design does not just change the technical aspect of energy consumption. It changes the occupant's behavior toward more sustainability. In order to make people change their energy-consuming behavior, we have to find out what the needs are. We have to find out what the occupants want and what makes them behave in a certain way. We have to find out what the motives for the choice of how they satisfy their needs are. We have to find out the root causes of their energy consumption. At the same time, we have to offer smart design solutions on how to satisfy those needs with less energy input (Fig. 17.1).

Smart Building Design Is About Thinking Outside the Box

For the occupant, all that matters is the satisfaction of their needs. For them, it does not matter how they achieve it. It does not matter if they need energy input for it or not. All that counts is that the needs are satisfied. The architect's and engineer's goal has to be to understand those needs and satisfy them in more resource-efficient ways. Maybe the occupant will not even need light at all. The solution could be somewhere completely else.

Energy consumption does not just depend on technical features. In fact, that is just a very small portion. Energy consumption, and consumption of all other resources, depends on the behavior and the lifestyles of the people who live, work, or play in the building. Comparing, for example, the energy consumption per capita in the United States with Europe, it becomes very obvious that behavioral and cultural aspects have a tremendous influence on the amounts of energy we consume. We all use similar building technologies and the same high-efficiency building from Europe could be built somewhere in the United States. However, the US version

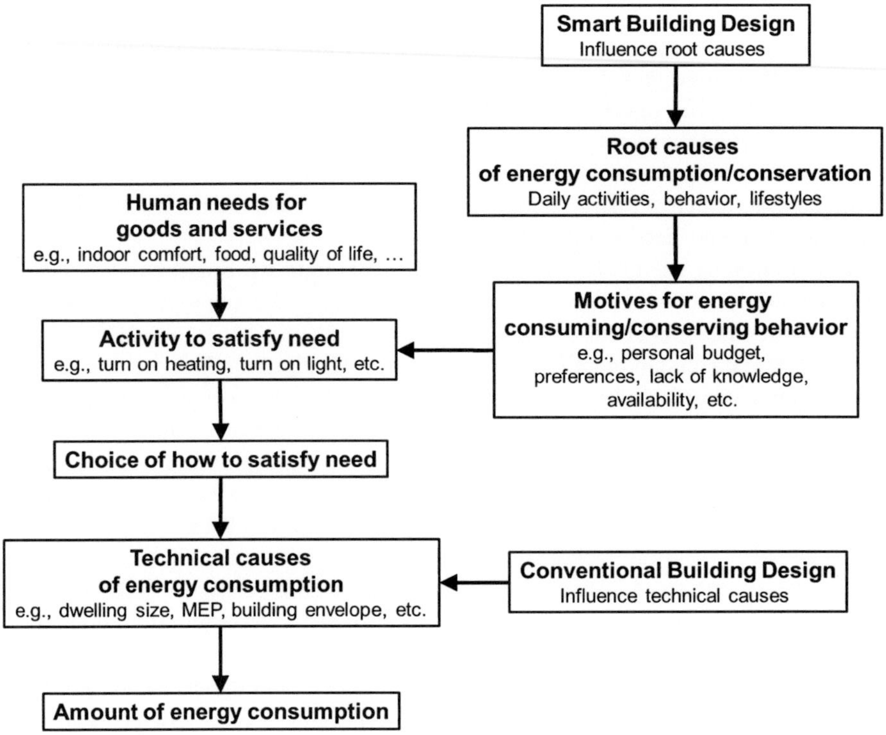

Fig. 17.1 Technical causes and root causes of energy consumption (Source: Stieninger (2013); adapted to building scale)

would probably still consume more than the European version just due to the ways people live, work, or play in that building.

The average US American consumes more than twice the amount of energy of the average European. Obviously, energy is much more expensive in Europe than in the United States, which gives a great incentive for Europeans to save energy. At the same time, there are many cultural aspects and differences in lifestyles and the way Europeans live their lives compared to Americans that result in less energy consumption in Europe (Fig. 17.2).

Both Americans and Europeans spend up to 90 % of their days indoors. However, in Europe, for example, it is still more common to open a window in order to increase the ventilation in a room (also called natural ventilation) instead of turning on the air conditioning or using mechanical ventilation systems. Many people in Europe do not like being in buildings that do not allow to open the windows. In the United States, almost 10 % of the total energy consumption in buildings is used for air conditioning, whereas in Europe, less than 5 % are being used for cooling.

Of course, climatic aspects play a very important role when it comes to sustainability in buildings. There is definitely a higher need for cooling in the United

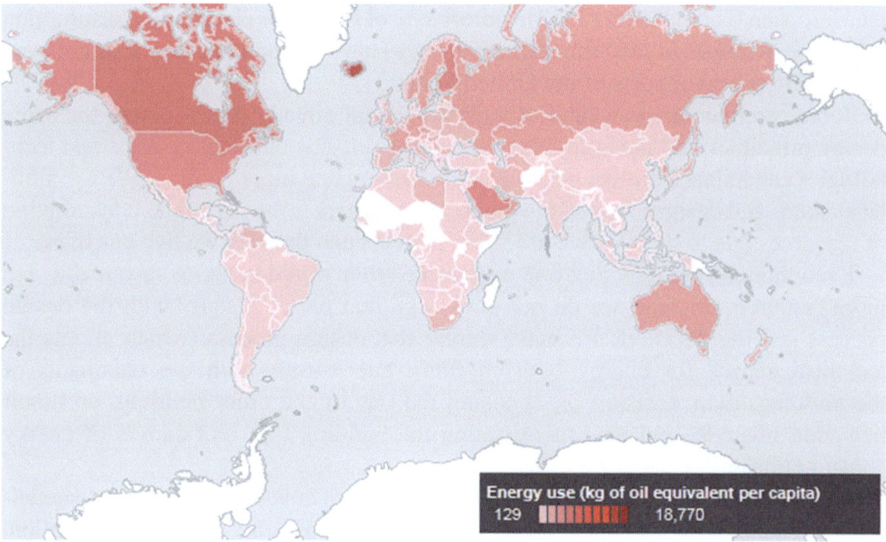

Fig. 17.2 Worldwide energy use per capita (Source: http://data.worldbank.org/indicator/EG.USE.PCAP.KG.OE/countries/1W?display=map)

States than in Europe due to climatic reasons. However, additionally, in certain cultures, the connection between the inside and the outside of a building plays a much more important role than in others. Europeans, in general, have a very strong connection to the outside. The importance of having the ability to open a window in a building, the option of offering outdoor seating in a restaurant, and the design of open public spaces is obvious when you look at European cities and their architecture and urban design. The high urban density which is also very common for European cities is balanced out with natural, open features that allow for healthy, high-quality indoor and outdoor environment and to connect with nature.[1]

Looking at the Middle East, buildings and entire cities used to be designed according to natural cooling elements. What is in some literature called "Islamic architecture" or "Islamic design" may coincide with Islamic principles such as the design focus on the inside of a building and focusing on the importance of family and water elements for the ritual washing. However, these design aspects totally make sense from a climatic point of view in that region as well. Clay structures with little connection to the outside creating shady courtyards and with water places on the inside are definitely more energy efficient in a desert environment and provide better natural cooling than transparent steel and glass structures the way they are being built almost everywhere nowadays.

On the other hand, apparently, Europeans like their morning coffee which requires energy input as well. According to the Energy Efficiency Status Report of

[1] Stieninger (2013).

the European Commission (2012), almost 2 % of the entire electricity consumption in European homes is used for coffeemakers, a statistical number that does not even show up in energy reports of the United States.[2]

Today, architecture can (and to a certain extent already does) feature the same design principles and ideas anywhere in the world. The building systems and technologies can balance out the need for more heating and more cooling. We use similar systems and designs around the globe. What differs is the per capita consumption of resources due to the way we use those systems and the way we live our lives.

Even though we know that one way or the other would be more sustainable and more resource efficient, we do not always go that route. It starts with the design process and the decisions we make during that design process (which affects the technical causes for energy consumption), but it ends with the occupants of the building, their decision on choosing the one or the other building, and their behavior, lifestyle, and way of operating the building (the root causes of energy consumption).

When designing a smart building, we must not just concentrate on energy modeling and calculations on paper. Those are technical causes for energy consumption. When designing a smart building, we have to focus on the different needs of the occupants and how to satisfy those needs in more sustainable ways. Thus, what we really want to find out are the root causes for the energy consumption or rather for behavior and lifestyles that require energy input.

17.2 The Five Root Causes of Energy Consumption: The Five A's

Research that I have done in the urban planning field resulted in five categories of root causes that can result in behavior change. Those five categories can also be applied at the building level. We choose certain products, services, or behavior and entire lifestyles because of their **affordability**, their **availability**, their **accessibility**, their **attractiveness**, and our **awareness** of the fact that they are affordable, available, accessible, and attractive.[3]

These five factors (the five A's) are not only the root causes of our sustainable or unsustainable behavior; in a design process of a smart building, they can become the tools for the architect and the design team. If the design team focuses on these five categories of root causes, they can turn root causes of unsustainable behavior into motivators for sustainable behavior. Designing a building in a smart way can make sustainable living affordable, accessible, available, and attractive. In addition, smart design can raise the awareness of the advantages of sustainable living (Fig. 17.3).

---

[2] https://ec.europa.eu/jrc/sites/default/files/energy-efficiency-status-report-2012.pdf.

[3] Stieninger (2013).

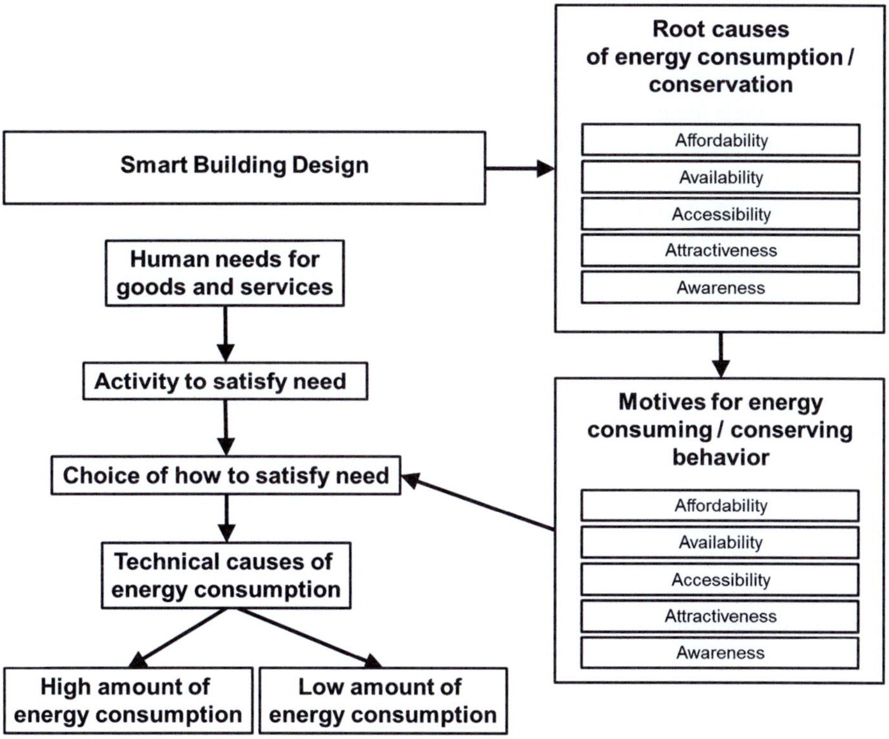

Fig. 17.3 The five root causes of energy consumption (Source: Stieninger (2013) – adapted to building scale)

We have to be able to afford something so we can consume it. High energy prices in Europe are one of the factors why Europeans consume less than half the energy per capita than Americans. However, either way, saving energy means saving money anywhere in the world. It is not just the affordability of the resource itself. It is also the affordability of actions that help conserve it and the affordability of adapting behavior and lifestyles to a more sustainable way.

In most cases, sustainability investments are not much more expensive than the unsustainable alternative (according to the US Green Building Council, investment costs for green buildings are only 2 % higher than conventional buildings).[4] Calculating the payoff, sustainability actually helps saving money on the long run. A lot of people are not aware of that or just do not want to think long-term. Therefore, creating awareness of sustainable options that are affordable is crucial as well. It is not just about making sustainable behavior affordable but it is even more about making people aware of the fact that it is affordable.

[4] U.S. Green Building Council: LEED® Core Concepts Guide. An Introduction to LEED and Green Building. Third Edition.

For instance, a lot of people perceive the energy-wasting alternative as cheaper than the sustainable, energy-efficient alternative. They see the difference in the investment costs the moment they make the purchase decision on their household appliances, their windows, or their HVAC systems. However, long-term, they would save a lot of money by choosing the higher investment cost which results in much lower operation costs. In most cases, operation costs are not perceived directly. They are hidden costs we are not directly aware of when choosing one or the other product.

Receiving an energy bill at the end of each month shows your consumption over a 1-month period but you are not aware of your specific consumption and when and why you consumed more or less energy. So there is no direct incentive to choose the energy-efficient solution.

Affordability is of course directly connected to availability. If certain services or product features are not available, I will not be able to implement them. Products that can be installed for more energy efficiency or programs from utility companies or building management companies have to be available. The availability can be influenced by the occupants to a certain extent only. And again, the occupants of a building have to be aware of the availability of programs, products, and services that help them to conserve energy. Information, promotion, and education are therefore imperative. Building occupants have to get informed and educated about their options and sustainable alternatives that are available.

Furthermore, buildings, infrastructure, and services that support sustainability have to be accessible for everyone. A lot of high-performance buildings or smart buildings are still standalone pilot projects with only an elite group of occupants that enjoys the green branding and is able to pay for it. High-performance, smart buildings should be state of the art and accessible for everyone, everywhere.

Sustainable, Smart Buildings Should Not Be the Exception—They Should Be State of the Art

Furthermore, sustainability also has to be attractive. Green buildings used to be considered ugly. They provided the required functionality in order to meet the efficiency goals but they did not offer beauty. This has changed drastically since the 1980s. In fact, sustainable building rating systems such as the Living Building Challenge evaluate the beauty of the building and the happiness of its occupants in addition to its sustainability.

Attractive design can change ways we perceive things and can therefore have a very strong influence on our behavior. Design has to be obvious in order to direct people toward a certain behavior. For example, the design of a staircase can influence our choice of using the staircase or the elevator. If we find a beautifully designed, light and open, safe staircase in the entrance area of our building, we will be more likely to use it than if it is just an emergency staircase somewhere in the back of the building. Walking up the stairs is much more energy efficient than using the elevator.

The most important factor is the awareness. Raising the awareness of the fact that people are using a lot of energy for ordinary, everyday life activities can change their energy consumption tremendously. When it comes to everyday activities, we are not always aware of our constant demand for energy. Information and education are crucial to make people aware of their wasteful behavior and teach them sustainable behavior.

It Is Not Just a Lack of Knowledge—It Is a Lack of Will

In many cases, we know that our behavior is unsustainable. We know about the consequences of leaving our air conditioning on at home while we are at work. We know about greenhouse gas emissions, their impact on our personal health and on climate change, and their interrelation with extreme weather situations that are threatening our personal lives. We know that recycling our waste minimizes the need for landfills and can even contribute to alternative ways of generating energy through waste incineration or other waste-to-energy technologies. We know it but we do not do it.

Obviously, there is a disconnection between environmental knowledge and environmental behavior. The American psychologist and behaviorist B.F. Skinner claimed in his work that behavior is determined by its consequences. Thus, we cannot expect people to modify their behavior as a result of information they received. In addition, the activator–behavior–consequences scheme (or ABC model) claims that certain activators direct behaviors and consequences motivate behavior. A consequence can be feedback, reinforcement, or punishment. However, in order to change behavior, the consequence has to be directly connected to the behavior or, at least, as close as possible.[5]

That is, taking a look at a monthly energy bill and noticing that the energy consumption and therefore the energy expenses increased might not necessarily change our behavior. A smart meter that is constantly collecting data about our energy consumption can bring the awareness and the feedback or consequence to the occupants of a building directly and immediately. Directly connected rewards or penalties are very important. If the smart meter tells us how much energy (or money) we wasted (or lost) due to a certain activity (e.g., in monetary figures), it may even trigger a much stronger emotion than if we just get some general information on our energy consumption. If we let people know how much they lose, it will give them a motivation to change their behavior. A smart meter that tells you that you wasted $20 by leaving your air conditioning on while you were at work or a smartphone app that tells you that you just wasted $2.50 for taking the elevator instead of the stairs (the same amount of money you paid for your train home from work) might motivate you to change those behaviors the next time.

[5] Geller (1989).

17.3 Sustainability Is About Understanding Behavior

A smart building is a building where people automatically change their behavior toward sustainability due to the building's design and the way it is operated. The five categories of root causes can be turned into motives for energy-conserving behavior if they correlate with smart building design. The five A's are not just the reason for our unsustainable behavior; they can also be used as tools in order to change that behavior and make it more sustainable. In addition, changing the five A's can result in an automatic change of our behavior toward more sustainability without even noticing it. Making the sustainable operation of a building affordable, available, attractive, and accessible and making people aware of it can result in an automatic behavior change. Today, we have all the technologies and design features available that allow for a construction of a smart building whose occupants behave more sustainably no matter where in world they live.

The most important thing when designing a smart building is therefore to understand the people who live, work, and play in the building. We need to gather information on their behavior, their preferences, and their needs. In addition, we need to collect data on the root causes for their behavior, their preferences, and their needs. This is where it comes to the era of smart technologies and smart buildings. It is not enough to just install an energy-efficient envelope or to put a solar panel on a rooftop. We have to learn from the occupant's behavior, understand it, and design accordingly. What are the needs of the occupants and how can we meet them in more energy-efficient ways?

Smart technology helps us to understand the behavior of the building occupants. Technology allows us to compensate the failures and find the connection between behavior, appliances, and the real needs and causes of energy consumption. Smart meters enable us to measure the energy use and consumption of each and every single appliance in a building, be it a coffeemaker, an air-conditioning unit, or the TV. In addition, they inform us on when energy consumption takes place and enable us to find failures of systems and potential for change. It is a mixture of machine-to-machine, machine-to-people, and people-to-people communication.

Smart technologies enable us to collect data on every activity, analyze it, visualize it, evaluate it, and make suggestions on how to solve the problem, how to satisfy the root cause, in a more sustainable way. We can collect data on daily activities and find out what the root causes of energy consumption are no matter what part in the world we are in. The root causes bridge the gap between technical causes of energy consumption and human behavior. In addition, constant feedback loops enable us to enter into a learning cycle that allows us to optimize technical systems, behavior, and their interactions.

Smart technologies allow for direct and immediate feedback. The energy user gets hit by the consequence the second they consume energy. It is a constant educating and awareness building on how much energy people use for their everyday lives. For example, a smartphone application that tells us how much money (energy) we

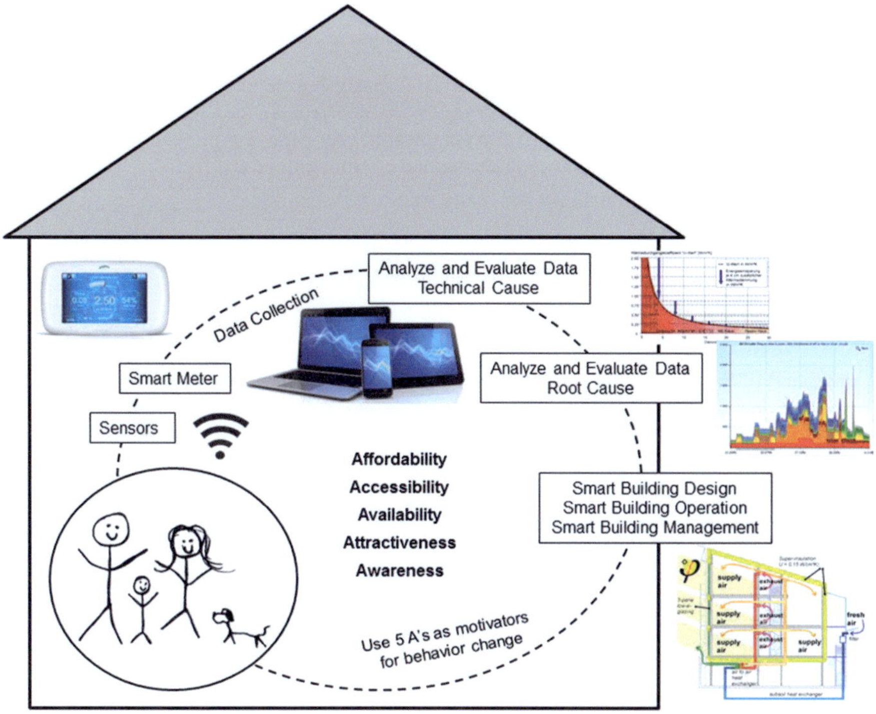

Fig. 17.4 Process from energy-efficient buildings to energy-efficient occupants (Source: author)

just wasted by turning on the air conditioning instead of opening the window will influence our behavior. Of course, the smartphone application can measure the outside and inside temperatures and will tell you how much money you save by opening the window (Fig. 17.4).

Sustainability is about understanding behavior. Designing a smart, resource-efficient building, the architect, the engineer, and the building manager have to understand what the needs of the occupants are and come up with solutions on how to satisfy them in sustainable ways. This is a constant process that starts during the design and continues throughout the operation of the building. It is a constant loop of data collection, analyzation, design and operation adaption, and lessons learning for everyone in the building.

Smart building design should be about making buildings more sustainable by enabling long-term behavior change. It is not about building energy-efficient, smart buildings. It is about creating energy-efficient, smart lifestyles in those buildings. The goal is to turn the root causes of unsustainable behavior and lifestyles into motivators for sustainable behavior and lifestyles.

References

Geller ES (1989) Applied behavior analysis and social marketing: an integration for environmental preservation. J Soc Issues 45(1):17–36

Stieninger P (2013) Changing human behavior towards energy saving through urban planning: creation of a new planning approach. Lessons learned from Europe and North America, Vienna/Chicago

Suggested Readings

Stieninger P (2015) Smart cities. Behavior change towards sustainability. An international perspective. Presentation at USGBC Spring Blitz, Chicago, 25 March 2015

http://data.worldbank.org/indicator/EG.USE.PCAP.KG.OE/countries/1W?display=map

http://living-future.org/lbc

Part III
The Smart Building in the Smart City

Chapter 18
Vision of Cities: From the Green City to the Smart City

Ingrid Kaltenegger and Helen Santiago Fink

Cities are complex, adaptive, social–ecological systems (SES)[1] "characterized by a particular human settlement pattern that associates with its functional or administrative region, a critical mass and density of people, man-made structures and activities" (UNEP 2011).

Cities house 60 % of the world's population and individuals spend 80–90 % of their time in buildings of one type or another. In contrary to common belief, urban systems can be more sustainable than rural or suburban areas. When a city is dense, people and resources are close to each other, and energy can be saved when ways are short. Densely populated cities can also be a source of innovation when human capital is concentrated on a relatively small area.

Through integrated planning, creative design, and innovative technologies, our built environments can reinforce the human–nature nexus to create smarter holistic communities that reinforce urban resiliency through addressing environmental and socioeconomic urban needs. Smart buildings can contribute to this aim by offering their surface areas for natural vegetation in support of heat and storm water mitigation, local food production, as well as energy conservation and define the sustainable cities of the present and future.

There have been several attempts to capture a "comprehensive and transferable model of sustainable urban development"[2] represented by terms like eco-cities and smart cities, yet there is no consensus as to what exactly these concepts mean and what criteria is necessary for classification. International cooperation is underway under the auspices of the French Standards Agency (AFNOR) with strong involvement from China to define smart communities with an inclination toward holistic environments that go beyond technological instrumentation to more fully address ecological and socioeconomic urban issues. But let us have a look at some definitions that are around for some years.

[1] Levin and Harvey (1999), Berkes et al. (2003), Gunderson and Holling (2001), Norberg and Cumming (2008), Evans (2008).

[2] Rapaport (2014).

A. Ahuja, *Integration of Nature and Technology for Smart Cities*,
DOI 10.1007/978-3-319-25715-0_18

The term "**eco-city**" was coined by Richard Register[3] for the first time in 1987. Originally described as "an urban environment system in which input (of resources) and output (of waste) are minimized" (Register 1987), there is still no one accepted definition or standard for an eco-city. The term is rather regarded as an "umbrella concept," or a collection of ideas of how to create more sustainable urban areas.[4] Nevertheless, there have been several attempts to define a set of normative criteria for eco-cities: Kenworthy in 2006 identified 10 eco-city dimensions, which emphasize:

1. A compact, mixed-use urban form
2. Public transit and non-motorized modes of transportation
3. A natural environment that is intact and helps sustain the city's food needs
4. The use of environmental technologies for resource and energy efficiency
5. Public culture
6. Community
7. Equity
8. A good governance
9. Economic performance and employment maximized by innovation
10. Creativity and uniqueness of the local environment

Examples for leading eco-cities in the world are Curitiba (Brazil), Freiburg (Germany), Stockholm (Sweden), and Adelaide (Australia) (Figs. 18.1 and 18.2).

As a synonym to eco-city, the term **sustainable city** is used. Also for a sustainable city, there is no completely agreed upon definition or on components that should be part of such a definition, but in general experts agree that a sustainable city should be linked to sustainability such that "a sustainable city should meet the needs of the present without sacrificing the ability of future generations to meet their own needs." This leaves a great variety and many possibilities for cities to find their very own way to become sustainable. According to the term sustainability, an ideal sustainable city is able to create an enduring way or life across the four dimensions: ecology, economy, politics, and culture.[5] Claiming to be sustainable, a city should at least be able to feed its citizens from the surrounding countryside, power itself from renewable energy sources, produce lowest pollution possible, efficiently use land, compost used materials and recycle them, or produce energy out of waste.

Three cities (of some more) across the globe are leading in all those actions mentioned above.

[3] Register: Ecocity Berkeley: Building Cities for a Healthy Future (1987).

[4] Rapaport (2014), Jabareen (2006).

[5] James, Paul; with Magee, Liam; Scerri, Andy; Steger, Manfred B (2015). *Urban Sustainability in Theory and Practice: Circles of Sustainability*. London: Routledge.

Fig. 18.1 Eco-city Adelaide/Australia (Source: http://www.travelblat.com/adelaide-is-the-ecocitizen's-city/)

18.1 Vancouver, Canada

Vancouver, already being on top in worldwide "livable city" rankings, has an ambitious goal of becoming the "greenest city in the world" by 2020.[6] Therefore, the city has developed a Greenest City Action Plan 2020.[7] The city gets already 90 % of its supply out of hydroelectric power but is also exploring wind, solar, and tidal as resources for power.

Vancouver also has the lowest carbon emission per capita[8] on the continent, thanks to mass transit, bike lanes (approx. 250 miles), ride-sharing programs, and greenways. In their action plan, an emission decrease of an additional 33 % is envisaged, while also enacting strict green building codes (all new buildings constructed from 2020 onward have to be carbon neutral in operations) and reduce the energy user and greenhouse gas emissions in existing buildings by 20 % over 2007 levels.

[6] http://vancouver.ca/green-vancouver.aspx.

[7] http://vancouver.ca/files/cov/Greenest-city-action-plan.pdf.

[8] http://vancouver.ca/green-vancouver.aspx.

Fig. 18.2 Eco-city Freiburg/Germany (Source:http://www.reisefuehrer-deutschland.de/baden-wuerttemberg/schwarzwald/freiburg/wandern-und-radfahren.htm)

The green mobility tasks are to make the majority (over 50 %) of the trips by foot, bicycle, and public transport and to reduce the average distance driven by the residents by 20 % from 2007 levels.

Overall, the goal of Vancouver is to reduce the city's ecological footprint by 33 % over 2006 levels (Fig. 18.3).

18.2 San Francisco, USA

One of the most densely populated metropolitan areas in the United States and a green city is San Francisco. San Francisco recently won the waste management category at the inaugural C40 and Siemens City Climate Leadership Awards, following a Zero Waste Program with the aim to achieve zero waste, or sending nothing to the landfill or incineration, by the year 2020.[9]

[9] http://sfenvironment.org/zero-waste/overview/zero-waste-by-2020.

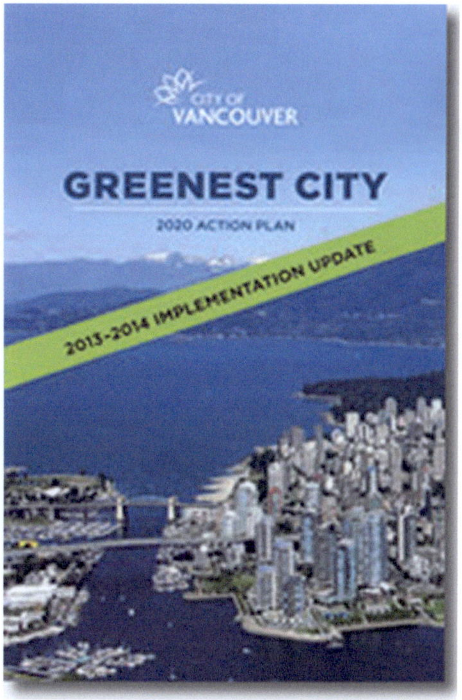

Fig. 18.3 Greenest City 2020 Action Plan, Vancouver (Source: http://vancouver.ca/greenest-city-2020-action-plan.aspx)

San Francisco is considered the electrical vehicle capital of the United States[10] with over 160 public charging stations and plans to install an additional 2750.[11] In April 2015, the city unveiled three installations of an off-the-grid solar-powered charging station, allowing electric vehicle owners to pull up and charge for free in a demonstration project (Fig. 18.4).[12]

Buildings shape the urban environment of San Francisco and much of the community's environmental impact. San Francisco is implementing a groundbreaking, comprehensive suite of policy initiatives and incentive programs to improve the performance of new and existing buildings. San Francisco has more than 497 LEED-certified green building projects.[13]

---

[10] http://sfenvironment.org/news?topic=details&ni=749.

[11] http://sfenvironment.org/news?topic=details&ni=749.

[12] http://www.cnet.com/news/san-francisco-introduces-free-solar-powered-electric-vehicle-charging/.

[13] http://sfenvironment.org/news?topic=details&ni=749.

Fig. 18.4 Easy-to-use three bin system in San Francisco (Source: http://www.c40.org/blog_posts/expert-voices-melanie-nutter-director-of-san-francisco-department-of-the-environment)

18.3 Oslo, Norway

The city of Oslo currently holds the record for the European continent's smallest per capita carbon footprint, around 2.3 tons of CO_2 equivalent per person in 2009. One of the reasons for this is the large proportion of hydropower in the energy supply. The city aims to cut carbon emissions by 50 % by 2030—Norway as a whole plans to be carbon neutral by 2050.[14]

The city of Oslo works to engage all of its 43,000 employees through eco-certification of all workplaces, departments, companies, schools, kindergartens, and other units. Over 350 municipal units and over 250 businesses in Oslo are already eco-certified. The eco-certification systems in use are ISO14001 for large departments and companies and the Norwegian Eco-Lighthouse for small- and medium-sized units/businesses (Fig. 18.5).[15]

Other examples of sustainable practices are intelligent lighting that adjusts intensity depending on traffic conditions and weather, bio-methane from waste to power mass transit and heating, and a good car- and bike-sharing program.[16]

---

[14] http://www.climateactionprogramme.org/climate-leader-papers/urban_ecology_in_oslo.

[15] http://www.iclei-europe.org/fileadmin/templates/iclei-europe/files/content/Membership/MUTS/Oslo/Oslo_brochure.PDF.

[16] http://www.ecomagination.com/top-five-most-sustainable-cities-in-the-world.

Fig. 18.5 Oslo, Norway (Source: http://www.climateactionprogramme.org/climate-leader-papers/urban_ecology_in_oslo)

Another often-used term is "**smart city**." Smart cities are characterized as cities that use innovation solutions mainly linked to the investment in technology to address burgeoning municipal problems. The term "smart" is usually a shorthand reference to public or private investment in information communication technology (ICT) or "smart" industry and other industries implying ICT in their production processes.

Examples of smart city investments include retrofitting of building stock, smart energy grids and broadband access, electric vehicle charging infrastructure, installation of heat networks, on-site renewable energy generation and involvement in more general adaptation and mitigation.[17]

It should be acknowledged that "smart" cities do not only imply technological investment or engagement in municipal spaces; rather, the term can also imply innovation and the improvement of cities' spaces in other "soft" industry areas. For example, the European smart city model considers a "smart city" to be a city which performs well in six distinct categories: smart economy, smart mobility, smart environment, smart people, smart living, and smart governance (Rudolf 2007).

The precedence of economic considerations is often cited as a reason for the limited success and impact of projects as well as the more recent emphasis on large top-down flagship sustainable urban development projects instead of the bottom-up initiatives that characterizes many of the early developments. The challenges facing

[17] Ogorkiewicz and Falconer (2013).

the implementation of sustainable cities highlight the need for a new approach, a shift in mindset, which recognizes that there may be significant tradeoffs among—and within—environmental, social, and economic goals, making the assessment and decision-making process more complex. Research that focuses on the experience and decision making of stakeholders when confronted with these complexities would reflect the realities facing local actors and enhance the transferability of solutions.

Numerous cities are actively pursuing a smart city strategy, one of them being Amsterdam. The city of Amsterdam started their initiative "Amsterdam Smart City" in 2009 and currently includes more than 70 projects, which are collaboratively developed by local residents, government, and businesses.

Another term was defined within the POCACITO project (Post-Carbon Cities of Tomorrow—EU FP 7), the **post carbon city**: The concept of "post-carbon cities" signifies a rupture in the carbon-dependent urban system, which has led to high levels of anthropogenic greenhouse gases, and the establishment of new types of cities that are low-carbon as well as environmentally, socially and economically sustainable. The term post-carbon emphasizes the process of transformation, a shift in paradigm, which is necessary to respond to the multiple challenges of climate change, ecosystem degradation, social equity and economic pressures. Through their adaptive capacity, post-carbon cities use the threat of climate change as an opportunity to reduce vulnerability as they restructure human–ecological and human–human relationships toward ecosystem health and a clean energy economy.[18]

The **biophilic city** is another term more recently coined by University of Virginia Professor Timothy Beatley that attempts to frame urban spaces in relation to their natural environments with a view toward holistically addressing current global trends. Biophilic cities are more than administrative jurisdictions with lots of greenery and natural spaces but also communities that contain governance structures, educational facilities, vested private sector, and committed civil society to communicating, investing, and connecting nature to all aspects of the urban fabric and local quality of life (see Fig. 18.6 and 18.7). They are also cities where concepts such as sustainable consumption and production and circular economy are understood and practiced, and local resources, including water, energy, waste, land, and food, are efficiently sourced and repurposed. Moreover, its inhabitants have a heightened sense of environmental consciousness that defines their behaviors, lifestyles, and political affiliations.

Why is it that technology has already achieved that coveted place connected to our bodies and minds and that which is our architect and benefactor—nature—has yet to be (fully) ingrained in our lives?

Biophilic cities present a fascinating opportunity to foster this nature–human nexus and exploit the marvels of technology to broadcast the multifaceted benefits of nature. Analogous to the fitness wristband on the market that monitors human body functions, technology can aid cities to measure, report, and verify local air and water quality, temperatures, precipitation levels, flora and fauna populations, and

[18] Evans (2008), p.3; based on Adger (2006), Neil Adger, Arnell, and Tompkins (2005).

Fig. 18.6 Singapore

Fig. 18.7 New York City

recreational location capacities as examples. ICT technologies and big data can also play an instrumental role in making the business case for biophilia by measuring the reactions of the citizens in unnatural/natural environments as well as the operations of urban systems working in coordination with green infrastructure applications to determine the socioeconomic and environmental impacts.

Cities can harness the power of its citizens and capacities for immediate action, to advance the local public health, climate change, and education agendas by incorporating urban ecosystems into their respective protocols. And local authorities and stakeholders can mandate biophilic principles and design in urban development plans, infrastructure investments, and policy measures. Ensuring all urban inhabitants share in the benefits of the local nature and its biodiversity requires political leadership, private sector engagement and citizenry support for multi-sectoral polices and diverse investments in low-carbon/impact development, including public/green/blue spaces, mixed-use built environment, education, multi modes of transport, and renewable energy among others.

Various demonstration cities within Tim Beatley's Biophilic Cities Project Network are pursuing aggressive agendas to address public concerns in the near as well as the long term often against tremendous odds. Birmingham, UK, a once industrial center, is committed to becoming a sustainable and a natural capital city both through reductions in carbon emissions and enhancing public health by capitalizing on the services, benefits, and economic assets of its natural environment. The city's Green Commission has developed a Green Living Spaces Plan targeted at generating economic revenues in part from saved health costs and rewarding residents for increased health and well-being. The city's 1000-hectare Sutton Park (UK's only urban reserve) and network of rivers and trails will be complemented with additional green infrastructure within the built environment and connected with alternative modes of mobility to optimize the public health potential and provide a new implementation interpretation to the UK's decentralization of national health policy to local levels.

One of the world's most dense cities, Singapore, has aggressively embraced biophilic principles and redesigned its buildings and neighborhoods to promote a human–nature coexistence. For example, the city uses park connectors and green corridors to broaden access to the natural environments both for citizens and local flora and fauna. By making use of innovative design and layering of both physical and natural infrastructure, such as elevated walks to access tree canopies, the city facilitates the preservation and regeneration of its biodiversity and strengthens its urban ecosystem services for its inhabitants. Political commitment is strong and fiscal incentives such as the Skyrise Greenery scheme encourages developers to incorporate green walls, facades, and other forms of green infrastructure into the built environment. The Singapore Biodiversity Index is used to monitor and measure the status of the local biodiversity and the impact on urban resources and human well-being. Such performance indicators are of growing interest among cities and governments for their contribution to global sustainability and climate agendas.

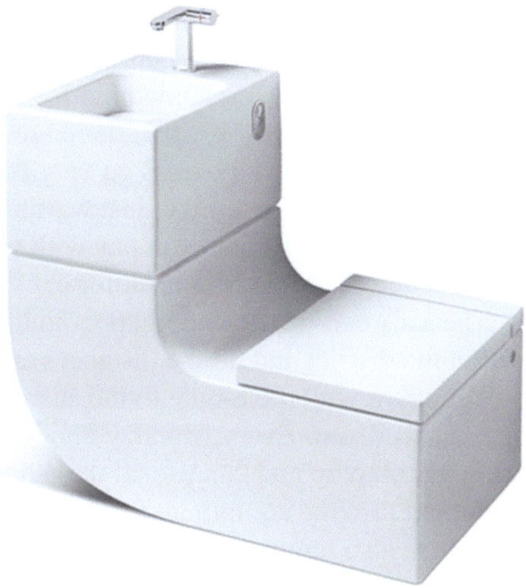

Fig. 18.8 All-in-one washbasin and toilet (Source: http://bathroom-kitchen-faucets.com/de/water-saving-all-in-one--washbasin-and-wc-to-maximise-space-and-conserve-water/all-in-one-washbasin-and-wc-1/)

18.4 A Walk Through a Sustainable, Eco, Green, Post-Carbon City District: Experiences of an Inhabitant

If you get up early in the morning, the first way is to the toilet. It still looks a bit weird but it is very functional: the water you use for washing your hands is used as flushing water for the toilet! (Fig. 18.8).

Taking a shower, the water is fresh, but only half of the amount of the average is used, thanks to the eco showerhead you have installed. The towels as well as the clothes you want to wear today are ready to be taken from the delivery box from the laundry. The laundry has a closed water cycle and the heat supply comes from solar collectors. They have a pickup and delivery service that also works when nobody is home—in the box outside of your apartment.

In the same box, you will also find the things you ordered for breakfast over the Internet—today that is some fresh muesli—made according to you own choice. All those services work on electro-mobility, transport bikes, or lorries.

There is a market hall right in the center of your district where you can buy all the things that you still want to choose on your own and see before you buy. All the food is organic; of course, most of it is grown on the roof gardens. Some of the vegetables are still for free; growing in the urban gardening places throughout the city, you just have to go there and pick!

The hallway looks very bright although it is right before dawn. This is due to the natural daylight that is directed into your apartment by special channels. Natural light is collected on the roofs or at the facades and directed to the rooms where it is needed, distributed by a diffuser. The blender for your morning juice runs on renewable energy sources, of course!

If you leave your apartment, you walk down the stairs (there is an elevator, but in the back of the building, behind the stairs, so it is much easier to take the stairs, and at the same time, do something for your fitness). On the board, just before you leave the building, you see that there are still 5 min left until the next tram arrives. Although the trams and busses go very frequently, the weather today is so nice that you decide to take the e-bike, just waiting there in front of your building. The key-card to open the key chain you pay with your rent, which includes also the ticket for the tram and all things you need for car sharing (electric vehicles only).

The kids can go to kindergarten and school on their own, as there is no traffic within the district and all the distances are very short.

You drive with the e-bike to the shared office within the district; you share with one other citizen. Every Monday and Wednesday, you can use one desk; you just bring your laptop and can use all the other equipment, like the printer, scanner, and copier but also the equipment for conferences and the coffee machine and the stove (of course also running on renewables). All the electronic equipment is very energy efficient, and if their lifetime has ended, they all get disassembled (which is very easy now due to new regulations according to eco-design guidelines) and separated into their components and reused or recycled.

Some nice contacts have been established during the last month your have been working there and there might also be a joint project with two other coworkers in the near future!

Just in case you need a car, there is a website and app available which offers all the available places on a platform where it is possible to see real-time availability and book these parking spots ahead. As a result, drivers spend less time searching for a single spot and reduce CO_2 emissions. Only a few people still have their own car. It is very inconvenient, as the cars have to be parked on the outside of the district, in shared garages, and as almost everything is available and reachable within walking or biking, a car is not really necessary.

On the way back, you pop by the local food store to get a refill on your favorite wine in the glass bottle you bring. Due to climate change, there are new vineyards now growing outside the city, not far from the city center.

Another stop is at the "do-it-yourself" shop, where you cannot only rent equipment but also a place at the workbench if you have some bigger projects to manage. You take home the newest drilling machine that is on the market as you have bought some new lights that need to be installed tonight. The shop is nearby the "Test Living Lab," which focuses on testing a range of innovative products related to smart grids, district heating networks, and improving the sustainability of houses. Such innovations include a hybrid heat pump that uses gas as well as sustainable energy, a virtual power plant that allows city residents to trade energy and an electric car that can load and unload itself. Experiments into how to encourage people to make "green" choices can also be conducted in the Test Living Lab, such as the use of energy displays.

On the way home, it gets dark already and you realize the smart lightening system the city has installed during the last years: the lighting can be adjusted for a range of situations via remote operation or sensors, helping to improve security and

Fig. 18.9 Smart street lights (Source: http://www.designboom.com/technology/tvilight-smart-streetlights/)

save energy. Lighting can be dimmed or adjusted according to the weather and according to number of pedestrians and cyclist on the roads. Movement sensors can also be used to register if the lighting needs to be brighter. Saved energy can then be used for other functions, such as powering the Wi-Fi network or measuring air quality (Fig. 18.9).

Back home, when entering your apartment, the smart air condition and the smart lighting system start to work: both systems realize when there is someone in the room and choose the conditions in the room according to your personal preferences. In general, the indoor climate is very nice throughout the year: big plants (that lose their leaves in winter) are placed on the outside of the building, regulating light, humidity, and oxygen coming in and keeping air pollutants from getting in. A look at your smart meter shows you.

Before you can sit down and finally relax, you have to bring out the garbage you forgot this morning. Most of the biodegradable waste goes into the biogas plant, where it gets converted into electricity and heat. The remaining substrate is being composted nearby the urban gardening beds, where it is used as fertilizer.

References

Adger WN (2006) Vulnerability. Glob Environ Chang 16(3):268–281

Beatley, T. (2011). Biophilic cities: Integrating nature into urban design and planning. Washington, DC: Island Press. http://biophiliccities.org/

Berkes F, Colding J, Folke C (2003) Navigating social-ecological systems: building resilience for complexity and change. Cambridge University Press. http://books.google.de/books?hl=en&lr=&id=Y5FnAq9kjxgC&oi=fnd&pg=PP1&dq=Berkes,+F.,+J.+Colding,+and+C.+Folke,+editors.+2003.+Navigating+social%E2%80%93ecological+systems:+building+resilience+for+complexity+and+change.&ots=-s94bblE5U&sig=vYRMZtAcIRMne4oJmIJbPBzRXk8

Evans G (2008) Transformation from 'Carbon Valley' to a 'Post-Carbon Society' in a climate change hot spot: the coalfields of the Hunter Valley, New South Wales, Australia. Ecol Soc 13(1):39

Gunderson LH, Holling CS (2001) Panarchy: understanding transformations in human and natural systems. Island Press. http://books.google.de/books?hl=en&lr=&id=o4u89akUhJMC&oi=fnd&pg=PR7&dq=Gunderson,+L.+H.,+and+C.+S.+Holling,+editors.+2002.+Panarchy:+understanding+transformations+in+human+and+natural+systems&ots=ec8AarlOXI&sig=M2yXwOisrMQmbBU-H1k53yWQWgU

Jabareen YR (2006) Sustainable urban forms their typologies, models, and concepts. J Plan Educ Res 26(1):38–52

James P, Magee L, Scerri A, Steger MB (2015) Urban sustainability in theory and practice: circles of sustainability. Routledge, London

Kenworthy JR (2006) The eco-city: ten key transport and planning dimensions for sustainable city development. Environ Urban 18(1):67–85

Levin SA, Harvey JA (1999) Fragile dominion: complexity and the commons. Perseus, Reading, http://osyka.pixub.com/data/fragile-dominion-complexity-and-the-commons-by-simon-leviin.pdf

Norberg J, Cumming G (2008) Complexity theory for a sustainable future. Columbia University Press, New York

Ogorkiewicz AM, Falconer G (2013) Smart Cities Stakeholder Platform: Finance Working Group – guidance document integrated action plan – report process & guidelines for smart cities. http://eu-smartcities.eu/sites/all/files/Integrated%20Action%20Plan.pdf

POCACITO. http://pocacito.eu/

Rapoport E (2014) Utopian visions and real estate dreams: the eco-city past, present and future. Geogr Compass 8(2):137–149. doi:10.1111/gec3.12113

Register R (1987) Ecocity Berkeley: building cities for a healthy future. North Atlantic Books. http://books.google.de/books?hl=en&lr=&id=QYE-Q8MAF3MC&oi=fnd&pg=PA3&dq=Register,+R.+(1987).+Ecocity+Berkeley.+Berkeley,+CA:+North+Atlantic+Books&ots=HsmssJ5wQh&sig=YdggnebXNwgVuPlGAA5mPQf6zU

Register, R. (2006). Ecocities rebuilding cities in balance with nature (Rev. ed.). Gabriola, B.C.: New Society

UNEP (2011) Cities: investing in energy and resource efficiency. http://www.unep.org/greeneconomy/Portals/88/documents/ger/GER_12_Cities.pdf

UNEP (2011) Towards a green economy: pathways to sustainable development and poverty eradication – a synthesis for policy makers. www.unep.org/greeneconomy

ZEUS 2020 (Zero Emission Urban Study): final report, the ZEUS Consortium. Blue Globe Report, Studie 27/2011. editor: Klima- und Energiefonds, Vienna, 2011

http://vancouver.ca/green-vancouver.aspx

http://vancouver.ca/files/cov/Greenest-city-action-plan.pdf

http://sfenvironment.org/news?topic=details&ni=749

http://www.cnet.com/news/san-francisco-introduces-free-solar-powered-electric-vehicle-charging/

http://sfenvironment.org/news?topic=details&ni=749

http://www.climateactionprogramme.org/climate-leader-papers/urban_ecology_in_oslo

http://www.iclei-europe.org/fileadmin/templates/icleieurope/files/content/Membership/MUTS/Oslo/Oslo_brochure.PDF

http://www.ecomagination.com/top-five-most-sustainable-cities-in-the-world

Chapter 19
The Smart Building in the Smart City

Petra Stieninger

19.1 Smart Buildings and the Urban System

Technologies enable us to make buildings more energy-efficient, more water-efficient, or in general more resource-efficient and sustainable. We design buildings with high-performance appliances that reduce the use of resources tremendously. We incorporate features that allow the building to produce more energy than it uses. And we install smart meters and integrate the Internet of Things to analyze big data on the efficiency of the building operation. Today, we create so-called smart buildings.

Technically, the operation of a building can give back more resources than it actually uses. However, this equation only takes the inside of the building into account. It only considers the building's needs for input resources, but it doesn't consider the building's surroundings, other systems the building is connected to, or the sources of its input resources.

How smart is a building whose occupants consume more energy to drive to the building than they actually use for its operation? How smart is a building that loses a large amount of the resources it needs for its operation (such as energy and water) before they actually get to the building? How smart is a building that operates like an island disconnected from people, surrounding systems, and nature? How smart is a building that saves resources and guarantees comfort, health, and happiness for its occupants in the inside but wastes resources and makes us sick on the outside?

We create fancy, net-zero energy, replicable, island-like buildings that meet all the calculation requirements to call them "green" and "smart" and that incorporate all the technology and widgets one can dream of not taking into account that the building and its operation don't end at its entrance door.

A. Ahuja, *Integration of Nature and Technology for Smart Cities*,
DOI 10.1007/978-3-319-25715-0_19

19.2 A Building Is Not an Island

A building is part of a larger urban system. Urban systems are very complex systems with the following, constantly interacting system components: the man-made, built environment (buildings, infrastructure, urban design, etc.); its societal environment based on paradigms, values, policies, and available technologies; and the natural environment providing resources the system needs for its existence. These urban system components are in a constant exchange with each other influencing the operation of the city, its neighborhoods, the operation of the buildings in the city, and the people who live, work, and play in the city (Stieninger 2013).

The building is part of that constant exchange between all system components. It is part of an exchange between goods, services, and people that want to access the building and live, work, or play in that building. It is part of an exchange of resources the building needs for its operation and waste and other output that result from its operation. It is part of a constant exchange of input and output (Table 19.1).

Urban systems are becoming bigger and bigger and therefore, more and more complex. The UN estimates the amount of people living in cities at almost 70 percent by 2050 (United Nations 2014). Cities are facing numerous challenges in combating negative impacts on human health and the environment due to the growing population numbers and the increasing demand for infrastructure, resources, and services.

The building is a part of the larger urban system and it is also responsible to contribute to a healthier and environmentally more responsible operation of that system. Sometimes we don't see the forest for the trees or rather the city for the buildings. When we claim we are building "smart" buildings, we must also understand the interconnections of the building within this larger system.

How is the building connected to its surroundings? How does it interrelate with the built environment it is located in? What is its relation to its natural environment? How do people get to and from the building? How are resources that the building needs for its operation being transported to the building? How is the waste the building generates being discharged and transported from the building? How is the inside of the building connected to the outside?

A smart building has to be smart inside and outside. The way the smart building is interconnected with the urban system influences its accessibility for people, its connection to goods, services, resources, and infrastructure, the efficiency of its operation, and its impact on the environment and the society.

A smart building has to provide sustainable accessibility. It has to be connected to sustainable systems that feed the building with the needed input and support the use, reuse, and recycling of the building's output. It has to balance what it takes from nature and what it can give back to nature (Fig. 19.1).

Since the building is a part of a larger urban system, it can have positive impacts on its surroundings and it can contribute to a more sustainable, smart outside in addition to its sustainable, smart inside. Smart building design considers the characteristics of the building's location and its potential connections with its surroundings. It is the building's responsibility to contribute to a more sustainable urban system.

Table 19.1 The complexity of the urban system

Urban system		
Societal environment	Built environment	Natural environment
Technology	Buildings and facilities	Resources
Social economy	Infrastructure (transportation, energy, water, waste, information, and communication)	Nature
Policy, laws, regulations		Geography
Values	Landscaping and urban design	Climate

Source: Stieninger (2013)

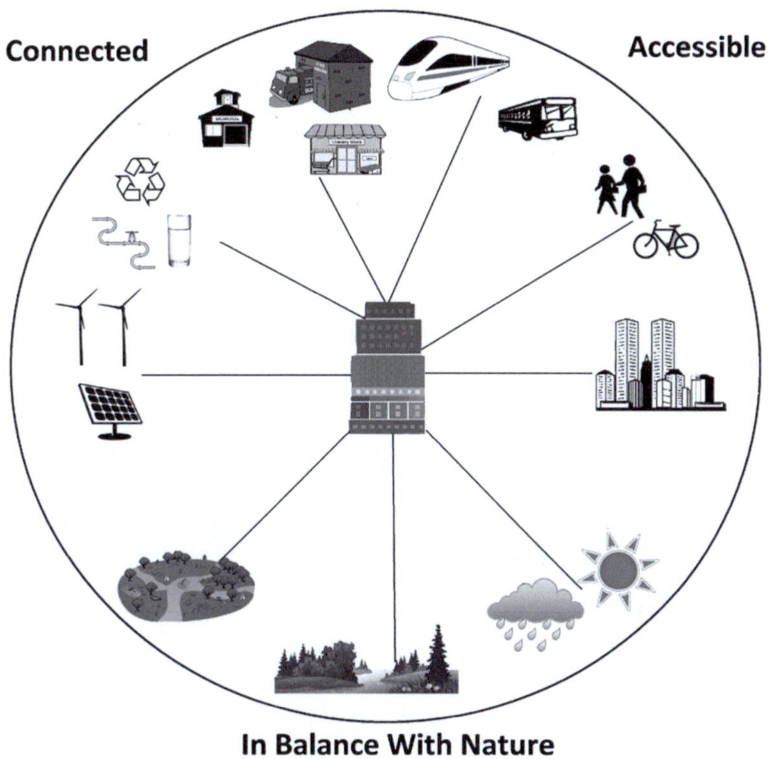

Fig. 19.1 A smart building is accessible, connected, and in balance with nature (Source: author)

19.3 The Accessible Building

Accessibility Is a Matter of Proximity and Mobility

According to the US Green Building Council (USGBC), in the United States, people use about 30 percent more energy in order to commute to an average office building and to access it (energy consumption for transportation) than what is actually needed to operate the building (energy consumption for heating, cooling, lighting, etc.) (U.S. Green Building Council 2015).

The smart building has to offer proximity to variety, that is, proximity to a variety of alternative transportation options and proximity to a variety of uses, people, and places. The way we access a building depends on the distance to the building and the available means of transportation. The location of the building plays the most important role when it comes to accessibility. Locating a building in the proximity of existing services, infrastructure, people, and places where activities take place makes its operation much more efficient than a building outside the city in the middle of nowhere. A building within a dense, urban context creates many more options for its accessibility. Its location within the city offers multiple alternatives of how to access it (walking, biking, public transit, driving). In addition, a larger number of people can access it from a shorter distance. At the same time, a larger number of people make the operation of infrastructure that is needed to access the building much more efficient.

A building located outside the city, in the middle of nowhere, surrounded by forests and farmland is more difficult to access than a building inside the city, in a dense, mixed-use neighborhood, close to a public transit station. The location in the middle of nowhere forces people to use their automobiles and drive there. We are not able to walk or bike there as it is too far from everything else. Public transit would not be economically feasible, as a single building does not meet the required demand to make a public transit system economical.

Cars are the most energy-consuming means of transportation. Buses consume less than 30 percent of the amount of energy a car uses per passenger kilometer and commuter rails or light rail systems consume less than 20 percent (Stieninger 2013). The building in the middle of nowhere outside the city forces its occupants to drive. The building in the dense, urban neighborhood offers multiple options of sustainable transportation.

A smart building that is resource efficient and sustainable inside should also support sustainability on the outside. Therefore, a smart building should be accessible by the most sustainable means of transportation. A smart building offers sustainable options for its occupants' choices of means of transportation such as walking, biking, or public transit. A smart building doesn't force its occupants to drive their cars. Dense, urban neighborhoods offer proximity and variety. Urban density moves things closer together, it goes hand in hand with mixed uses and variety, and it makes the installation and operation of public transit systems cost-effective.

There are many factors that influence our choice of means of transportation. In order to make it happen, we have to find the right mix of rewards and penalties, a balance between carrot and stick. The core idea is to make driving unattractive and alternative means of transportation attractive, not just from a design perspective but also in our minds. The smart building's occupants have to be aware of the benefits of using alternative means of transportation, and at the same time, they have to be aware of the disadvantages of driving a car.

Urban growth and development have always taken place along and around transportation options. What used to be rivers and oceans, at some point, became railroads and later roads, streets, and highways. The average daily commute time to work has stayed more or less the same over time; only the daily commute distance changed. While people were walking, cities were much smaller. Later, new means of transportation allowed for longer distances (Cerwenka et al. 2007). Railways and automobiles made distances less important, and the telephone and the Internet made them almost irrelevant.

However, cities that realized that public transit is a much more resource-efficient option compared to driving the car also realized that in order to make public transit cost-effective a certain number of passengers per station per day are a key. These numbers can oftentimes not be met in sprawled areas. Cities such as Vancouver, Canada, and Chicago, Illinois, realized that in order to make their transit systems operable they needed to increase the density around transit stations. The so-called transit-oriented development (TOD) ordinance requires higher densities around public transit stops and supports minimized numbers or the complete omission of parking spots per dwelling unit in those areas. Thus, people who live and work close to transit stations have more incentives to take the train or bus for their daily commutes. At the same time, the transit company serves a sufficient number of passengers in order to create revenue that can be used to make the system more attractive and keep it clean and safe.

We cannot influence the density of a neighborhood with a single building, but we can influence proximity and variety by choosing the right location of the building.

There are many actions we can take in order to make driving unattractive and alternative transportation attractive. However, not all of these actions are within the power of a building. Many decisions have to come from the municipality or other authorities. Nevertheless, the building itself has a very strong influence on our transportation behavior.

Obviously, decisions regarding the transit system itself, the city-wide bike infrastructure, and the layout and design of the city have to be made by the city. But the building itself can contribute to the quality of the neighborhood it is located in by providing attractive design, landscaping features, and a mix of uses to the neighborhood. The building is a part of the built environment with all its other buildings, infrastructure, and facilities. Every single piece of the neighborhood has to contribute to its walkability, density, and quality. A neighborhood can be very dense and can have lots of different uses close to each other. Yet, narrow sidewalks and a lack of landscaping can make it unattractive for walking. On the other hand, a neighborhood can have beautiful building designs, landscaping features, and wide sidewalks, but if it is homogenous (e.g., only residential uses), there is nothing to walk to. The right balance of building uses and designs combined with density makes a neighborhood walkable, and all of this can be done on a building level.

The most important factors are the location and the purpose of the building. Combining different uses within a building already adds more variety to the building and the rest of the neighborhood. For example, use the ground floor space for retail, the first couple of floors for office space, and the higher floors for residential use. Build "a city in the city" the way Bertrand Goldberg did it in 1964 when he built the Marina City Towers in Chicago or the way almost every historical city center was originally built in Europe.

Greenery, wide sidewalks, and life on the street make a neighborhood safe and walkable. Attractive public spaces and restaurant gardens can bring life to the street. As the architect Jan Gehl proved with the research he conducted, the more life on the street, the more people get attracted to spending time outside, in public spaces. The feeling of place and community in public spaces improves their safety and quality. Gehl called it the 5 km/h architecture designed for walking, sojourning, and enjoying versus the 60 km/h architecture designed for driving by and missing all the details and beauty (Gehl 2011).

The following table shows the variety of actions that can be taken in order to make driving unattractive and alternative transportation attractive and what can be influenced by the building directly (Table 19.2).

The right mix of incentives and penalties, carrot (attractive sustainable means of transportation) and stick (unattractive driving), is crucial in order to change people's choices on their means of transportation. The sustainable means of transportation such as public transit, biking, and walking have to be attractive, affordable, accessible, and available for the building's occupants. That is, the building has to be close to a public transit stop and within a bikeable and walkable environment. At the same time, the building has to provide the necessary infrastructure on site that is needed to make the use of these means of transportation more attractive. For instance, a lot of people would argue that riding the bike to work is impossible as they would have to change clothes and take a shower after their bike ride. Therefore, the installation of locker rooms and showers in the building can make the access by bike more attractive. In addition, the occupants have to get informed and educated on sustainable alternatives and specific commuter programs that support the use of alternative transportation.

The easiest way to make access by car unattractive is to minimize parking on site. Of course, a smart building won't necessarily be able to increase the cost of driving in terms of gas prices, taxes, etc., but it can still make its accessibility by car unattractive (if local building codes allow).

Looking at building codes and how we access our buildings, it becomes obvious that the priority is still to provide a minimum number of parking spaces so at least a certain number of building occupants can access the building by car. Where are the building codes that require a maximum number of parking spaces? And where are the building codes that require a maximum distance to a public transit station, a maximum distance to bike lanes, and a minimum requirement of walkability?

Table 19.2 The building's influence on the choice of means of transportation

Action	Within influence of building	Outside of influence of building
Make access by public transit attractive	– Location of building close to transit station – Commuter incentive programs for occupants – Public transit real-time displays in building to inform occupants on next available train – Provide information and educate occupants on transportation options – Provide information on commuter programs	– Provide attractive and safe public transit – Provide affordable public transit
Make access by bike attractive	– Location of building connected to bike lane network – Weather-protected bike racks in building – Showers and locker rooms in building – Bike share station on site – Commuter incentive programs – Provide information and educate occupants on biking options – Provide information on commuter programs	– Provide attractive and safe bike lane network – Provide bike share program
Make access by walking attractive	– Location of building in dense, walkable neighborhood – Attractive and walkable entrance area of building – Natural elements, landscaping, park area around entrance area of building – Contribute to walkability of neighborhood with building design and entrance area connection – Provide mix of uses in building	– Create walkable and safe neighborhoods – Increase urban density – Mixed-use neighborhoods
Make access by driving unattractive	– Limit or omit parking spots – Prioritize pedestrian and bike access design on property – Provide information and educate occupants on alternatives and on negative impacts of driving	– Change parking requirements of local building codes – Increase gas tax – Road pricing – Pedestrian zones – Transportation priorities: walking and biking > public transit > driving

Source: author

Factors such as the location or the mix of uses have to be considered at the beginning of the planning phase of a project. Once a building is built, the operation itself has to influence what occupants do and how they choose one or the other means of transportation to get there. Information and awareness of alternatives play an important role.

Smart buildings use information and communication technologies (ICT) to inform and educate in order to change their occupants' behavior. ICT are tools for metering, simulation, modeling, analysis, monitoring, and visualization of energy use or water use in the building. ICT inform the occupants about their energy or water consumption, they educate them on how they can reduce their use, and they change their behavior by raising the awareness or adapting through building automation.

On the outside of the building, ICT can inform about public transit options in real time, can educate about alternatives and the negative impacts of driving, and can change behavior by raising awareness and spread information about alternatives. It can be a single smartphone application that everyone can download on their phones. It can be a building website that informs about options and that is set as a default website for all computers. Or it can be an information and media center in the lobby such as a simple TV screen in the lobby that has information on public transit lines as well as real-time information on arrivals and departures, availability of bikes from the bike share program, incentives for the commuter program, etc.

For the occupants who cannot use public transit, bike, or walk to work for whatever reasons, there are also more sustainable options to get to work when having to use their cars. New technologies such as electric cars enable us to also make this option more sustainable. Stefan Melzer elaborated this topic in more detail in Chap. 21.

The smart building offers proximity and variety and supports smart transportation.

19.4 The Connected Building

Technically, a building could be an island. We can build a building in the middle of the desert; put some solar panels on the roof; treat the gray, black, and storm water to a drinkable quality level; compost organic waste and use it as fertilizer for landscaping; and generate energy out of the waste that is left. Thus, a building can technically be a closed system; nothing goes in (despite solar radiation and rain water) and nothing comes out. But do we really want to completely close that system? Do we want buildings to be islands? Do we want cities to be collections of disconnected islands? We don't. Even if it might be technically doable, it would be very inefficient. A building that is imbedded in its surroundings and connected to other systems can take advantage of shared infrastructure assets and achieve efficiency through exchange and interconnection. Even if our calculations confirm that the building is completely self-sufficient, what happens if there is an outage of one of the systems? What happens when the sun is not shining? And what happens if our peak demand doesn't correlate with our peak production of "in-house resources"? We can't have a building where every system is redundant. The building needs exchange with other systems. The building needs exchange with the larger urban system. The inside of the building needs exchange with the outside of the building.

The accessible building was about how people can access it. The connected building seeks to connect with the systems that provide input resources it needs for its operation and with systems that provide options for the use, reuse, and recycle of output from the building's operation.

The building's operation requires energy for lighting when the sun is not shining, for heating in the winter, and for cooling when it gets hot in the summer. It needs energy to transport other resources from the outside to the building such as water and to circle it around in the inside. It needs energy to transport waste from the building, and at the same time, it can generate energy from its waste and wastewater. The smart building can take advantage of sharing input resources and output resources with other surrounding systems.

19.4.1 The Connected Building Is a Power Plant in a Larger Urban System

Technically, we could build a building in the middle of nowhere and operate it completely independently from the power grid. Passive design such as natural ventilation or passive heating and energy-efficient appliances reduces the need for energy to a minimum. The remaining need for energy can be generated by renewable sources on site. Most commonly, on a building level, solar energy is used for electricity production (photovoltaic (PV) systems) and solar water heating systems. PV technology has dramatically improved in the last couple of years but is still a very volatile market heavily depending on subsidies and monetary incentives. Geothermal energy can be used almost everywhere in the world for heating and/or cooling. Geothermal heating and cooling systems take advantage of the constant temperature found within the earth's surface. By installing a ground source heat pump, in the summer, the cooler ground temperature can be used for cooling, and in the winter, the warmer ground temperature can be used for heating. Recently, buildings also started to install wind turbines in order to use wind energy at a building scale. However, this technology is not yet at a level where it would be economically or technically efficient.

There is one major issue about renewable energy (and especially solar energy) though. You need sunshine in order to produce energy. On a sunny day, you might be able to produce more energy than you actually need. But what happens with the energy you don't need? And what happens at night, once the sun sets and everyone wants to turn on their TVs, starts cooking their dinners, or needs to simply turn on the light? Especially in residential buildings, the peak hours of demand don't correlate with the times of energy generation from renewable energy sources (Fig. 19.2).

The main problem with renewable energy is the missing correlation of peak demand and peak generation and the ability to store energy. This is not only a problem on the building level but a general problem that utility companies have to deal with every day. However, being part of a larger grid makes it easier to balance demand and supply as well as overcapacities. Smart grids allow the needed flexibility to balance renewable energy generation with energy storage, distribution, and generation from nonrenewable alternatives. Smart grids can store energy that is produced during sunny and windy daytimes and can supply the generated renewable energy at a later time.

In addition, smart grids offer demand response programs in order to balance demand and supply. These types of programs can offer time-based rates such as time-of-use pricing, critical peak pricing, variable peak pricing, real-time pricing, and critical peak rebates. Other forms of demand response programs can be direct

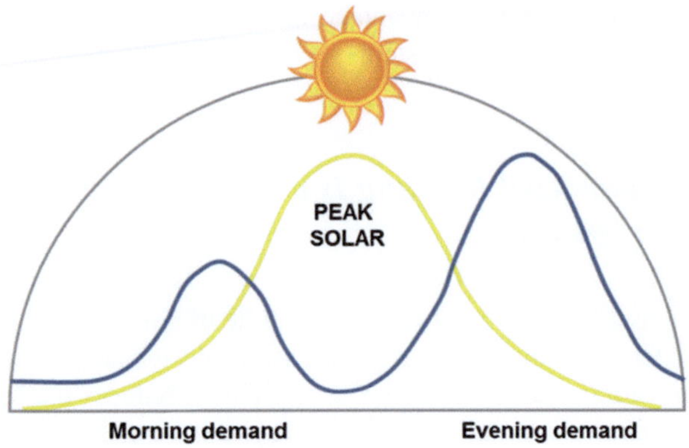

Fig. 19.2 Disconnection between solar energy production and peak demand (Source: author)

load control programs where utilities have direct control on air conditioners and water heaters to turn them on or off during periods of peak demand. The owner or building management receives either a financial incentive or lower electric bills in exchange.[1]

Smart grids are taking advantage of this constant exchange of information on demand and supply between the energy provider and the customers. They balance demand-side management (DSM) and supply-side management (SSM). On the demand side, they aim at reducing the energy demand and shifting times of peak demands without reducing the energy service quality. On the supply side, they increase generation capacities by offering more alternatives and shifting supply times, improve generation efficiency, and reduce losses from transportation and distribution.

This is also crucial in cases where the customers themselves produce energy that they can use for their own demand or feed back into the grid. Smart meters and other control systems are needed to keep track of this constant exchange. It is a constant give and take between the building and the utility, a constant exchange between the building inside operation and its outside surrounding systems.

Additionally, energy storage on site is becoming more and more important. The problem of energy storage at a building level has gained a lot of attention, and a lot of research has been done in recent years. Batteries that promise to solve that problem at the building scale have been developed in order to provide a backup system in case of power outages and to become more independent from the utility grid. Theoretically, we could now produce energy during the day as long as the sun is shining and then consume it at night when we turn on the light and the TV. At the utility scale, however, solutions for battery storage are not yet advanced enough.

The smart building produces more energy than it actually needs. The US Green Building Council's Leadership in Energy and Environmental Design (LEED) pro-

[1] http://energy.gov/oe/technology-development/smart-grid/demand-response

gram and other rating systems require an energy efficiency of more than 30 percent of conventional buildings. Knowing the technologies that are out there, this is a pretty conservative number. The Living Building Challenge, on the other hand, requires the on-site energy generation of 105 percent of energy what is being used in the building, which is a 5 percent plus energy. Buildings that achieved a Living Building Challenge certification, which requires a verification of an entire year of operation, prove that it is doable.

The smart building is not just a consumer of energy but also a producer. The smart building is a power plant within a larger system of energy demanding and producing entities. The conventional idea of utilities supplying and the building demanding is outdated in the era of smart buildings. In the era of smart buildings, a new model for utilities is needed. The utilities are no longer the main producer of energy. Their main task is to manage the multiple sources of energy production and to distribute properly. Everyone in the system becomes a micro power plant and the utility is responsible to manage it (Fig. 19.3).

The management of such a complex system of producers and users can be overwhelming. Therefore, the energy grid has to be within a certain size boundary. We call it micro grids. Cities don't need one big power plant that supplies the energy for the entire city and that is managed by one company. Smart cities are a system of multiple sources of energy produced by the users themselves as well as the utility whose main task is to manage and distribute. In order to manage a smart grid, it has to be split into a number of decentralized micro grids.

Decentralized micro grids will also improve the transmission and distribution losses (T&D losses). About a quarter of all energy generated is being lost on the way from the power plant to the consumer. The smaller the system is, the shorter is the distance between supply and demand, and the fewer losses can occur during distribution. The grid itself will become a means of energy storage for everyone who has plus energy to feed in.

In addition, decentralized systems are much more reliable and resilient. In case of outages, the system can take advantage of adjacent systems that can serve as backup. Big, centralized systems are dependent on one main system and have to install multiple, additional backup systems that are operated redundantly.

19.4.2 The Connected Building Turns Waste into Valuable Resources

The smart building reduces, reuses, and recycles waste. And not only that, it also uses waste as an additional source for energy production. Other than solar panels and heat pumps, the building has many more options to produce energy that it can feed back into the micro grid. Waste can be turned into energy as well. However, for an island building, it would be pretty inefficient to handle waste-to-energy on site

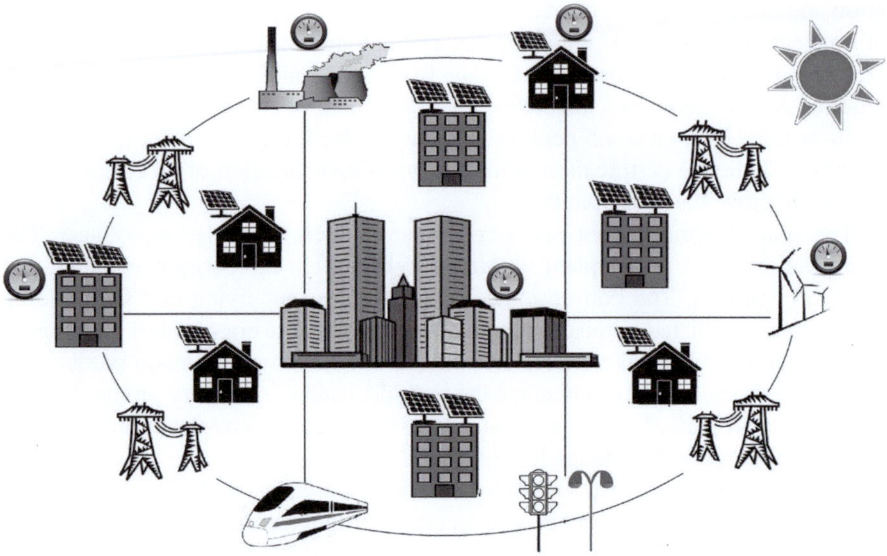

Fig. 19.3 The connected building as micro power plant in a smart grid (Source: author)

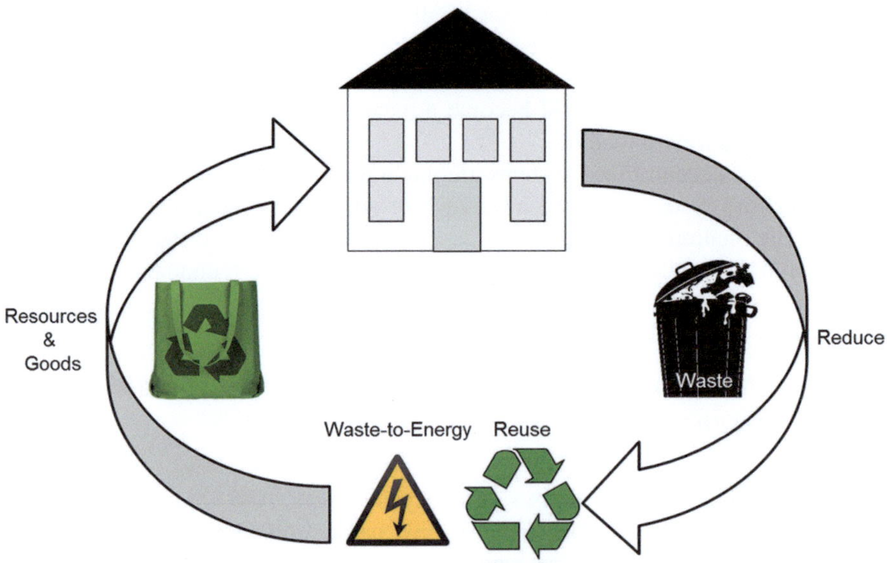

Fig. 19.4 The connected building turns waste into valuable resources (Source: author)

without collaboration with other system components. For certain operations, the building has to take advantage of the system components surrounding it (Fig. 19.4).

Solid waste can be incinerated or used as pallets. Waste heat can be recovered. Wastewater sewage sludge can be turned into biogas. All types of waste (liquid, solid, and gaseous) can be used for district heating or cooling and/or electricity production. Obviously, all these options can't be handled efficiently at a building scale. Decentralized waste-to-energy "power plants" must support the energy production at a district scale supporting exchange between multiple micro grids.

Thus, also waste management companies and wastewater treatment plants become energy utilities. Together with conventional energy utilities and the building owners, they feed energy made from waste, waste heat, and sewage into their local micro grid. It is obvious that a building needs to be connected to other systems in order to manage its waste outputs in sustainable ways. Chapter 15 provides more details on recycling and waste-to-energy.

19.4.3 The Connected Building Manages Water Resources Within Micro Grids

The smart building is not just a power plant though. In order to operate a building, it also needs water and has to find ways to treat and manage the wastewater it produces and the stormwater that falls onto the site.

Obviously, freshwater has to come from somewhere outside the building and can be treated and reused within the building. Technologies are out there to not just treat and reuse stormwater or gray water; nowadays, we are able to even turn black water into potable water. The main barriers to do so, however, are our minds that don't like the idea of drinking what we flushed down our toilets a couple of hours ago. In addition, in most countries and states, laws don't allow the reuse of black water for potable water purposes.

Similar like waste and energy, for water, it also makes sense to treat and manage it at a district scale. Even though we do have the technology to treat water on site within the building, it is still more efficient to share systems with others and have an exchange. One single large wastewater treatment plant for the entire city requires a huge network of pipes and a stormwater management system that is interconnected with it. In the United States, a lot of cities struggle with sewer overflows during heavy rain events, as they are handling stormwater and wastewater in combined sewer systems that overflow and pollute rivers, lakes, and oceans when they are over capacity (Fig. 19.5).

Water and especially wastewater management at a micro scale can be handled easier with a more efficient and smarter grid solution. Furthermore, stormwater management should be done on site. The smart building doesn't rely on a city-wide stormwater system, and it won't cause sewer overflows. Green infrastructure is a

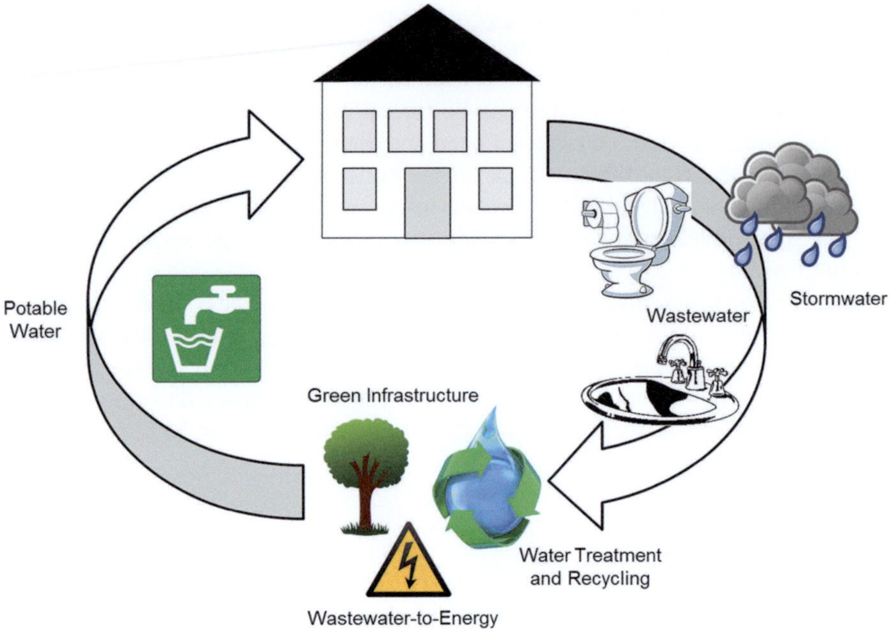

Fig. 19.5 The connected building manages water resources within micro grids (Source: author)

solution for sustainable water management on site "that protects, restores, or mimics the natural water cycle. Green infrastructure is effective, economical, and enhances community safety and quality of life. It means planting trees and restoring wetlands, rather than building a costly new water treatment plant. [...] Green infrastructure incorporates both the natural environment and engineered systems to provide clean water, conserve ecosystem values and functions, and provide a wide array of benefits to people and wildlife. Green infrastructure solutions can be applied on different scales, from the house or building level, to the broader landscape level. On the local level, green infrastructure practices include rain gardens, permeable pavements, green roofs, infiltration planters, trees and tree boxes, and rainwater harvesting systems."[2]

Also for water as an important input resource in a building, it becomes obvious that a connection to the surrounding systems of the building is imperative. The building can help to manage local stormwater, and the micro grid supplies potable water to the building and treats and reuses wastewater that comes from the building.

---

[2] http://www.americanrivers.org/initiatives/pollution/green-infrastructure/what-is-green-infrastructure/

The water topic and its connection to the building and the entire urban system are addressed in more detail in Chap. 20.

19.4.4 The Connected Building Is in Constant Exchange with Its Surroundings and the Larger Urban System

Last, but not least, the smart building doesn't just exchange people and resources with its surroundings, it also exchanges information and communication. The building communicates with its surroundings. It collects, analyzes, and evaluates data and adapts its systems accordingly in order to provide an efficient and smart operation.

Information and communication technologies (ICT) support the building's operation and its interrelations with the urban system and its surrounding micro grids. ICT inform about demand and supply. ICT educate about consumption behavior and supply capacities. ICT inform about options and alternatives. ICT support the connections of the building's inside with its outside and the connection between the building systems and its occupants (Fig. 19.6).

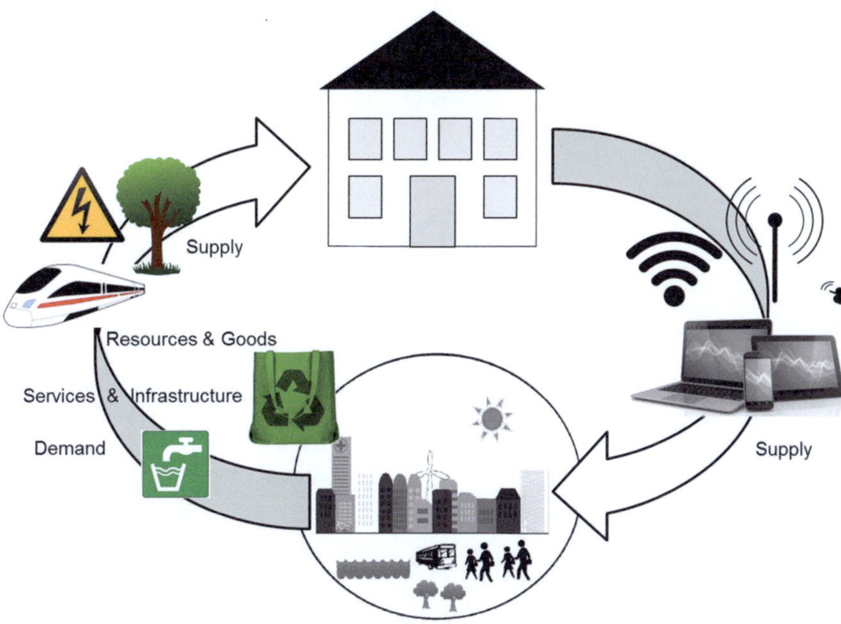

Fig. 19.6 The connected building is in constant exchange with its surroundings and the larger urban system

19.5 The Building in Balance with Nature

Previous parts of this chapter explained the importance of the accessibility of the building for its occupants and the connectivity to other system components. In order to enhance the quality of life, the occupant's experience, and the happiness and health of the people who live, work, or play in the smart building, there is one more important factor—the occupant's connection to nature, to beauty, and to aesthetics. A lot of high efficiency buildings have the technology to make them accessible and to guarantee an efficient supply and discharge or reuse of resources. The calculation can look wonderful on paper, but how wonderful is a high efficiency building with unhappy, unhealthy people?

The phenomenon of human beings and their connection to nature is called biophilia. Various studies have shown that nature and natural elements in the built environment have positive impacts on human health, well-being, productivity, and happiness. In addition, more and more engineers and architects try to learn from nature and implement technologies that are mimicking nature and processes that occur in the natural environment. This is called biomimicry. These two phenomena are described in more detail in Chap. 13.

Nature gives well-being to the occupant of the smart building and the smart building gives back and protects nature from harm and negative impacts. The smart building supports accessibility through environmentally friendly means of transportation and reduces the consumption of resources in order to minimize greenhouse gas emissions. It treats and reuses its waste, wastewater, polluted air, and waste heat before they leave the building to enter a new cycle of exchange between the inside and the outside of the building (Fig. 19.7).

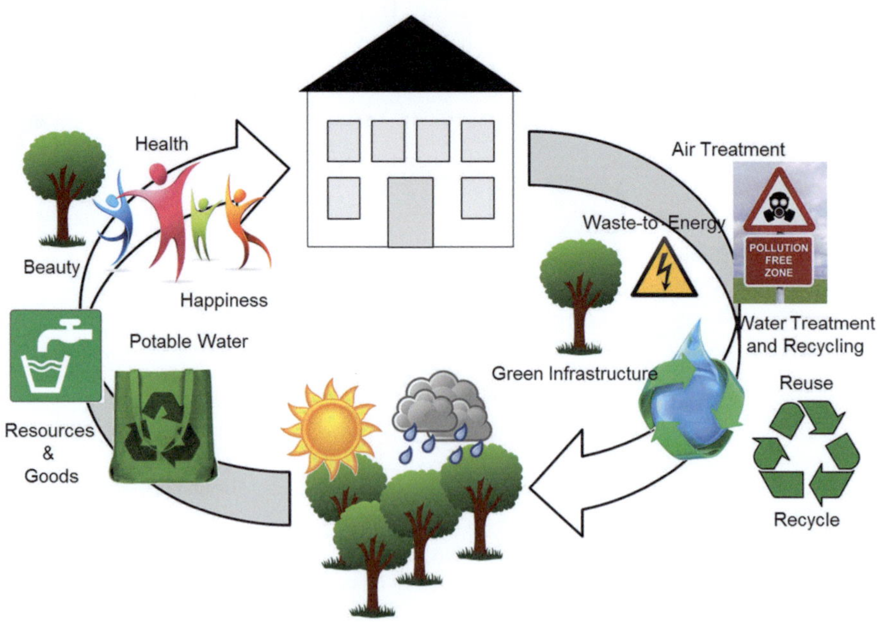

Fig. 19.7 The building in balance with nature

19.6 The Smart Building in the Smart City

It is technically doable to build an island-like building and operate it as a closed system; however, connecting it with other systems within a larger urban context is crucial to guarantee efficiency. A smart building has to be accessible for its occupants by sustainable means of transportation. It has to be connected with other systems and resources off-site in order to take advantage of the masses and to share infrastructure. And it has to be in balance with nature using and reusing natural resources while protecting the environment. The smart building has to provide a constant exchange between its inside and its outside. The smart city operates decentralized micro systems that are connected with ICT in order to understand and manage them correctly. The smart building is embedded in a smart urban system.

A smart city is not just a collection of smart buildings that are connected with other smart buildings and decentralized micro systems though. A smart city is a very complex system that manages constant demand and supply, constant exchange between give and take, and a constant exchange between its built environment, its societal environment, and its natural environment. A smart city can only operate if all its system components take advantage of this constant flow of exchange while providing information and feedback loops on the needed inputs and the resulting outputs.

References

Cerwenka P et al (2007) Handbuch der Verkehrsystemplanung. Österreichischer Kunst- und Kulturverlag, Vienna

Gehl J (2011) Life between buildings. Using public space. Island Press, Washington D.C

Stieninger P (2013) Changing human behavior towards energy saving through urban planning: creation of a new planning approach. Lessons learned from Europe and North America. LAP LAMBERT Academic Publishing, Vienna/Chicago

United Nations (2014) World urbanization prospects [Highlights], 2014 Revision. http://esa.un.org/unpd/wup/Publications/Files/WUP2014-Highlights.pdf

U.S. Green Building Council (2015) LEED® core concepts guide. An introduction to LEED and green building, 3rd edn. U.S. Green Building Council, Washington, DC

Chapter 20
The Journey of Water

Peter Mulvaney

20.1 Background

The world has a finite volume of water, and only 1 % of it is fresh. Of that freshwater, only a small fraction is available for human consumption (see Fig. 20.1, tiny ball over Atlanta). This is all the water we have. This text will discuss how we should manage this tiny volume of water for a sustainable way of life.

Most of us understand that liquid water evaporates and then condenses in the sky as clouds, which return the water back to earth as rain. And most of us are aware that the rain which falls on land seeps into the ground fills our aquifers and rivers and returns water back to the ocean. Thus, the cycle is a closed loop. But not many have considered how this water gets to our homes much less what happens during the remarkable journey of water through time and space.

Yes, space. We should appreciate that all of our water comes from outer space. Water, like everything else on the planet, was pulled from space into the earth's gravity. In fact, many planets have an abundance of water, and it is earth's remarkable orbit relative to the sun which produces a temperature where water is present in all three phases—ice, liquid, and vapor—and allows the simple water cycle we began to understand in grade school. The importance of the position of earth's orbit cannot be overstated; it allows us an atmosphere, not too close to burn off the water and not too far to be an icy plant. We are in the unique place that allows all three phases of water. Liquid, ice, and vapor each is critical to all life on the planet. Perhaps that is obvious, yet it is worth stating here, as water in all its phases is what allows life, and those phases are a result of our orbit. The point here is not just a gee-whiz moment but also to emphasize that we are unlikely to gain any more water on earth. The water here on planet earth (not planet water) is what we have to share for all needs. So let's take care of it.

While thinking about our simplified water cycle, let's consider that the vast majority of water—about 98 %, give or take—is saline, meaning it has a salt concentration greater than three percent. However, it was not always that way. Water,

A. Ahuja, *Integration of Nature and Technology for Smart Cities*,
DOI 10.1007/978-3-319-25715-0_20

Fig. 20.1 The largest sphere (860 miles in diameter) represents *all* of earth's water (40 % size of the moon). The smaller sphere (170 miles in diameter) represents all of *earth's liquid freshwater (including frozen/glaciers)*. The tiny sphere over Atlanta (35 miles in diameter) *represents all fresh surface water (could call this lakes and rivers)*—on which human and other land-based life depend. The data comes from Igor Shiklomanov's estimate of global water distribution and is referenced by both US Geological Society (source of image) and the United Nations. The reason this visual is so hard to believe is the volume of earth and the relative thinness of our oceans, rivers, and lakes compared to the diameter of earth. We are wisely called planet earth, not planet water. We may look *blue* from afar, but the amount of water on earth is remarkably scarce

on planet earth, started out pretty fresh. Consider that as rain lands on earth, it interacts with the soils and dissolves minerals and salts that are essential to life. The rain percolates through the ground, fills aquifers, and becomes the base flows of rivers. These rivers make their way down to the oceans, carrying the minerals and salts of the earth toward the oceans. As water evaporates, the salts are left behind. This process of making oceans of water saline has taken millennia and is still going on today.

What does this have to do with smart city technology? Well, a smart city understands the global scarcity of water and its impact to the natural cycles that make life possible. A smart city understands its ecological footprint through monitoring and data analysis and strives for balance with the larger geological trends. Fortunately, water has a few remarkable qualities that allow us to use, reuse, and share it infinitely with generations to come.

Water is the epitome of being small but strong—of the three atoms that make up water, two of them (hydrogen) are the smallest elements in the periodic table. Its bonding strength makes water one of the most unusual substances on the planet, with many qualities that are exactly what is needed for life, such as how water organizes into crystalline structure as it freezes, becomes less dense than liquid water, and floats. Also, water does not easily compress and thus is used in hydraulic lifts and presses. Water has a high capacity to absorb heat (think how long it takes to boil a pot of water) and thus moderates climate and weather. In all, water is a wonderful carrier of all sorts of materials, from ions and organic compounds to minerals and salts. Yet despite all this aqueous chemistry, the water molecule remains resilient to change, presenting many iterations of itself as a vapor, a liquid and ice, but always water. All these characteristics and more make water extraordinarily resilient.

Further, it is important to realize that the pollutants in water do not change the water itself, and water remains two hydrogen atoms bonded to an oxygen atom. Although it may appear muddy, or oily, or it may smell badly, the water is not the problem; it is all the other stuff that comes and goes into it that makes water either healthy or toxic, replenishing or dehydrating.

This resilience is exemplified in the facts that the water we have today is the original water that came to earth from space. It is the same water the dinosaurs walked through and is the same water our ancestors used to wash and drink. Water has taken care of itself, pollutants have come and gone, and the water cycle is still functioning. It has provided for all manner of life, and it still does. So where is the problem?

The simple fact that water is unevenly distributed across the globe creates locations of extreme abundance and deficit. A water deficit is characterized as scarcity. There are two types of scarcity that human populations face (Fig. 20.2).

One is a physical scarcity. This is where the demand for water has exceeded its supply. In this definition, there may be lots of water, just not enough to meet the demand. Typical solutions associated with resource scarcity are to either reduce your demands (through efficiency and/or changing the uses) or find new water supplies (drill wells to aquifers, dam rivers for storage). To elucidate this idea of resource scarcity, think of a city that has grown alongside a river. At first the river supplies all the needs of a small city. But rivers do not grow with population, and as the population and economies grow, they start to demand more water than the river can supply. So people find other sources and reach ever further up the landscape to supply water to the city.

The second type of scarcity, termed economic scarcity, occurs where water resources are sufficient, but the management, governance, or institutional know-how is not able to deliver the resource to the population. In this case, a scarcity of water is created if the city may not have the technology to clean the water, the governance to stop pollution, or the capital to build and maintain an appropriate pipe network. This is typified in many sub-Saharan African nations and in parts of India.

It is simple to pick up a newspaper today and see that cities around the globe are all facing both resource and economic scarcity. Modern cities from all populated continents such as Los Angeles, Beijing, Sydney, and Sao Paolo are making headline news as they face severe water scarcity.

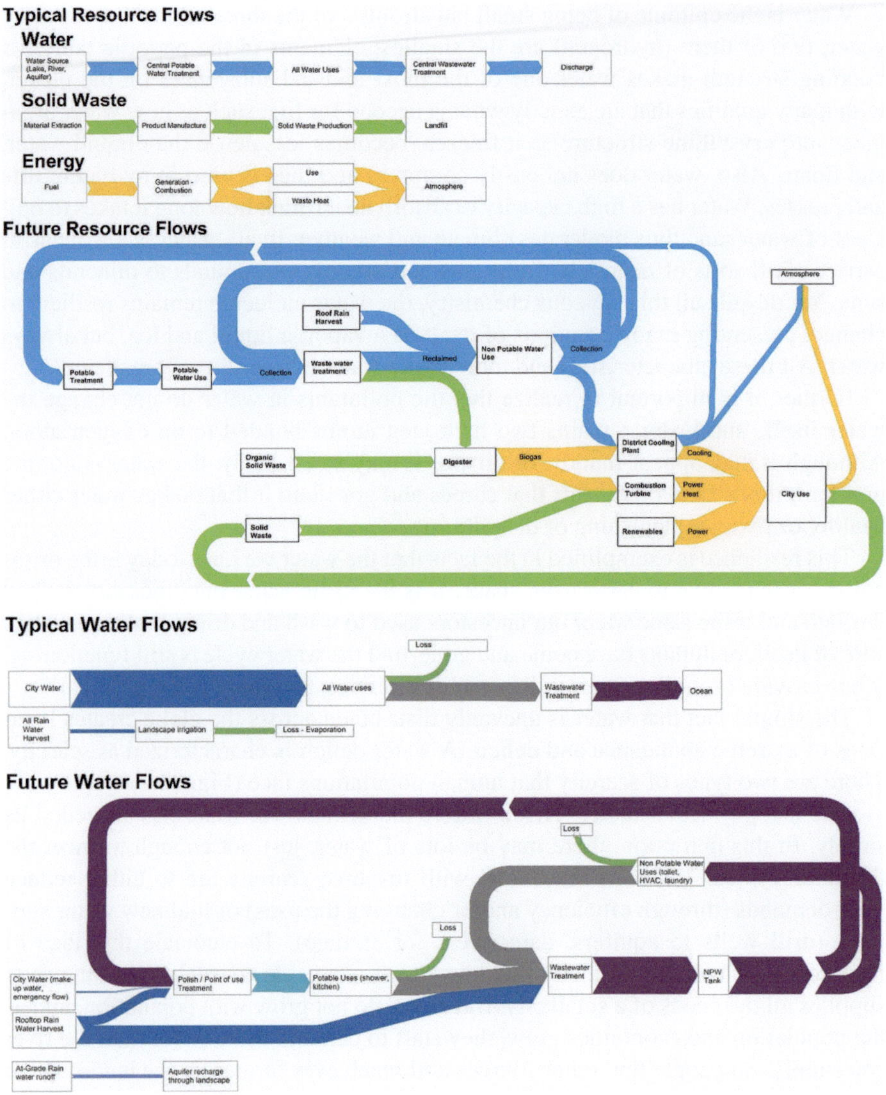

Fig. 20.2 Uneven distribution of water: For example, about 20 % of the earth's freshwater is deposited in North America's Great Lakes in amazing abundance; on the other hand, water in the White Desert of Egypt is extremely scarce, with less than 1 in. of rainfall a year

Yet water gives us great hope. We must be reminded that the problem is not the water but the stuff in the water. And once that stuff is removed, the water is as good as new. We must learn to stop polluting and to apply the treatment technologies we already have that can remove the current contaminants for a healthier ecosystem for all of Mother Nature's cycles.

This chapter will talk about the amazing journey of water in three parts. First we will further explore hydrology—the movement of freshwater through our landscapes. Next we will explore the movement of freshwater through our cities considering the smart water management systems of today and tomorrow. Finally we will explore urban flooding and the future of storm water management.

20.2 Hydrology

As freshwater journeys toward the oceans, its shear force and relentless drive to lower ground erode stone and move the earth. Along the way it forms our aquifers, rivers, ponds, and lakes as a function of topography, geology, and climate. Thus, when we interfere with hydrology for the gain of industry (irrigation, power generation, drinking water reservoirs), there are impacts both large and small, obvious and subtle. The more technology can make us aware of these impacts, the more we can choose to live, reduce, and minimize our footprints.

"Water footprinting" differentiates between water withdrawals—which extracts water for a use and returns it to the source—and water consumption which extracts water and evaporates it as part of the use, e.g., evaporative cooling.

A smart city monitors its outside environment and brings this information into the decision-making process. In fact, this has been happening for decades between energy and agriculture, the two industries with the largest withdrawals of freshwater. This dialog has evolved with the slow realization that there are direct and irreversible implications to misallocating freshwater. An unfortunate manifestation of allocation on the horizon is a growing conflict between municipal and industrial demands for water. For example, DuPont has had to address its employees regarding the water demand for its production as being a cause of water scarcity at their homes. But technology such as real-time monitoring and control, groundwater well monitoring, remote-controlled valves, and hydrologic modeling are all available to us today. We have the ability to take this outside data and internalize it. The implications of this outside–inside data are tremendous. We could, and will in the future, understand our impacts on hydrology, communicate this information to consumers, and price the use of water accordingly. We will be able to predict and manage floods, use water when it is available, and conserve in advance of scarcity.

One example of an emerging approach that alters the paradigm of existing storm water detention is to consider that there are tens of thousands of storm detention facilities, all of which are dumb. Dumb in the sense that they have passive controls. But consider how these facilities would be enhanced with a data feed from a weather service to determine the volume, time, and intensity of an anticipated rainfall. This technology is available today by the company OptiRTC which transforms passive systems into smart detention systems, which use active controls to hold water when the capacity is truly needed and to pass water when the natural and man-made systems have the capacity to convey it. Now extend this idea to rain harvest, and detention systems are no longer only managing flood risks but also serve as a water source.

Now consider the value of extending this type of information to into our ecological systems where we track and understand the movement of water through our rivers, lakes, and groundwater, and we begin to manage the ecological flows in conjunction with forecasted water and energy demands. To do this would take sensor technology that can survive a wide range of environmental conditions, run on years with minimal energy draw, send information wirelessly, and be controlled from a great distance—a challenge, but well within our means.

The US National Science Foundation has begun to stimulate exactly this effort with the Water Environment Research System (WATERS), which is intended to monitor hydrologic systems and to develop the monitoring technology that is akin to the Internet of Things for ecology. The relationship between river levels, rainfall, temperature, wind, and the moon is absolute. We must begin to understand how these parameters interact so that our cities can be designed in such a way to harmonize with the outside world. Bringing data from outside into the decision-making process and into our consciousness is a first step to a smart city and to managing water in a sustainable manner.

It is telling that the more progressive corporations are looking more and more to nature for models of reuse, interdependence, and value of diversity. The smart city will be plugged into the outside world such that decisions inside every home will have a known impact on the ecology of the watershed.

20.3 Dams/Water/Energy

Much is written about the nexus between water and energy. In the United States, about 39 % of our water withdrawals is used in the generation of electricity. Wind and solar electricty use almost no water in the generation of electricity (in some cases water is used to wash solar panels). Whereas, nuclear power withdraws approximately 42,000 gallons of water per megawattt hour of electricty, and coal approximately 35,000 gallons of water per megawatt hour. If we want to move the needle in how we manage water, we need to shift from thermal combustion electricity generation to renewable wind and solar energy. That is an understated yet powerful reality.

As urban water scarcity becomes increasingly present, the shift in dialog of power generation competing on cost will shift to one that competes on water consumption and withdrawal. When this shift happens, renewables will become more valuable for water conservation than for carbon reduction. A smart city will understand this and start to secure its water supply by shifting to renewable energy other than hydropower.

20.3.1 Water for Food

Globally, about 70 % of human water use is close to 70 %. In the United States, the withdrawal rate is closer to 32 %. In Western United States, this water is largely withdrawn from aquifers, whereas in Eastern United States, the water is largely from surface rivers and lakes. But as pumps become stronger, wells get deeper. After World War II, the United States had developed and installed enough pumping capacity to withdraw more water from the ground than fell from the sky, and our aquifers have been depleting ever since. Similarly, as our rivers are dammed, water is lost to evaporation and sediments are trapped at the bottom of the lake instead of replenishing the floodplains. Once again the technology and know-how to reduce our water usage for agriculture is available to us.

The most appropriate approach is simply to grow the right crop at the right time in the right place. There is no viable, sustainable reason to grow rice in the deserts of Southwestern United States. Second, we need to monitor our aquifers, with meters, data, and models with complete hydrology. Groundwater does not stop at stateliness, nor should our tools to help us manage it.

Moreover, we need to monitor the landscape for salinity. As freshwater passing over the earth will carry some salts, and as more water is pumped into fields for flood irrigation and evaporates or transpires, it leaves behind traces of salt. This salt builds up with the increased flood frequency, and begins to impair the soil quality. The salt reduces yields, limits the viable varieties, and ultimately leads to the failure of the land to supply food. Ancient and modern societies alike, from Mesopotamia to Afghanistan, India, and China, have failed to manage their soils and lost entire civilizations as a result. It is time to apply our technology to monitor and manage this very predictable and devastating effect of mismanaging our water resources.

Water salinity based on dissolved salts			
Freshwater	Brackish water	Saline water	Brine
<0.05 %	0.05–3 %	3–5 %	> 5 %

A smart city will have their farm fields monitored and collect the data needed to avoid the deterioration of the land, and farms can be restorative and live in harmony with its urban markets by being the sink to the wastes that the urban environment creates. But first comes the monitoring and management of the key drivers. Next-generation sensors will need to be deployed to track the soil conditions, identify where the soil is deficient, and signal to the land manager where, how much, and when to deliver a solution.

But there is also a tremendous influx if emerging technologies are available to farmers. Drones are flying over fields to pinpoint where to apply water and fertilizer or monitor for insects. They can even send information real time to farmers, as they fly autonomously over the fields to identify any particular threat to the crops. Robots are roving the rows applying only the quantities of water, fertilizer, and insecticides that the individual plant needs. The next generation of meters and monitors will be tracking the soil conditions and the hydrology of not only the field but the up- and downstream conditions of the watershed, so that the farmer can integrate the information regarding crop health. The data will be collected by drones, connecting ground meters, with weather data and crop information. This can result in tremendous productivity output as well as a higher water efficiency and allow lands to be farmed in sync with natural cycles. These are much needed technologies as the world population grows and the demand for food and water increases. At a minimum, the future city will be able to make informed decisions regarding its resource depletion and have an opportunity to respond to restore a balance.

Concurrent to smart agriculture, more food will be grown in the city centers. Both urban farming and indoor hydroponic systems can be amplified to supply significant amounts of food, grown near population centers with a far smaller land and water footprint. Ultimately, the smart city will lead the movement to smaller, more productive and diverse farming, and water will be a significant driver of that movement. Cities will increasingly support local food supplies and gardening, especially fresh fruits and vegetables where access to food is limited. Urban hydroponics will also start to play an increasing role in the urban food supply, as it does not only bring the benefits of being local but is far less vulnerable to flood, drought, hail, insects, and climate change. It will be organic, water efficient, nutrient efficient, and in a much smaller footprint. It is unlikely that urban hydroponics will replace outdoor agriculture, and certain crops are not yet ready for indoor growth. But it will prove out a market that supports healthy food, and it will reduce the agricultural footprint and our vulnerabilities to drought. Out of necessity of water conservation, the future city will need more food grown with less water, and this will likely take farming indoors.

20.4 Smarter Cities

20.4.1 Know Your Source Water

Data has long been used to manage products and goods and to develop global logistics such that your every desire is delivered within days. Yet water, the foundation of life, is only just beginning to be well monitored. Over the past decade or so, water utilities have experimented with enterprise data management leveraging sensors, telecommunication networks, and software to improve the management of pipes, pumps, and plants—the fundamentals of urban hydrology. With some lessons learned, these vanguard utilities experienced improved efficiencies and safety while lowering their regulatory risks. But few used data collection beyond their assets of pipes and facilities to think about the natural systems of forests, wetland rivers, and lakes. The smart water city will be wired to their customers and also to the rivers, streams, and earth which are the raw materials for healthy water. They will gather data from the natural world and bring it into their systems so that the water can be managed to serve both populations and the ecologies of the watershed. This is where utilities need to head, to begin to monitor and manage the source of the water in this changing climate.

20.4.2 *Know Your Plumbing Water for Reuse*

As mentioned, smart water cities will know their customer demands through the use of smart meters. Surprisingly, most water meters in use today require a human to look at and read the meter. These meters are dependent upon moving parts and thus are susceptible to wear and tear, eroding accuracy of a marginally accurate system, especially in low and high flow conditions. Fortunately, the past decade (2005–2015) has seen tremendous increase in meter technology being applied to water systems. The improvement in water meter technology has moved to the point where more accurate reads can be made without moving parts and data collected in real time. The benefits are remarkable. Instead of collecting data on a monthly or quarterly basis, it can be collected at any time point. Given the ability to measure the flow of water through a magnetic field allows water to be measured in extremely low flows, without moving parts or any physical barrier to the flow of water. This hydrologic improvement reduces the pressure losses across the meter and in effect throughout the entire system. This pressure reduction allows a lower system pressure (the EPA was mandated in the United States to be ~20 psi) which reduces water loss from leakage and reduces the energy needed in pumping and treating water.

The plumbing systems are becoming more and more equipped with sensors, and management is developing insights from data. At the time of writing, there are approximately 100 million water meters installed in the United States, of which about 20 % are considered “smart meters” with an accuracy of approximately 0.02 gallons per minute and the ability to send data directly to the utility about water consumption. Every day more smart meters are being deployed by utilities, as collecting the data is a critical means of making intelligent decisions about how water is used and managed. Similarly, utilities with smart meters can make data available on an hourly basis or even real-time basis, so that a customer can understand which activities utilize the most water, allowing them to change their behavior in effective ways. This technology is beginning to be rolled out at large scale. Given the installation rate of smart meters increasing about 25 % in this decade, it is possible to see a future where consumers know their consumption patterns and alter their behavior and that utilities can implement pricing strategies to drive changes that reduce the demands and costs of infrastructure.

Insurance agencies have begun to notice this trend in smart meters and are evaluating ways to price their policies if the meters are outfitted with technology that will have early detection of leaks or even predict if a leak is likely. The data will change the way water is managed, from source to consumption to reclamation, and all the associated supporting industries.

Another means to smarten up our cities is to design and plan the city to leverage its physical assets and development plan. Currently, modern homes are tied to collection systems that aggregate and convey wastes to a centralized facility (think sewer/wastewater treatment for liquid waste, trucks to landfills for solid wastes) and distribution systems that take from a distant central facility and distribute to the disparate end users (think potable water facility to homes, gas storage to furnaces, or power plant to end user). Now imagine a scenario where each building or clusters of buildings treats their own wastes and extract the resources such as biogas to produce heat or electricity and soil for gardens to grow food or landscapes. The smart city will consist of clustered and nested networks that greatly minimize the reliance on single-purpose large-scale infrastructure. Now, let's focus this concept on water systems.

It is possible with today's membrane technology to convert wastewater into drinking water. In fact, some cities in the United States have begun to plumb their wastewater effluent directly into the potable water plant, a true recycling of its water, what is known as a closed loop. It goes from drinking, to sewer, to treatment, and back to drinking, like nature, a cycle of water. Now, as we improve technology and reduce its size, and energy demands, we can see how this might be possible at the scale of a home. But this leads to a problem. Although the technology can be reduced, a gallon of water will occupy about 231 cubic inches. Water cannot be compressed into a smaller volume the way gases, like CO_2, is compressed and sold in pressurized vessels. So we need to figure out how to deal with the storage of water. This is why a mixed use is important as the "storage" can occur as a use that is set aside in time. For example, most offices have relatively consistent demands for water Monday–Friday from 8:00 A.M. to 6:00 P.M. This kind of profile can be marred with a use that has high demands at night and little in the daytime, such as a hotel or a restaurant district. So the same water that serves the offices in the day can be treated and serve the hotel in the evening. By designing the city to be mixed use, we enable the sharing of local resources in a smaller, tighter network. To sweeten the story, urban mixed-use communities are also more walkable, livable, and resilient to changes, all things we hope to achieve in a smart city.

As discussed, water can be recycled infinitely; we just need to get the contaminating stuff out of it. Of course, the most compelling way to make this a reality is to stop making toxic substances in the first place. But in the meantime, there are a lot of chemical and biological contaminations to deal with. And many are developing technological solutions to help us.

One of the current challenges to the water industry is the ever-improving detection technology. Our ability to detect contaminants in water is extraordinary; we are now measuring in the parts per trillion, one contaminant molecule for every trillion water molecules. This visibility of contaminants has raised many challenges. It raises the very complex question of how much contamination is safe for human health. Is it the concentration or the interaction of contaminants that matters? How different are people's tolerance to contamination? There are very complex questions to answer, and there is likely no right answer. The obvious thing is to not make or use toxic compounds, but many of the pollutants (such as nitrates, radon, and bacteria) are naturally occurring and a concern to many generations.

Humans have been trying to purify water for thousands of years. There are clear indications that in 1500 the Egyptians were treating their water with alum to remove sediment from their drinking water.

But it wasn't until the late 1800s and early 1900s that scientists figured out that microbes caused illnesses and that water could be treated with chlorine to eliminate them. Fortunately, both large firms and individual entrepreneurs are working to both make water available and to purify water at all scales, from low-tech solar distillers producing to meet daily needs to large-scale microfiltration where millions of gallons are drawn through micropores, exposed to ultraviolet radiation, settled in tanks, and treated to prevent microbial growth. The innovation is astounding.

Fortunately, the technology of efficiency is reaching the point that leaks could become a thing of the past (globally, roughly 30 % of the treated water supply leaks into the ground). Leaking water mains are a good place to make conservation investments. New technologies are available that make pipes self-healing, meaning they will reseal small leaks and extend the life of a pipe. They also sense leaks and notify the owners and can even record and transmit data directly to field crews and utility managers.

Here is where the relationship between energy and water becomes evident. Prior to megacities of 10 million people, and prior to global trade of agriculture, cities first relied on surface water, sometimes creating small dams or cisterns for storage and redirecting water for flood irrigation. But as these resources became insufficient, and as people spread to new lands, people started to tap groundwater. So long as less groundwater was taken out than rain could replenish, then all was well. But as we increased our pumping capacity, and our ability to drill ever deeper, we can now pump more water out of the ground than falls from the sky. This is non-sustainable.

So we turn to the seas.

20.4.3 Water from the Ocean

One of the mega trends of our time is the migration of human population to the coasts. Our largest cities are coastal and often located on river deltas, the incredibly diverse and dynamic world of estuaries, deltas, and marshes. This readily available abundance comes with the risks of sea storms and riparian floods. Today, our freshwater rivers have been either dammed to the maximum extent, polluted beyond safe use, or dried up; likewise our aquifers rapidly depleting, we look to the sea for our water, the most abundant source of water, and we are learning to remove the salts ever more efficiently. The Middle East is most dependent upon desalination, with Israel being a leader in the technology. Israel desalination capacity is over 130 billion gallons a year, and they are planning for a capacity of over 200 billion gallons a year by 2020. Desalination is a technology that will save many cities from dehydration, but it comes with a high energy cost and an underestimated brine waste impact. Ironically, life first began in the salty seas, where a single-celled organism differentiated itself from the salty environment by evolving a sodium pump, where it could pump sodium out of this interior and thus actively maintain itself as different from its world. The first desalination systems were biological and the start of life itself. Today, we use either distillation processes (boiling seawater and collecting the water vapor) or membranes which separate the water from the salts. Either way, the salt is separated from the water, but the salts did not disappear; they are concentrated as the water is removed, and this concentrated salt, called brine, must be disposed of, often back to the sea where it creates a variety of proximal damage as a result of the changed chemistry. But the real barrier to date has been energy costs.

Desalination plants are becoming more energy efficient, and in the Middle East, much of the desalination is powered by waste heat from the production of electricity ("free" power for the desalination, but of course the origin of the heat is fossil fuel combustion).

As water is foundational to life, and as water scarcity begins to threaten life, livelihoods, and economies, we will see dramatic shifts in humanities relationship with water. With the recent headlines about the California drought and the Toledo water ban, people are beginning to understand. People are not only hearing about it, but it is starting to impact their lives and the jobs.

20.4.4 Know Your Behavior

A smart city will look to reduce its demands, invest in behavior changes and efficiency, and even plan its industry and economy around its water resources. Yet, we live with the paradigm where capital projects are easier to manage than behavior change. As such, utility managers are good at building infrastructure; it is a tangible way to reach voters, raise capital, and increase a central authority over water. What we need more are smart economics—creative pricing schemes that allocate water to the highest values. We need education from our utilites regarding water use relative to neighbors, or relative to comparable households, and incentives to conserve and to invest in water-efficient fixtures. Simple policy changes that limit irrigation are an easy way to reduce the largest percentage of residential water use. Cities can follow the many examples set by leaders who have curtailed the growth of lawns, promoted xeriscaping, and limited irrigation of certain days and times, even ticketing for water that runs off your property and into a public drain. Second is the bathrooms—and toilets in particular. The New York Department of Environmental Protection has launched a 125 vouchers to eligible residents for the purchase an EPA-certified high efficiency toilet from approved vendors. On average, an American flushes approximately 24 gallons a day down the toilet.

20.4.5 Know Your Storm Water

While having enough water for our cities is a major growing concern globally, we must also design our cities for too much water. As the climate warms up, it can hold more water (which it pulls from the surface, making the surface drier). Thus, when it rains, it rains with more intensity. Add to this the ever-increasing amount of impervious areas (the hardscape of our cities: roads, roofs, driveways, patios, etc.), and it is easy to see that storms can quickly create urban flooding. Further, the world's sewers are aging. Just consider how much your nearest city has changed over the past three decades. Now consider how much the sewer system has changed. It is extremely unlikely that the sewer has kept pace with the changes on the landscape.

When sewers were designed in most American cities, few engineers could have conceived the amount of impervious areas to come. Simply put, the sewers were not designing to manage cities that sit above them. Add to that the complexities of climate change which fundamentally changes when, where, and how much water is allocated, and we are truly facing the need to rethink our infrastructure.

Sewers come in two different schemes, and we will see how a smart city will add a third element into its design. Early American sewers were designed to carry both sanitary waste (from the internal plumbing of homes and industry) and storm water runoff from roads and rooftops in the same pipe. These sewers are referred to as combined sewers, as it combines sanitary flow and storm flows in the same pipe. Oftentimes the "combining" happens underneath the home where the rooftop drains connect to the sanitary line from the internal plumbing, so that only one drain line goes out from the home into the public sewer in the right of way. By the way, sewers are the deepest of the utilities, as few want leaking sewage to drain onto their underground assets. As such, sewers are often the most expensive utility as much of the cost is in the trenching of the road and the subsequent rebuilding of the road. Combined sewers also have the added cost of treating the combination of storm and sanitary flows, with the benefit that all the water gets treated.

The other type of sewer scheme is called a separated sewer for the sewer consists of two independent systems: one is for storm water and the other for sanitary. The key thing to note here is that the sanitary line is carried to a wastewater treatment plant, while the storm flows (larger pipe) are directed to a water body such as a river or lake. Until recent regulatory changes, the storm water went untreated. It was thought, rather foolishly, that urban runoff was clean enough to discharge straight into our rivers. Today, regulators are wrangling over new regulations that require the retrofit of storm sewers with some treatment and that brings us to the third scheme of sewers and to the notion of inside/outside.

As many cities of the future are yet to be built or are in the process of being built, better urban planning with efficient, integrated systems for urban water management and the conservation of water resources can go a long way in reducing both water and energy demand in cities.

Cities will include more natural systems and less impervious areas. This is happening in many cities across the globe. Philadelphia was one of the first to push the US Environmental Protection Agency (USEPA) for this type of natural solution, where their combined sewer system is augmented with landscape changes to reduce the burden on the sewer. So water is diverted from roads into planters where it can percolate into the earth. Parking lots and lanes are being converted into porous materials to allow water to pass through the pavement instead of over it. In China, where they are building new cities for tens of millions of people, they are integrating large open spaces for storm water infiltration. And of course Germany has been building their cities with green roofs (roofs with soil and vegetation) to reduce and clean urban runoff for decades. All these strategies have ancillary benefits over the business as usual.

Infiltration recharges ground aquifers with rain, which improves vegetation, which benefits wildlife and lowers the urban heat island all while creating a more diverse aesthetic and lower costs in treating water. Infiltration, a simple, natural and ancient way to manage stormater is in many ways smarter than our modern machinery. It is hard to argue against these benefits, and although many engineers and regulators have resisted for years, the benefits are now clearly outweighing the costs. To be sure, the traditional systems are still adding the fundamental value, be they combined or separated, but the inclusion of the landscape into the system is simply smart and including what should have been put in place from the outset (Figs. 20.3, 20.4, and 20.5).

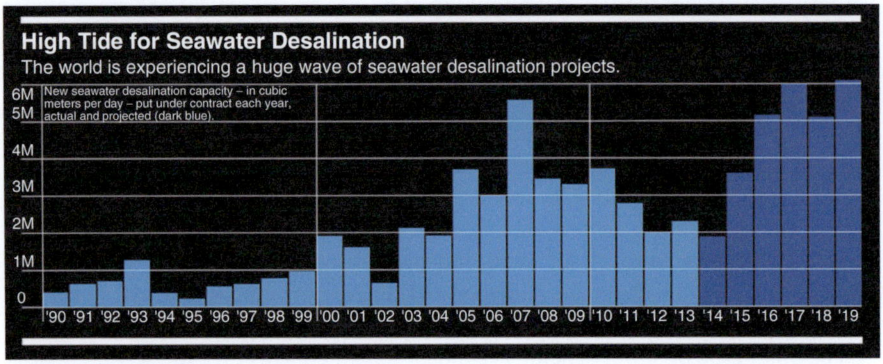

Fig. 20.3 High tide for seawater desalination (Source: http://www.technologyreview.com/featuredstory/534996/megascale-desalination/)

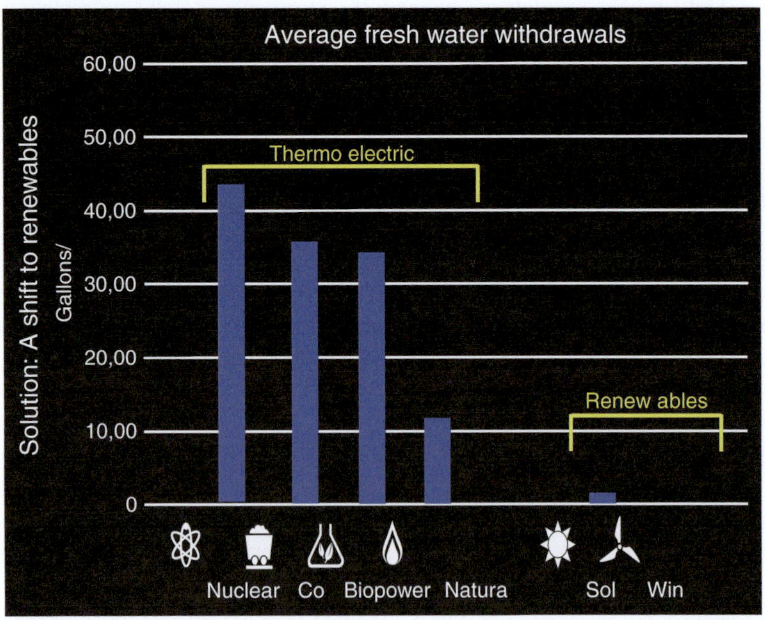

Fig. 20.4 Data from Consumptive Water Use for U.S. Power Production; December 2003, NREL/TP-550-33905; P. Torcellini, N. Long, and R. Judkoff National Renewable Energy Laboratory 1617 Cole Boulevard; Golden, Colorado 80401-3393; NREL is a U.S. Department of Energy Laboratory Operated by Midwest Research Institute the graphic was developed by Skidmore Owings and Merrill, LLP

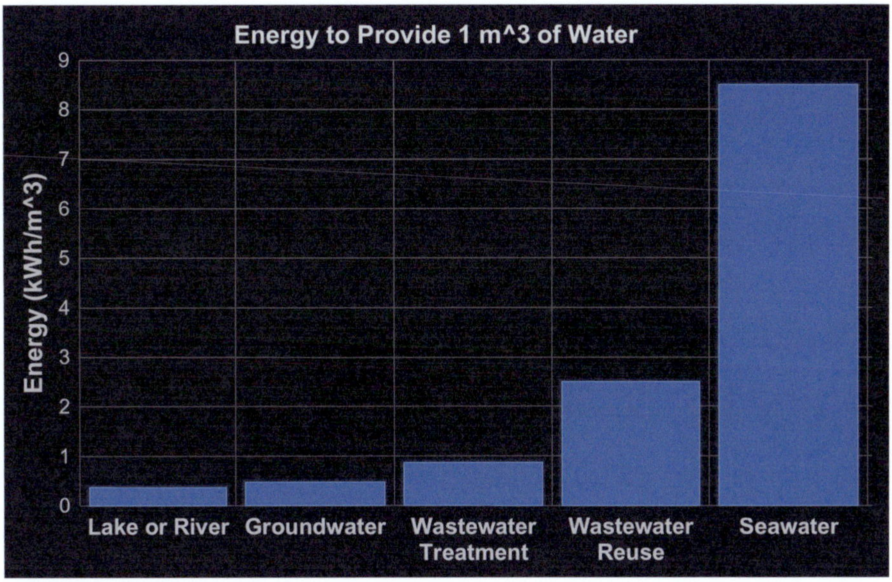

Fig. 20.5 Range of water withdraws for typical electricity generation in the United States

Chapter 21
Integrated Affordable Mobility Solutions in a Smart Neighborhood

Stefan Melzer

Climate change is evident—on a global level, there is an international congress setting up goals in order to reduce emissions along with the reduction of energy consumption nearly every year. Seventy-five percent of the CO_2 emissions throughout the world are currently caused by burning fossil fuels in cities.[1] From an institutional point of view, cities are obliged to create a legal framework which supports planners and citizens in fighting climate change or switch to renewable energy sources.

As cities and agglomerations are growing and the gathering pace of urbanization is increasing, the potential for smart solutions in smart cities is very high. The mobility sector is a key factor for achieving climate goals.

Mode shares in Chicago identify private transport at 63 %.[2] Cities with sustainable development strategies like Vienna show that an investment in smart activities is not only an ecological asset but also an economic and social one. Vienna has invested a lot in eco-friendly means of transport so that the share of motorized individual traffic was reduced from 40 % [3] in 1993 to 27 % [4] in 2014. Attractive alternatives where created and even the numbers of inhabitants who owned a car decreased. The city administration started with the development and implementation of innovative smart mobility initiatives in selected pilot areas. Urbanization is growing fast; more infrastructure is needed for more and more inhabitants. Urban mobility is going to be one of the hardest challenges.

As a matter of fact, the transport sector is one of the biggest carbon-emitting sectors, and one of the major goals of every community is to reduce traffic-

[1] Smart City Vienna Agency, The Initiative, https://smartcity.wien.gv.at/site/en/initiative/, 2015.03.16.

[2] Chicago regional household travel inventory: mode choice and trip purpose, Chicago Metropolitan Agency for Planning, Chicago 2008.

[3] STEP2025, Vienna City Administration, Municipal Department 18, Wien 2014, page 106.

[4] OTS-Press-release, vice-mayor Maria Vassilakou, 2015.02.10 http://www.ots.at/presseaussendung/OTS_20150210_OTS0091/modal-split-2014-radfahren-in-wien-immer-beliebter, 2015.03.12.

A. Ahuja, *Integration of Nature and Technology for Smart Cities*,
DOI 10.1007/978-3-319-25715-0_21

based emissions, so it is necessary to implement sustainable mobility solutions not only on the city level but also to the apartment complex level. That sounds compulsory, but nearly 80 % of all trips by citizens per day start or end at home,[5] the place of living is therefore essential for a sustainable mobility management. The choice of your personal mean of transport does not only influence your personal health; it has an impact on the living standards and carbon footprint of the whole city too.

When we want to reduce CO_2 emissions, we have to reflect the actual situation and focus on the key transportation systems in a city. Public transport systems consume far less amount of energy per RPK (*r*evenue *p*assenger *k*ilometers) and CO_2 emissions are only a small fragment of emissions from individual motorized vehicles [6] (Fig. 21.1).

In a smart city, intelligent mobility is eco-friendly and has no need for individual use of automobiles. Individual car use consumes the most energy and produces the highest number of emissions.

So we have to get rid of a vast number of cars. How can we do that?

Many new mobility providers have been emerging during the last years. Sharing initiatives entered the market, starting with car-sharing-providers; nowadays we can find bike-sharing, ride-sharing, and even tool-sharing platforms. In a technological era, digital platforms where demand and provision meet are getting more and more important. But first we have to think about solutions to reduce the vast number of privately owned cars in the cities and then we have to create or enhance innovative eco-friendly means of transport and invent smart user-friendly information systems.

Other important steps are the improvements for bicycle users and pedestrians. As a cheap alternative to automobiles and public transit, the public authorities should reinvest funds which were originally allocated for motorized transport. The quality of footpaths and bike lanes are a major infrastructural factor. Mobility infrastructure is often seen as the most important factor for potential investors. In a smart city mobility infrastructure is not a simple road for cars; in a smart city mobility infrastructure is seen as a holistic answer where we find various sustainable means of transport.

Individual motorized traffic causes the highest emissions per RPK and consumes most of the space available. The majority of the consumed space is space used by private cars for parking. On average a car is only used 1 h a day; in the remaining 23 h, this vehicle is just using public or in less cases private space.[7] The following graph[8] shows the space needed by the different means of transport (Fig. 21.2).

---

[5] Mobilität in der Stadt Graz, Graz City Administration, Municipal Department Traffic Planing, Graz 2014, page 2.

[6] VCÖ-factsheet: Wie Mobilität in einer Smart City aussieht, VCÖ, Wien 2013, page 3.

[7] VCÖ-factsheet: Carsharing verringert Verkehrsprobleme, VCÖ, Wien 2011, page 2.

[8] Road Space Allocation: Roadway Design and Management to support transportation alternatives, Victoria Transport Policy Institute, URL http://www.vtpi.org/tdm/tdm56.htm, 2015.07.27.

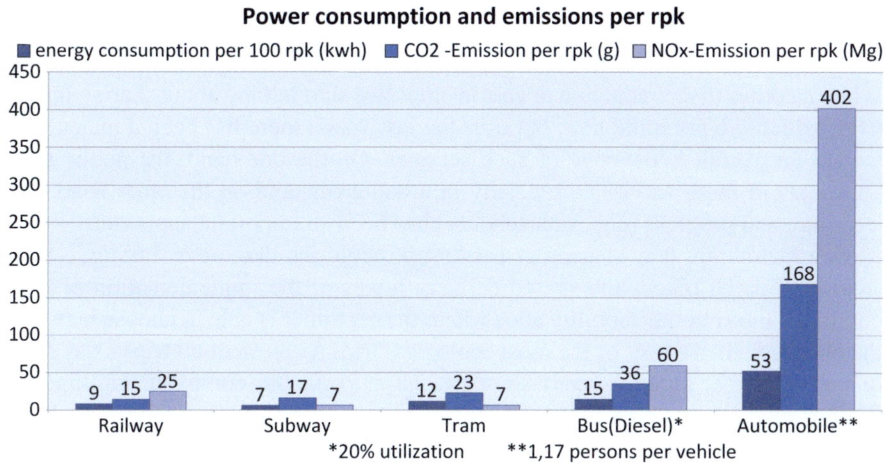

Fig. 21.1 Power consumption and emissions per rpk

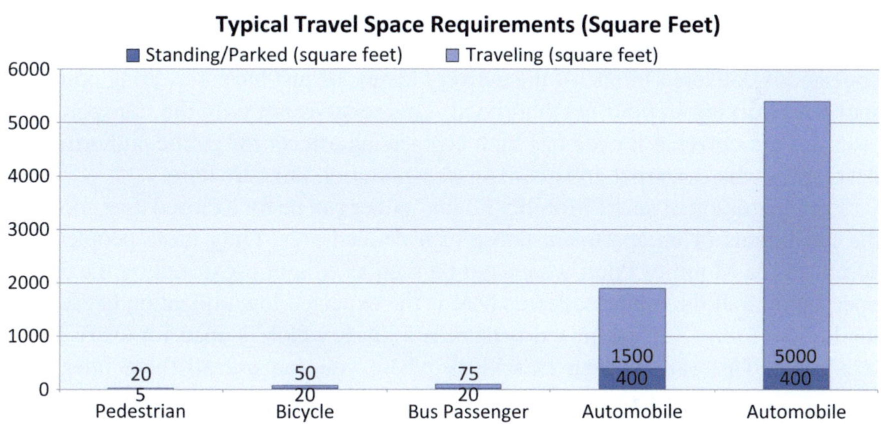

Fig. 21.2 Typical travel space requirements (square feet)

Spaces taken by the different means are calculated by vehicle length, width of necessary travel track, and space needed for braking distance. We can see that private cars, solo driven, take up the most space and are therefore highly inefficient. In a smart city these inefficient means of transport should be replaced with investments in space savers like bike lanes, pedestrian walks, or other carbon footprint reducers.

So, to summarize this, the first step is we have to get rid of privately owned cars in cities, which goes along with the prioritization of public transport. The second step is to search for new innovative smart mobility solutions.

21.1 Sharing Initiatives

When it comes to the reduction of cars in cities, we start talking about sharing initiatives, which are not quite new, but over the last years, there has been a major step toward an overall knowledge of such services. On the one hand, the public now knows about these services, especially in urban areas, and on the other hand, the technological progress (e.g., smartphones) has been an important supporter.

Key factors are new sharing or pooling possibilities like e-car sharing, e-bike sharing, e-cargo bikes, and shared delivery boxes or the implementation of new combined multimodal-mobility apps where the costumer is able to choose between the cheapest, the fastest, or the most ecological trip. As 80 % of all trips a day start or end at home, modern smart strategies have to dictate combined multimodal sharing solutions not only for new houses but for existing areas as well. So we have to start thinking about sharing initiatives right at the place where we are living (Fig. 21.3).

Convenience for users is essential here. Easy access, attractive affordable offers, good integration, and low-threshold networking opportunities are the other factors which will lead to success and high utilization. One strategy for new city development areas is the integrated multimodal "Mobility Point" where you can share different cars for different purposes, rent a bike for a short trip to the supermarket, get your goods delivered to one of the delivery boxes, or just have a meeting point for social networking in your neighborhood. This reduces not only the transportation costs for the citizens; it even has high cost saving effects for public authorities or developers who construct and maintain transportation infrastructure.

The integration of such "Mobility Points" either can be for a closed user pool like the inhabitants of an apartment house or a defined area. Only these people have access to the Mobility Point which can be seen as benefit organized by the developer, but one of the major concerns here is the expected low utilization because of the limited users. The optimized version is a kiosk which is open for every registered user. This can be seen as a kiosk where you can use all these integrated mobility services and in addition to that a local supplier acts as a café or small supermarket.

Ideally this kiosk is not only a local supplier or cafe but a social meeting point in that area too. In addition to that, shop owners can provide mobility information; rent e-bikes, e-cars, or e-cargo bikes; and even repair bikes.

In addition to these services, loads of new technologies were born in the last years. For instance e-cargo bikes are said to be able to handle up to 51 % [9] of the urban motorized trips with goods transport in a city. Transported goods in European cities tend to be suitable for cargo bikes—on average goods have less than 100 kg and less than 1 m^3 volume.[10] Especially in cities with hills an electrified engine

[9] Cyclelogistic – moving Europe forward, Austrian Mobility Research, Graz 2013, page 7.

[10] Article City-Logistics, Maria Randelhoff, URL http://www.zukunft-mobilitaet.net/thema/city-logistik, 2015.05.04.

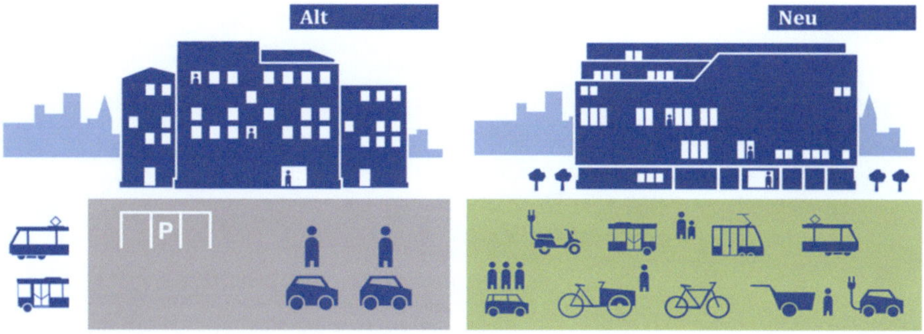

Fig. 21.3 WOMO enhances multimodal transport (Source: WOMO – Wohnen und Moblität AP2, Wien 2014, page 8)

could save here a lot of human energy and does not frighten potential users. In more and more cities, people are allowed to ride bikes in shared spaces and pedestrians zones. Maximum walking speed is allowed but you may use these zones.

In the late 1880s, people started to transport goods and luggage by cargo bikes because it was a safe and easy way to transport things in a city. Today the cargo bike is going through a renaissance, starting in the northern countries like Denmark or the Netherlands. When urban logistics is discussed among experts, private trips are often not considered. The biggest share of convertible trips is private shopping trips starting from home. Numbers are changing because of online shopping but there is still the largest potential for saving trips.

Cargo bikes take up more space than ordinary bikes—there is a need for adequate and safe parking in cities. Municipalities have to create new pavement markings and parking spots for these bikes. Apartment complexes have to be built with extra-large bike rooms and upgraded elevators where there is space for more than one bike. The following picture shows you a design study for a cargo bike parking spot, unfortunately without a charging station (Fig. 21.4).

So the demand for these cargo bikes is getting bigger and bigger, but how can common people or developers share or provide these bikes?

During the last months, several companies invented a new locking system for bikes. Keyless systems with GPS modules, siren warnings, and the possibility to set up your own sharing business are now available. The first prototypes are being transported to interested customers at the moment (Fig. 21.5).

Keyless locks look like usual locks but they are smart. It is very simple. You only have to walk to your bike and press a button to lock or unlock your vehicle. You can search and manage your drives via the smartphone app and it is very easy to use. On bigger scale, companies can start now their own bike-sharing business. It seems there is no more need for big and expensive terminals and docking stations. Most of these classic bike-sharing providers offer only a limited range of available bikes. That system claims that any bike can be used, no matter if it is a cargo bike or an e-bike; even scooters are possible. For business users, fleet management is essential.

Fig. 21.4 Cargo bike station (Source: Design Company Cyclehoop (The Copenhagenize Bar, Cyclehoop Design Study, URL http://copenhagenize.eu/portfolio/project01.html, 2015.07.27))

Fig. 21.5 Variety of keyless locks

The payment process, access policy, service plans, geolocation, maintenance, alerts, and statistics should be implemented right from the beginning. Besides that, social media appearance, insurance issues, and a state-of-the-art website design will play a major role for a successful start. Some companies even claim that they detect many types of serious crashes using its built-in accelerometer and notify your supervisor.

21.1.1 *How Does It Work?*

BitLock uses Bluetooth 4.0 technology is used to replace your bike key with your smartphone. It detects your presence as you come within 3 ft of your bike and without the need to interact with your phone. Sure you have to make sure you have Bluetooth activated and the app is running in the background—so it sends a message to your home screen to lock or unlock by simply pressing the button. The battery is said to provide a 5-year-long life without charging, and once the level is low, costumers get notified. The lock itself does not provide a GPS module; the application on your smartphone uses the integrated GPS module to save the location where the bike is parked. Keyless locks are made of heat-treated and cut-resistant steel and use security protocols often used in online banking systems with encryption based on open and published standards. Other systems use their built-in 3-axis accelerometer to notify you that the bike is getting moved or shaked. This can be used too as a crash alert. One company has a built-in solar panel that is able to recharge the battery. Although these systems are rather expensive devices at the moment, there is definitely a potential for future use, property developers who plan to set up easy to use bike- or e-bike-sharing systems, for instance.

21.2 The Next Big Thing?

Now we want to go one step further and have a look at new improved abilities to locate the best available parking place right near your target area. How can we do that? How can we use new technologies to achieve that? The next big thing for cities could be smart parking solutions which were installed in Buckinghamshire, UK, early this year (Fig. 21.6).[11]

These 300 Smart Parking RFID vehicle detection sensors enable drivers to check the number of free spaces and provide analysis to the Council in order to manage their strategies when it comes to parking spaces in public areas. On the one hand, it is a big advantage for individual drivers, who can save time for not searching for free spaces and do not have to return to their car for changing tickets, and on the other hand, public authorities gain better data concerning parking issues and the whole parking experience will improve because of the reduced searching time of cruising vehicles through the area. And in the end congestion and emissions will reduce significantly. At the start infrastructure costs are quite high, but studies show that revenue from parking infringements boosts like occupancy levels and satisfied customers who use this ticketless, cashless pay-by-phone systems which saves up 15 min of average search time in dense urban areas. This reduces time spent on the parking process and reduces parking costs to the parking provider and enhances penalty charge revenue because wardens now know where vehicles park which have no valid ticket. This software collects and analyzes live information on how parking

[11] Smart Parking Homepage, URL http://www.smartparking.com, 2015.07.27.

Fig. 21.6 Smart parking

space is used. Real-time data gives us information about daily management and future planning. It is simple to use and compatible to most of the existing systems. Combined with low costs for maintenance, it will make parking quicker and easier for customers and providers.

21.2.1 How Does It Work?

SmartEye, the technology system behind this smart parking company is based on infrared sensor technology. Infrared sensor technology comes with long-life batteries, and because of its wireless technology, it is quite easy to install. Now that SmartEye registers all the vehicles, the information on space occupancy is gathered and fed to the management software in the back. The data is managed and forwarded to the used app in order to get the information about the next available parking space.

Other services like Mobypark from the Netherlands set up shared parking platforms via smartphone apps which provide available spaces in private parking lots, public parking garages, hotels, or hospitals. They claim to reduce CO_2 emissions because drivers spend less time searching for parking spots.

Others criticize such services because they claim that making free parking spaces more visible will make it more attractive for car users to drive in these areas because now they can reserve a parking space. They are not producing more parking spaces but they are definitely producing more traffic.

It is not that technology driven, but combined mobility platforms as a platform to connect multimodal mobility systems are the right solution when it comes to gathering information about the mobility products which are available.

21.3 Combined Integrated Mobility Platforms

Such services can easily be integrated into digital mobility platforms. They represent the digital layer between the potential users and the mobility services. At the moment, there are several research programs working on such integrated mobility platforms. One of them is the project "SMILE" (*S*mart *M*obility *I*nfo and Ticketing System *L*eading the Way for Effective *E*-Mobility Services) led by the OEBB (Austrian federal railway company) which started a pilot only a few months ago. Depending on the purpose of a trip, this smartphone app chooses the right means of transport for the right time. It encompasses a route planner, shows you which means of transport are available in your area, and informs you about the expected price of that trip. At the end of the month, you receive a summary of all your trips and a combined invoice. Project partners are car-sharing companies, public transport providers, parking garage operators, research institutes, and charging station providers. With this smartphone app, you are informed about the available transport, the optimal route to reach your goal, and the actual costs of that trip. Furthermore it keeps you informed about travel time and CO_2 emissions.[12] Combined with the integrated mobility solutions mentioned in this chapter before, it is a powerful tool to keep you informed and no need to carry around a handful of member cards.

Researchers from PNAS (*P*roceedings of the *N*ational *A*cademy of *S*ciences) found out after analyzing more than 150 million taxi trips in New York City that taxi-sharing models could save up to 40 % of all the trips. In this theoretical survey, the starting and ending point were known so it was an optimal analysis situation. In reality the survey showed that there is still a potential of 16–25 % of the taxi trips if customers are willing to drive two minutes longer but with shared costs. So it takes a little bit longer but it is a lot cheaper. It is another program which shows how to save trips, save emissions, and save money if we are able to use real-time data for route-optimizing applications.

One of the most important questions here is the optimal operator model. Do public institutions have an obligation to implement such services or is this a new chance for public private partnership models?

21.4 Delivery Solutions: Drones

Another interesting part of smart city logistics is drone technology. Right now drone technology is being tested in the United States and in Europe.

At a first look, it seems that drone technology has a large potential in reducing CO_2 emissions—empowering them should not be a big problem too. Greenwire estimates that there would only be an increase in electricity consumption by a hundredth of a percent in the United States. On the other hand, the costs for such a

[12] Smile Project Homepage, URL http://www.smile-einfachmobil.at, 2015.07.07.

delivery could rise up to +100 % because of a single delivery instead of bulk deliveries organized via trucks.[13]

At the time when drone aircraft is no longer only used by military operations, there is a need for strict regulations when it comes to data security and security at all. A drone is an unmanned aircraft that can fly more or less automatically. In the last years, there are more and more fields in which drones become more and more important, like observing nature, filming sport events and commercials, and package delivery. This year the FAA (*F*ederal *A*viation *A*dministration) and the Department of Transportation proposed new regulations concerning commercial use of drones for applications like photography and surveying. With this announcement the FAA is opening a 2-month public comment period before that proposal can become law.

Drone technology is now cleared for testing but not yet ready for takeoff. Amazon's goal is the usage of drones to deliver small packages to people's doorsteps, fully automated. Amazon Prime Air is able to test its drones at 400 ft or below and under other restrictions. In Germany, DHL researches the transport of goods to more or less isolated places. The main goal is to test emergency transports (pharmaceutical goods) to the island of Juist in the northern sea. The island Juist is not reachable via streets and the sea is sometimes that rough that no ships can land there. So this seems to be a perfect spot for research.

In Vienna a research project from Siemens operates in Vienna's urban lakeside in order to overview the construction sites progress and produce certain 3D pictures. It is funded by the Austrian Ministry for Transportation.

21.5 Automated Vehicles

As per definition these cars use technologies like radar, GPS, or, for example, computer vision in order to find the correct path. The very first prototypes were tested in the early 1920s, but the biggest steps toward a driverless system have been invented during the last years. Every major car-manufacturing company has developed their own vehicles and even Google have started their tests. As it is a future technology, there are many potential advantages such as lower emissions or less traffic congestion due to optimized distance control. One or in my opinion the most important advantage is the estimated reduction of traffic collisions. Ninety-seven percent of all traffic accidents are based on human factors. By reducing the number of rides and the number of cars, automated vehicles could be another major factor in reducing the ecological footprint. Analysis estimates that within automated taxis the amount of greenhouse gas could be reduced by 90 % compared to current petrol-driven taxis.[14]

[13] E&E Publishing, Technology – Amazon's drones are coming, URL http://www.eenews.net/stories/1059991434, 2015.27.07.

[14] The Independent: Driverless cars and robot taxis could cut carbon emissions by 90%, URL http://www.independent.co.uk/news/science/driverless-cars-and-superuber-robot-taxis-could-cut-carbon-emissions-by-90-10369503.html, 2015.07.27.

Chapter 22
International Rating Systems for Smart Buildings and Smart Cities

Rahil Kaka and Petra Stieninger

22.1 Rating Systems for Buildings

"Buildings consume about 40 percent of global energy and 25 percent of global water." In addition, they emit over 30 percent of greenhouse gas emissions.[1] Thus, the way we operate buildings nowadays has substantial, negative impacts on the environment and our health. Stakeholders in the building sector such as building owners, building management companies, architects, and engineers are aware of this threat to human health and nature and have been trying to make buildings more resource-efficient and sustainable. In recent years, buildings have become more and more efficient, and some recent developments have made even net zero-energy or net zero-water buildings possible. Organizations such as Building Research Establishment in the UK, the US Green Building Council (USGBC), the International Living Future Institute, and Green Globes have established rating systems for buildings in order to measure, evaluate, and certify their sustainability. With rating systems such as USGBC's Leadership in Energy and Environmental Design (LEED), the Living Building Challenge, or the British Building Research Establishment Environmental Assessment Methodology (BREEAM), buildings can achieve a certain number of points according to their efficiency and sustainability and receive a certain level of certification. In some countries, the building owner receives funding or tax incentives for achieving a certain level of sustainability. In the following, the most popular ratings systems will be described in more detail.

[1] http://www.unep.org/sbci/AboutSBCI/Background.asp.

A. Ahuja, *Integration of Nature and Technology for Smart Cities*,
DOI 10.1007/978-3-319-25715-0_22

22.1.1 BREEAM

Building Research Establishment Environmental Assessment Methodology (BREEAM), founded in 1990, is the longest established assessment method for the sustainability of buildings. While most prominent in the UK, BREEAM has certified over 250,000 buildings worldwide. BREEAM works with buildings at any stage of their "life cycle," whether it is new buildings, in-use buildings, or refurbishing. BREEAM utilizes several categories for sustainability such as management, health and well-being, energy, transport, water, materials, land use, or pollution. Depending on where a building is in its life cycle, BREEAM can evaluate buildings on whether or not they pass certain protocols and, if not, what they can do to improve their sustainability (Fig. 22.1).

22.1.2 LEED

The Leadership in Energy and Environmental Design (LEED) certification was developed by the US Green Building Council in 1993 and has become one of the most popular building certification programs in the world. LEED is a certification that promotes green or environmentally friendly building. It stresses efficient and cost-effective building while also having a positive impact on the environment with the use of renewable energy. LEED uses a point system as shown in Fig. 22.2, in which different certification levels can be reached based on various eco-friendly features for new or existing buildings.

The point system shown in Fig. 22.2 is used to assess a LEED level of certification of buildings. The levels are as follows:

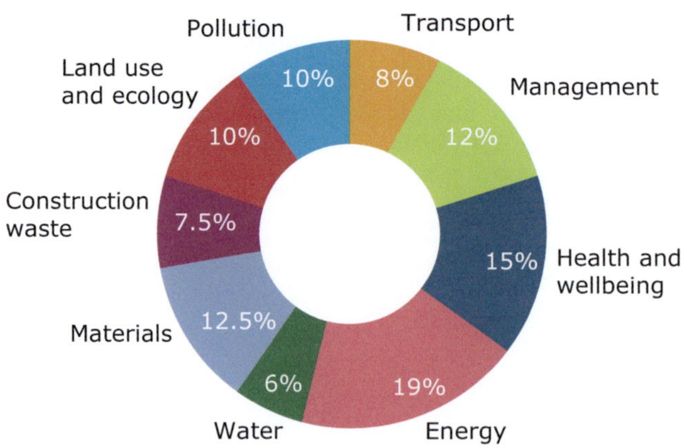

Fig. 22.1 The BREEAM rating system with each category and their respective weighting (Source: www.pmarchitects.com)

LEED v4 for BD+C: New Construction and Major Renovation
Project Checklist

Project Name:
Date:

Y	?	N			
			Credit	Integrative Process	1
0	0	0		**Location and Transportation**	**16**
			Credit	LEED for Neighborhood Development Location	16
			Credit	Sensitive Land Protection	1
			Credit	High Priority Site	2
			Credit	Surrounding Density and Diverse Uses	5
			Credit	Access to Quality Transit	5
			Credit	Bicycle Facilities	1
			Credit	Reduced Parking Footprint	1
			Credit	Green Vehicles	1
0	0	0		**Sustainable Sites**	**10**
Y			Prereq	Construction Activity Pollution Prevention	Required
			Credit	Site Assessment	1
			Credit	Site Development - Protect or Restore Habitat	2
			Credit	Open Space	1
			Credit	Rainwater Management	3
			Credit	Heat Island Reduction	2
			Credit	Light Pollution Reduction	1
0	0	0		**Water Efficiency**	**11**
Y			Prereq	Outdoor Water Use Reduction	Required
Y			Prereq	Indoor Water Use Reduction	Required
Y			Prereq	Building-Level Water Metering	Required
			Credit	Outdoor Water Use Reduction	2
			Credit	Indoor Water Use Reduction	6
			Credit	Cooling Tower Water Use	2
			Credit	Water Metering	1
0	0	0		**Energy and Atmosphere**	**33**
Y			Prereq	Fundamental Commissioning and Verification	Required
Y			Prereq	Minimum Energy Performance	Required
Y			Prereq	Building-Level Energy Metering	Required
Y			Prereq	Fundamental Refrigerant Management	Required
			Credit	Enhanced Commissioning	6
			Credit	Optimize Energy Performance	18
			Credit	Advanced Energy Metering	1
			Credit	Demand Response	2
			Credit	Renewable Energy Production	3
			Credit	Enhanced Refrigerant Management	1
			Credit	Green Power and Carbon Offsets	2
0	0	0		**Materials and Resources**	**13**
Y			Prereq	Storage and Collection of Recyclables	Required
Y			Prereq	Construction and Demolition Waste Management Planning	Required
			Credit	Building Life-Cycle Impact Reduction	5
			Credit	Building Product Disclosure and Optimization - Environmental Product Declarations	2
			Credit	Building Product Disclosure and Optimization - Sourcing of Raw Materials	2
			Credit	Building Product Disclosure and Optimization - Material Ingredients	2
			Credit	Construction and Demolition Waste Management	2
0	0	0		**Indoor Environmental Quality**	**16**
Y			Prereq	Minimum Indoor Air Quality Performance	Required
Y			Prereq	Environmental Tobacco Smoke Control	Required
			Credit	Enhanced Indoor Air Quality Strategies	2
			Credit	Low-Emitting Materials	3
			Credit	Construction Indoor Air Quality Management Plan	1
			Credit	Indoor Air Quality Assessment	2
			Credit	Thermal Comfort	1
			Credit	Interior Lighting	2
			Credit	Daylight	3
			Credit	Quality Views	1
			Credit	Acoustic Performance	1
0	0	0		**Innovation**	**6**
			Credit	Innovation	5
			Credit	LEED Accredited Professional	1
0	0	0		**Regional Priority**	**4**
			Credit	Regional Priority: Specific Credit	1
			Credit	Regional Priority: Specific Credit	1
			Credit	Regional Priority: Specific Credit	1
			Credit	Regional Priority: Specific Credit	1
0	0	0		**TOTALS** Possible Points:	**110**

Certified: 40 to 49 points, **Silver:** 50 to 59 points, **Gold:** 60 to 79 points, **Platinum:** 80 to 110

Fig. 22.2 The LEED certification point system for buildings (Source: http://www.usgbc.org/resources/leed-v4-building-design-and-construction-checklist)

- Certified: 40–49 points
- Silver: 50–59 points
- Gold: 60–79 points
- Platinum: 80+ points

22.1.3 Green Globes

Green Globes is an environmentally oriented building certification that was founded in 2000. Green Globes issues certifications to buildings primarily in the United States and Canada. Green Globes is often used as an alternative to LEED, as it offers an online assessment that may be more flexible and affordable than LEED. Similarly to LEED, Green Globes uses a point system to evaluate a building on its environmental impact. However, unlike LEED, the pre-assessment can be done by a project manager or design team, instead of requiring a consultant.

The Green Globes point allocation system is shown in Figs 22.3.

22.1.4 Living Building Challenge

The Living Building Challenge was created in 2006 by the International Living Future Institute. It works to achieve the most advanced sustainability measures in buildings. It strives to achieve futuristic goals for building performance and issues certifications based on seven categories, or "Petals," which are place, water, energy, health and happiness, materials, and equity and beauty. Each petal is further divided into subcategories with specific focuses. These categories and subcategories can be

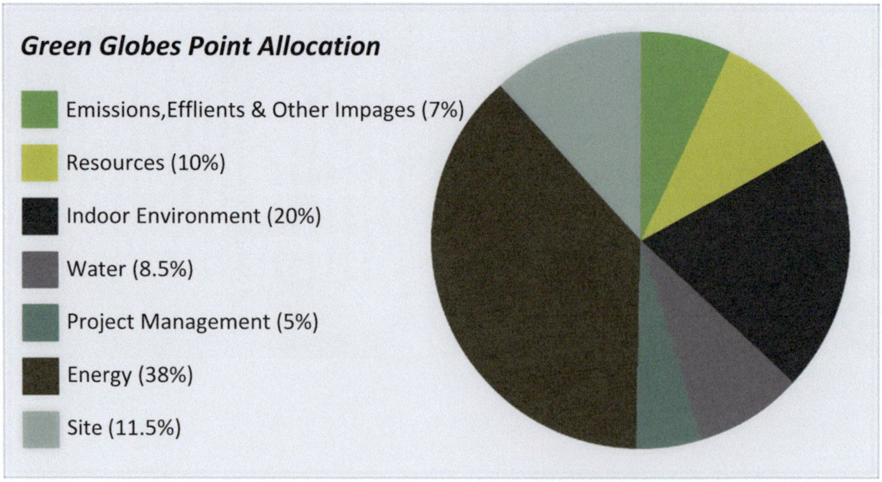

Fig. 22.3 The Green Globes weighting system for certifications (Source: http://www.mcgill-smithpunshon.com/contents/company/2014Fall.html)

used to certify developments of all scales and can often be more rigorous than LEED or BREEAM. Unlike other certification organizations, the Living Building Challenge is a "challenge" or something that new buildings attempt to pursue.

A living building is able to generate its own energy through renewable processes or process its own water. In order to achieve many of its futuristic goals, the Living Building Challenge works with politicians or government officials to ease the limitations on building sustainability and make it more beneficial to build with sustainability in mind. The end goal is to change buildings from being a negative environmental impact to a regenerative and environmentally conscious system.

The following table shows the Living Building Challenge criteria for its water petal (Table 22.1).

Table 22.1 Comparison of different building rating systems

	BREEAM	LEED	Green Globes	Living Building Challenge
Organization	Building Research Establishment	US Green Building Council	Green Building Initiative	International Living Future Institute
Type	Rating system	Rating system	Rating system	Rating system
Type of projects	New or existing buildings	New or existing buildings	New or existing buildings	New or existing buildings
Topics or categories	9 categories Pollution Management Health and well-being Energy Transport Water Materials Land use and ecology Waste +1 innovation category	8 categories for scoring Location and transport Sustainable sites Water efficiency Materials and resources Indoor environmental quality Regional priority Air quality Energy and atmosphere	7 categories with different weightings Emissions Indoor environment Resources Water Project management Energy Site	7 Petals Place Water Energy Health and happiness Materials Equity Beauty and spirit
Indicators	40 individual assessment issues 5 core technical categories + innovation category	12 prerequisites (mandatory) 40 credits (for point allocation)	Various sub-criteria per category	20 imperatives in 7 Petal areas
Levels of certification	Unclassified (<30 %) Pass (≥30 %) Good (≥45 %) Very good (≥55 %) Excellent (≥70 %) Outstanding (≥85 %)	Certified (40–49 pts) Silver (50–59 pts) Gold (60–79 pts) Platinum (80+ pts)	1 Globe (35–54 %) 2 Globes (55–69 %) 3 Globes (70–84 %) 4 Globes (85–100 %)	Meet all subcategories to earn each Petal

22.2 Rating Systems for Cities

Over 50 percent of the world's population lives in cities. By 2050, this number is estimated to increase to 70 percent. City life is considered the most sustainable form of living due to the urban density that allows short distances for transportation, more efficient operation of infrastructure, and a more resource-efficient operation of buildings. Dense urban settlement patterns can be operated more efficiently than sprawled areas. However, cities today face a lot of different challenges related to density and population growth such as air pollution, health issues, and waste and resource management.

With increasing population numbers in cities, the role of cities is becoming more and more important when it comes to sustainability. Therefore, it is also important to make the city's sustainability measurable and create certain indicators in order to compare and evaluate the strengths and weaknesses of different cities.

Buildings account for roughly a third of the US energy consumption (about 40 percent globally) and are therefore responsible for a large amount of greenhouse gas emissions. However, transportation also accounts for a third of energy consumption in the United States (globally about 14 percent) and is also an important factor that should be taken into account when talking about sustainability in cities.

Just in recent years, we began to incorporate more holistic, whole-system-related approaches into existing rating systems and created new ones that focus on neighborhoods or cities as a whole system. We understood that the interrelations between the buildings are even more important than their single components when it comes to sustainability. In the United States, for example, people consume more energy for their commute to work than the amount of energy commercial buildings consume for their operation.

Several building rating systems such as the US Green Building Council's Leadership in Energy and Environmental Design (LEED) rating system or the Living Building Challenge have been expanded in scope for entire neighborhoods in order to address those interconnections (LEED Neighborhood Development, Living Community Challenge).

In addition, international city organizations such as ICLEI (International Council for Local Environmental Initiatives) collaborated with other organizations in order to create rating systems for neighborhoods and communities including topic-related rating systems such as climate change, greenhouse gas emissions, and social equity. For instance, the C40 Cities Climate Leadership Group (C40) was formed to combat climate change issues and CO_2 emissions in cities.

For a city-wide approach, in 2009, nine cities together with the University of Toronto and the World Bank started an effort to make the sustainability of entire cities measurable and comparable. Out of this effort, they created the ISO 37120 standard for sustainable development communities.

22.2.1 *ISO 37120:2014 Sustainable Development of Communities: Indicators for City Services and Quality of Life*

With this international standard, definitions and methodologies for a set of indicators to measure the performance of city services and quality of life have been developed. The ISO 37120 standard defines 100 city performance indicators that can be used as tools to ensure that policies are put into practice that promote liveable, tolerant, inclusive, sustainable, resilient, economically attractive, and prosperous cities. Forty-six of these city performance indicators are defined as core or "shall" indicators and 54 are defined as supporting or "should" indicators.

The standard covers all areas and activities that are taking place within a city:

- Economy
- Education
- Energy
- Environment
- Finance
- Fire and emergency response
- Governance
- Health
- Recreation
- Safety
- Shelter
- Solid waste
- Telecommunications and innovation
- Transportation
- Urban planning
- Wastewater
- Water and sanitation

The standard is not considered a rating system and has three main goals. It aims to help cities to[2]

(a) **Measure** performance management of city services and quality of life over time
(b) **Learn** from one another by allowing comparison across a wide range of performance measures
(c) **Share** best practices

The performance measuring and ability to compare between different cities will enable cities to provide more effective governance services and a better connection between the city population and the government. The transparency and data that can

[2] https://www.iso.org/obp/ui/#iso:std:iso:37120:ed-1:v1:en.

be gained through that will support local sustainability planning and will additionally leverage by international benchmarking and best practices. The standard aims at becoming a tool for planners, politicians, business leaders, and other decision-makers as well as for citizens and other stakeholders that are involved in sustainability improvement of cities worldwide.

It promotes a holistic and integrated approach to sustainable development and resilience. With this standard, cities will be able to compare where they stand in relation to other cities globally, identify best practices, and learn from each other.

Currently, the World Council on City Data (WCCD) is piloting ISO 37120 conformance with 17 cities.

22.2.2 STAR Community Rating System

The first certification program for sustainable communities in the United States was the STAR Community Rating System (Sustainability Tools for Assessing and Rating Communities). ICLEI–Local Governments for Sustainability, the US Green Building Council, and the Center for American Progress launched the concept for the STAR Community Rating System (STAR) at Greenbuild in Chicago in 2007.

Since 2012, STAR Communities is an independent nonprofit organization with the mission to evaluate, improve, and certify sustainable communities. The STAR Community Rating System is a US-wide certification program to recognize sustainable communities. It is not just a certification program though; it also serves as a framework for local governments who use it "as a roadmap, as a planning tool, to organize/guide public engagement processes, to aid in decision-making, for performance-based budgeting, and as a measuring stick to determine whether their investments are achieving outcomes."[3]

STAR aims at creating more livable communities that are more healthy, inclusive, and prosperous across seven goal areas:

- Built environment: quality, choice, and access where we live, work, and play
- Climate and energy: increase efficiency, reduce impact
- Economy and jobs: quality jobs, shared prosperity
- Education, art, and community: vibrant, connected, and diverse culture
- Equity and empowerment: inclusion and access for all community members
- Health and safety: strong, resilient, and safe
- Natural systems: protect and restore the resources of life

When evaluating communities, STAR is looking at two aspects: the community-level outcomes (measurable results that show community progress) and local actions that have to be done to achieve certain outcomes. Thus, the rating system can help to identify and evaluate as well as implement best practices.

---

[3] http://www.starcommunities.org/.

STAR defines a sustainable community as a community that "thinks and acts systematically, instills resiliency, fosters innovation, redefines progress, lives within means, cultivates collaboration, ensures equity, embraces diversity, inspires leadership, and continuously improves."[4]

22.2.3 BREEAM Communities

BREEAM stands for Building Research Establishment Environmental Assessment Methodology and was the first international building rating system for green, sustainable buildings established by Building Research Establishment (BRE) in the UK in 1990. BREEAM has been executed in over 50 countries. The main sustainability categories in BREEAM are management, energy, health and well-being, transport, water, materials, waste, land use and ecology, and pollution.

In order to also assess the sustainability of entire communities, BREEAM Communities was developed. This international assessment method aims at creating places where people want to live and work that are environmentally responsible and economically successful. It can be used for mixed-use community projects or single-use developments of a certain size. The projects can be either new construction or regeneration if there is a significant change to the urban fabric.

The overall performance of a BREEAM project is determined by the following elements:

- The mandatory BREEAM Communities standards
- The BREEAM Communities assessment issues and credits
- Awarding credits for innovation
- The category and assessment issue weightings
- The BREEAM rating level benchmarks

The system covers forty different topics such as housing provision, transportation networks, community facilities, and economic impact. All three core aspects, environment, society, and economy, are equally considered. In order to provide sustainable developments that function now and in the future, the engagement of the community in the projects is a crucial factor.

22.2.4 LEED Neighborhood Development

The LEED Neighborhood Development system (LEED-ND) is a joint effort between the US Green Building Council, the Congress for the New Urbanism, and the Natural Resources Defense Council based on the LEED rating system for green buildings. It expands the scope of LEED to a more holistic approach

[4] http://www.starcommunities.org/.

looking at the buildings' connections to their surroundings. It's a rating system at the neighborhood or community scale that certifies green (sustainable) neighborhoods combining the principles of green building, smart growth, and new urbanism.[5]

The system focuses on the following categories of a neighborhood: smart location and linkage, the neighborhood pattern and design, and green infrastructure and buildings. In addition, credits can be achieved for innovation and design process and regional priorities. Similar like the LEED systems for buildings, certain prerequisites have to be met in each category. In addition, a certain number of points can be achieved through credits in all categories. The levels of certification range from certified (the basic certification) to silver, gold, and platinum[6] (Fig. 22.4).

22.2.5 Living Community Challenge

The Living Community Challenge is a program formed and executed by the International Living Future Institute. Similar like USGBC, they also started with a building rating system (Living Building Challenge) which they extended to a community/neighborhood-scale rating system.

The program serves as a planning tool for planners and developers for community-scale projects at the master planning stage. Additionally, it is a rating system for campus-scale, neighborhood-scale, or simply street-, block-, or corridor-size projects. The program rates any type of project from residential to commercial to mixed-use neighborhoods.[7]

The main idea of the Living Building Challenge is to think of a building as a flower. Flowers need energy and water as well as other resources to be able to live. At the same time, they don't produce harmful emissions. The Living Community Challenge continues this thought on a community scale.

The Living Community Challenge consists of seven performance areas, also called "Petals," that are further split into 20 imperatives:

- Place
- Water
- Energy
- Health and happiness
- Materials
- Equity
- Beauty and spirit

[5] http://www.cnu.org/node/2964.

[6] http://www.usgbc.org/.

[7] http://living-future.org/lcc.

LEED v4 for Neighborhood Development Plan
Project Checklist

Project Name:
Date:

Yes	?	No		Smart Location & Linkage	28
0	0	0			
Y			Prereq	Smart Location	Required
Y			Prereq	Imperiled Species and Ecological Communities	Required
Y			Prereq	Wetland and Water Body Conservation	Required
Y			Prereq	Agricultural Land Conservation	Required
Y			Prereq	Floodplain Avoidance	Required
			Credit	Preferred Locations	10
			Credit	Brownfield Remediation	2
			Credit	Access to Quality Transit	7
			Credit	Bicycle Facilities	2
			Credit	Housing and Jobs Proximity	3
			Credit	Steep Slope Protection	1
			Credit	Site Design for Habitat or Wetland and Water Body Conservation	1
			Credit	Restoration of Habitat or Wetlands and Water Bodies	1
			Credit	Long-Term Conservation Management of Habitat or Wetlands and Water Bodies	1

Yes	?	No		Neighborhood Pattern & Design	41
0	0	0			
Y			Prereq	Walkable Streets	Required
Y			Prereq	Compact Development	Required
Y			Prereq	Connected and Open Community	Required
			Credit	Walkable Streets	9
			Credit	Compact Development	6
			Credit	Mixed-Use Neighborhoods	4
			Credit	Housing Types and Affordability	7
			Credit	Reduced Parking Footprint	1
			Credit	Connected and Open Community	2
			Credit	Transit Facilities	1
			Credit	Transportation Demand Management	2
			Credit	Access to Civic & Public Space	1
			Credit	Access to Recreation Facilities	1
			Credit	Visitability and Universal Design	1
			Credit	Community Outreach and Involvement	2
			Credit	Local Food Production	1
			Credit	Tree-Lined and Shaded Streetscapes	2
			Credit	Neighborhood Schools	1

Yes	?	No		Green Infrastructure & Buildings	31
0	0	0			
Y			Prereq	Certified Green Building	Required
Y			Prereq	Minimum Building Energy Performance	Required
Y			Prereq	Indoor Water Use Reduction	Required
Y			Prereq	Construction Activity Pollution Prevention	Required
			Credit	Certified Green Buildings	5
			Credit	Optimize Building Energy Performance	2
			Credit	Indoor Water Use Reduction	1
			Credit	Outdoor Water Use Reduction	2
			Credit	Building Reuse	1
			Credit	Historic Resource Preservation and Adaptive Reuse	2
			Credit	Minimized Site Disturbance	1
			Credit	Rainwater Management	4
			Credit	Heat Island Reduction	1
			Credit	Solar Orientation	1
			Credit	Renewable Energy Production	3
			Credit	District Heating and Cooling	2
			Credit	Infrastructure Energy Efficiency	1
			Credit	Wastewater Management	2
			Credit	Recycled and Reused Infrastructure	1
			Credit	Solid Waste Management	1
			Credit	Light Pollution Reduction	1

Yes	?	No		Innovation & Design Process	6
0	0	0			
			Credit	Innovation	5
			Credit	LEED® Accredited Professional	1

Yes	?	No		Regional Priority Credits	4
0	0	0			
			Credit	Regional Priority Credit: Region Defined	1
			Credit	Regional Priority Credit: Region Defined	1
			Credit	Regional Priority Credit: Region Defined	1
			Credit	Regional Priority Credit: Region Defined	1
0	0	0		**PROJECT TOTALS (Certification estimates)**	110

Certified: 40-49 points, **Silver:** 50-59 points, **Gold:** 60-79 points, **Platinum:** 80+ points

Fig. 22.4 LEED-ND checklist for prerequisites and available credits (http://www.usgbc.org/resources/leed-v4-neighborhood-development-checklist)

Two levels of certification can be achieved:

1. Living Community Certification: All 20 imperatives in all 7 areas must be met.
2. Petal Community Certification: All imperatives of at least three Petals have to be met. One of the three Petals has to be either water or energy or materials Petal.

The uniqueness of this rating system compared to others is that it measures the performance of a project based on 1 year of operation. That is, other rating systems such as BREEAM or LEED calculate efficiencies based on what is planned and the results of the commissioning process but not on the results during operation of the building or neighborhood. A project can only get Living Community Challenge certified if the 1-year performance meets all the required thresholds and goals.

In addition, the Living Community Challenge rates the happiness of the occupants (with an occupant survey) and the beauty and spirit of a project. The integration of community and public art into the project evaluation is another factor that the Living Community Challenge evaluates (Table 22.2).

Table 22.2 Comparison of different city and neighborhood rating systems

	ISO 37120:2014	STAR Community Rating System	BREEAM	LEED Neighborhood Development	Living Community Challenge
Organization	ISO (International Organization for Standardization)	STAR Communities	Building Research Establishment	US Green Building Council	International Living Future Institute
Type	ISO standard	Rating system and framework for local governments	Rating system	Rating system	Rating system
Scale	City	City/community	Communities	At least 2 habitable buildings and max. 1500 acres	Communities, neighborhoods, campus, street-, block-, or corridor-size projects
Project stage	Existing cities	Existing communities	Master plan stage or redevelopment of existing development	Any stage of the development process	Master plan stage or completed construction
Topics or categories	17 areas: • Economy • Education • Energy • Environment • Finance • Fire and emergency response • Governance • Health • Recreation • Safety • Shelter • Solid waste • Telecommunications and innovation • Transportation • Urban planning • Wastewater • Water and sanitation	7 goal areas: • Built environment • Climate and energy • Economy and jobs • Education, art, and community • Equity and empowerment • Health and safety • Natural systems	5 core technical categories • Governance • Land use and ecology • Resources and energy • Social and economic well-being • Transport and movement +1 Innovation category	5 credit categories • Smart location and linkage • Neighborhood pattern and design • Green infrastructure and buildings • Innovation and design process • Regional priorities	7 Petals: • Place • Water • Energy • Health and happiness • Materials • Equity • Beauty and spirit

(continued)

Table 22.2 (continued)

	ISO 37120:2014	STAR Community Rating System	BREEAM	LEED Neighborhood Development	Living Community Challenge
Indicators	100 city performance indicators 46 *core* or *shall* indicators 54 *supporting* or *should* indicators	44 sustainability objectives 526 measureable indicators	40 individual assessment issues 5 core technical categories + innovation category	12 prerequisites (mandatory) 47 credits (for point allocation)	20 imperatives in 7 Petal areas
Levels of certification	ISO certified	• Reporting STAR Community (50–199 pts) • 3-STAR Community = recognized for sustainability leadership (200–399 pts) • 4-STAR Community = recognized for national excellence (400–599 pts) • 5-STAR Community = recognized as top tier achiever in national sustainability (600+ pts)	• Unclassified (<30 %) • Pass (≥30 %) • Good (≥45 %) • Very good (≥55 %) • Excellent (≥70 %) • Outstanding (≥85 %)	• Certified (40–49 pts) • Silver (50–59 pts) • Gold (60–79 pts) • Platinum (80+ pts)	2 levels of certification: • Living Community Certification (meet all 20 imperatives across all 7 Petals) • Living Community Petal Certification (meet all the imperatives of at least three Petals)

ERRATUM

Integration of Nature and Technology for Smart Cities

Anil Ahuja

A. Ahuja, *Integration of Nature and Technology for Smart Cities*,
DOI 10.1007/978-3-319-25715-0

DOI 10.1007/978-3-319-25715-0_23

This book is third edition, but the preface mentions this book as a second edition. This is now updated correctly to Third edition.

The updated online version of the original book can be found at
http://dx.doi.org/10.1007/978-3-319-25715-0

A. Ahuja, *Integration of Nature and Technology for Smart Cities*,
DOI 10.1007/978-3-319-25715-0_23

References

http://smartcitiescouncil.com/article/dissecting-iso-37120-why-new-smart-city-standard-good-news-cities
http://www.iso.org/iso/home/news_index/news_archive/news.htm?refid=Ref1848
https://www.iso.org/obp/ui/#iso:std:iso:37120:ed-1:v1:en
http://www.usgbc.org/
http://www.cnu.org/node/2964
http://living-future.org/lcc
http://www.starcommunities.org/
http://www.breeam.org/filelibrary/BREEAM%20Communities/Introduction_to_BREEAM_Communities.pdf
http://www.pmarchitects.com
http://www.greenglobes.com/home.asp
http://www.mcgillsmithpunshon.com/contents/company/2014Fall.html
http://landscapemanagement.net/sustainability-standards-faceoff/

A. Ahuja, *Integration of Nature and Technology for Smart Cities*,
DOI 10.1007/978-3-319-25715-0

Index

A. Ahuja, *Integration of Nature and Technology for Smart Cities*,
DOI 10.1007/978-3-319-25715-0

G

H

I

MIX
Papier aus verantwortungsvollen Quellen
Paper from responsible sources
FSC® C105338
FSC
www.fsc.org

If you have any concerns about our products,
you can contact us on
ProductSafety@springernature.com

In case Publisher is established outside the EU,
the EU authorized representative is:
Springer Nature Customer Service Center GmbH
Europaplatz 3, 69115 Heidelberg, Germany

Printed by Libri Plureos GmbH
in Hamburg, Germany